The Roots of Logistics

Peter Klaus • Stefanie Müller
Editors

The Roots of Logistics

A Reader of Classical Contributions to the History and Conceptual Foundations of the Science of Logistics

Springer

Editors
Prof. (em.) Peter Klaus D.B.A. (Boston Univ.)
School of Business and Economics
Friedrich-Alexander University
Erlangen-Nuremberg
Germany

Prof. Dr. Stefanie Müller
Faculty of Business Administration
Georg-Simon-Ohm University of Applied Sciences, Nuremberg
Germany

ISBN 978-3-642-27921-8 ISBN 978-3-642-27922-5 (eBook)
DOI 10.1007/978-3-642-27922-5
Springer Heidelberg New York Dordrecht London

Library of Congress Control Number: 2012937223

Printed on acid-free paper

Springer is part of Springer Science+Business Media (www.springer.com)

Foreword

The story of modern logistics is one of success on many accounts: In the world of business, over the past 30+ years, companies in most industries and all over the world established logistics departments and employed logistics personnel. The fragmented markets of transportation, forwarding, warehousing, packaging and added services are all now part of a multi-billion "logistics service provider" industry. Professional associations focusing on logistics claim tens of thousands of members across the world. The impact of professional logistics on the overall competitiveness of companies and even nations is now widely understood. Last but not least: Logistics curricula, research institutes, chairs and professorships dedicated to logistics (and supply chain management) have been established at universities and colleges everywhere.

But one important question is still being debated in the academic world: The question of the identity of logistics as a science. This volume aims to make a contribution to this debate in two ways:

Firstly, it retraces the evolution of the field from its very early roots to recent "milestone" contributions through a selection of original papers and quotations, which – we believe – could be properly called the "classics" of the field. We hope to provide a service to logisticians interested in the history and evolution of logistics by making these – in some cases hard to access – contributions available in one convenient volume.

Secondly, by the arrangement of these materials in a structure along three major lines of development, supplemented by an introductory essay and brief comments on each of the selections, we offer our own interpretation of the evolution and evolving identity of the "Science of Logistics" to further stimulate the debate. Any suggestions and critique regarding our selection and comments will be highly appreciated.

For reading convenience, the references in our comments to those papers reprinted are highlighted in bold type face. Any other references are found at the end of the respective comment or orginal article.

We like to thank the many institutions and individuals who helped us in the preparation of this volume by providing copyrights. Special thanks go to Jennifer Pruden for contributing to the translation of those articles orginally published in German. Nicole Lubecki and Helen Rogers also offered their time in assisting us to improve the quality of the final draft. The publishing house of Springer and its editorial staff were patient, helpful and cooperative regarding the layout and completion of the volume. Last but not least, we are indebted to two sponsors who provided financial support to this book project: the University of Applied Sciences, Offenburg and DB Schenker AG (Stinnes Foundation), Berlin.

Nuremberg, February 2012

Peter Klaus
Stefanie Müller

Table of Contents

Part B: The Evolution of Business Logistics as an Autonomous Field of Science

Part C: The Instrumentation of Logistics: Terminology, Concepts and Methodological Approaches

V. Contributions from Quantitative and Engineering Sciences

VI. Contributions from Management Sciences

VII. Contributions from Economics and Organizational Sciences

INTRODUCTORY ESSAY

Peter Klaus/Stefanie Müller: Towards a Science of Logistics: Milestones along Converging Paths[1]

I. IN SEARCH OF A SCIENCE OF LOGISTICS

Is logistics a science? This essay is to discuss the claim that the field of logistics – over a time span of approximately 50 years – has evolved from a new label, which was pragmatically attached to a set of problems within the field of Marketing, towards a scientific discipline in its own right. The claim is not unchallenged.[2] And to the extent it is accepted among the members and observers of the large "community of logisticians", tacitly or openly, there are ongoing debates about the 'proper' definition, the boundaries, and the essential characteristics of the discipline.[3]

The following discussion intends to add to this debate in two ways:

- First: It retraces and documents the path of "milestone" contributions to the evolution of the discipline, including some of the outstanding older conceptual roots and ideas, on which modern logistics is founded. A selection of those contributions is reprinted and briefly commented on in this essay and in the following sections of this volume.[4] It is hoped that this collection of – in some cases hard to access – original documents will be useful to researchers and practitioners in the field of logistics, regardless whether they agree or do not agree with the "logistics science" claim.
- Second: The essay offers an interpretation of the identity of the emerging "science of logistics" in content. It argues for this interpretation being a basis for future research and progress in the achievement of economic, ecological and social objectives. It makes an effort to reflect the logic by which

---

[1] Parts of this essay are based on an earlier paper by Klaus (2009).

[2] Logistics, especially as seen from the outside perspective of other fields of science, is still seen as part of Marketing, Operations Management, Operations Research. See also Krampe/Lucke/Lüsch (1990).

[3] See for example Kirsch (1971), Ihde (1972), Novack/Rinehart/Wells (1992); Weber (1992); Klaus (1993); Göpfert (1996); Cooper/Lambert (1997); Kent/Flint (1997); Ballou (2007); Davis-Sramek/Fugate (2007); Frankel et al. (2008); Charvet/Cooper/Gardner (2008), Hopp (2008).

[4] On the following pages, the citations of those articles are **HIGHLIGHTED** that are reprinted in this volume.

P. Klaus and S. Müller (eds.), *The Roots of Logistics*, DOI 10.1007/978-3-642-27922-5_1,

researchers accumulated and integrated knowledge in the name of "logistics".

To start the discussion, the question about "what it is that makes a science?" will be asked. Looking back to the roots of logistics, the following sections of the essay retrace the path of the terminological, conceptual, and "instrumentation" developments. Based on this retrospective review, it is assumed, the intellectual roots of the field may be better understood, landmark contributions highlighted, and a vision of what course the field might take in the future may result. The review shows that there are several roots which grew in parallel, but have been converging gradually to support an entity which can now be called a science. Some thoughts about the future challenges and potentials of this development conclude the discussion.

1. What Makes a Science?

For the purpose of a discussion on "what makes a science", some thoughts from Thomas Kuhn's (1970) famous work on *The Structure of Scientific Revolutions* will be borrowed. He argues that a science is identified by a combination of external manifestations and some shared conceptual, intrinsic ties.

As to the external manifestations: There must a group of professionals who share an interest in a certain class of facts and problems. These persons share their observations and research through dedicated technical journals and conferences – and they are related through similar educational experiences and professional initiations. Finally they "see themselves responsible for the pursuit of a set of shared goals, including the training of their successors" and thus become a "scientific community" (Kuhn (1970), p. 177).

There is no question any more that these external manifestations of a science of logistics could be observed at least since the 1970s, when professional associations like the (then) "Council Logistics Management (CLM)" in the United States, the "Bundesvereinigung Logistik (BVL)" in Germany, and a growing number of comparable associations and groupings in many other countries were established. Dedicated university departments, professorships, and academic curricula, as well as academic journals were established,[5] and professional conferences both for practitioners and academics were routinely held.

But beyond the external manifestations of an academic discipline, Kuhn asks what it is that "accounts for the relative fullness of the professional communication and the relative unanimity of their professional judgement" (Kuhn (1970), p. 182). What are the conceptual, intrinsic commonalities which tie the community together? Kuhn refers to these as the "paradigm" of an established "normal" science. A paradigm, according to his argument, is characterized by a shared set of questions, and a family of shared facts and problems. He asks "on what aspects of nature do scientists ordinarily report? What determines their choice [of those aspects] and what motivates

[5] First university courses were established in the US in the 1960s. First academic journals were *Journal of Business Logistics* (since 1979), *The International Journal of Logistics Management* (since 1989).

the scientist to pursue that choice to a conclusion?" (Kuhn (1970), p. 25f.). Furthermore, he suggests that the members of a scientific community share a set of metaphors, models, techniques and rules, which help to interpret observations and derive predictions and recommendations on the basis of these in similar ways (Kuhn (1970), pp. 25f and 181ff.). What are the elements of that shared set of instruments, and how is the process of the "instrumentation" of a science taking place? If there is a "science of logistics", there must be answers to these questions about the paradigm – the intrinsic elements of the identity of the field. The question of whether all this applies to logistics and if logistics can hence be considered to be a science will be addressed in the following pages by way of a retrospective on the roots, the milestones, and the evolution of logistics.

2. Remarks on the Search Process

Looking back, now, the question about the roots and important milestones in the development of the science of logistics arises. Inspired by Kuhn's thoughts referred to in the previous section, answers are sought by tracing scientific logistical thinking in the literature from some of its very early predecessors through 50-years' history of logistics between the 1960s and today:

- What kinds of major facts and issues have been addressed over time?
- What kinds of models, metaphors, techniques, and instruments that help to develop answers and solutions have been suggested by different members and groups of the scientific community of logisticians?

And, to the extent this can be identified:

- What are the early scientific roots that inspired those groups and informed the answers that they were proposing?

The level of detail of the following discussion is intended at a middle range: It should say more about the substance of the ideas, research questions and answers proposed than the frequently suggested very broad categorizations of the scope of logistics from a (mono) "functional", "physical distribution", to a "supply chain/ cross-functional" and a "global supply network" perspective.[6] However it should also avoid the hard-to-structure 'jungle' of mostly technical concepts and terms that were discussed with varying intensity over time, such as "Materials Requirements Planning (MRP)", "Just-in-Time (JIT)", "Vendor Management Inventory (VMI)", "Collaborative Planning, Forecasting, and Replenishment (CPFR)", etc.

The primary criteria for clustering the terminological and conceptual contributions will be their historical context and association with certain scientific communities.[7]

---

[6] Most recent examples of this kind of categorizing the evolution of logistics are in Ballou (2007) and Baumgarten (2008).

[7] The choice of these criteria is inspired by Rheinberger's (2007) argument that the study of the development of scientific knowledge should reflect to historical context and the means and methodologies through which the process of knowledge generation takes place. The analysis is deliberately qualitative and interpretive rather than based on formal methodology (as, e.g., in Charvet/Cooper/Gardner's (2008) recent article).

Logistics science has evolved as a cross-disciplinary field, stimulated and informed by a wide range of other, older fields of science, ranging from Economics, Mathematics, to the younger fields of Business Administration, the Organization Sciences, and Engineering. Each of these has its own perspective and prefers its own set of methods, concepts, and instruments. The logisticians of the first generations – if they did not operate in purely pragmatic, intuitive ways – brought with them the perspectives and methods from those fields and institutions where they came from, since there were no indigenous logistics courses and institutions until quite recently. And even among the younger generation academics in logistics who received their education in contemporary logistics and supply chain management programs and departments, the style and direction of their work is usually influenced by roots in either Marketing Science, Economics, Operations Research, Management Studies, Engineering, Behavioural Science and Organization Theory, or other similar affiliations. Logistics, significantly more than other fields, is embedded in a diverse network of intellectual relationships – which also explains the difficulty for logisticians to establish their own, distinct scientific identity.

On the basis of these preliminary considerations about the appropriate level of detail and the logic of clustering materials by their disciplinary affinities, several development paths of idea development are traced, each starting from some roots in "classical" academic contributions. First, there is a line of developments starting from the early references to logistics in the military context, evolving through several stages towards the "flow systems" logistics paradigm. Then three other lines of development are traced which describe the broadening and deepening of logistics' set of metaphors, models, techniques and rules: The process of the "instrumentation" of logistics.

II. TRACING THE EARLIEST ROOTS OF LOGISTICS

1. Logistics and Military Science

The etymological and historical roots of the term "logistics" are often sought in the Greek word *λόγος* (logos, standing for "reason" or "rational"), in the Latin *locare* (standing for "arranging", "placing") and in the related French word *loger* (standing for the accommodation of guests or soldiers), and its uses in the military context.

The French general Antoine-Henri Jomini was the first to explicitly elaborate the idea of logistics as the military function of lodging soldiers in his classical works on military strategy *The Art of War* (**JOMINI (1838/1862)**). His explanation of logistics is based on a list of 18 principal duties of a military quartermaster (the "major général de logis") including "the preparation of all material necessary for setting the army in motion". This elementary idea of logistics as a function to assure "supply of men and materials" for an operation is present already in this very first documented use of the term.

At the time of publication Jomini's book didn't attract much interest. A few years later, however, the English translation found its way into the standard lectures and literature used in American military schools. Thus, "logistics" became a common term in military language.

The line of the development of logistics as a critical function of war preparation and the sustained maintenance of the ability of the military forces to fight continued into the 20th century. The next milestone contributions were by the American officer George Cyrus Thorpe (1917) in his book on *Pure Logistics: The Science of War Preparation* and discussions by Admiral Henry **Eccles (1954)** and the famous economist Oskar **Morgenstern (1955)** on the essence of logistics and the possibilities to develop logistics theory. They are milestones for more recent contributions to military logistics and examples of first incidences of the debate about the possibility of a science of logistics – albeit still strictly in the context of its military uses.

2. The Emergence of Logistics as a Discipline of Marketing

Quite apart from military applications, a different meaning of logistics emerged out of the earliest questions studied in the field of Economics: How is economic value and wealth created? What are the basic economic activities and mechanisms for growing the production and supply of goods, and how are people's needs and demands becoming satisfied? These considerations started in the 18th century with Adam Smith's (1776) famous work on the *Wealth of Nations*. With hindsight it has become clear that some of these classical works anticipated and laid important groundwork to the conceptions of modern logistics.

An outstanding early contribution on the question of how value can be created is the great British economist's Alfred **Marshall's (1890)** book on *Principles of Economics*. Marshall claimed that "man cannot create material things" and made clear that the economic activity of making things "serviceable" and valuable to people is about the "movement", "arrangement" and "rearrangement", not about the creation of things (Marshall (1890), p. 116f.). He had the foresight to see that it is not production in the usual sense, which should be the center of attention in considering the creation of economic value, but the activities of moving and arranging natural matter – activities which we now consider to be at the core of logistics.

Following on from Marshall, the American economist Louis **Weld (1919)** specified four fundamental utilities that economic activity can generate. He demanded that the systematic generation and enhancement of those utilities, at the business level, is the essential task of the Marketing function: There is the "utility of form" – which corresponds to Marshall's "arrangement" and forming of physical matter. In more modern terminology it is a result of "manufacturing". But then there are – no less important to meet customers' and markets' needs – the utilities of "time", of "place", and of "possession", which are created through the elementary logistical functions of storing, transporting, and bringing things into useful order by parcelling, sorting, consolidating, etc.

The German logistician Gösta B. **Ihde (1984)** took these thoughts still one step further to graphically show how the elementary economic activity of production as the "transformation of matter" (respectively creating "form utility", performing manufacturing operations) and the activities of the transfer of things in space ("place utility" through transportation), in time ("time utility" through storage, warehousing, buffering) and order ("possession utility" through parcelling, consolidating/deconsolidating, patterning,) are interrelated.

Through Marshall's and Weld's insights, the idea of the indispensability of logistical operations in the processes of economic value creation and their key role in the successful marketing of products was established very early. However it was not until the 1960s in the United States, when these insights – after some more scattered other references to this effect had appeared in the business and economics literature.[8] **CONVERSE'S (1954)** argument for the importance of physical distribution – *The Other Half of Marketing* – started to attract broad interest in the business and academic communities. The historical motivation for this happening in the 1960s was a "Marketing Revolution"[9] which began in the United States after World War II and soon spilled over to other parts of the world. No longer was the capacity to efficiently produce scarce goods the decisive factor for business success. Rather, the ability to attract and service customers who have many alternative choices to satisfy their needs became the key to economic success. Physical distribution and logistics – the systematic creation of utilities of "place", "time", and "possession" in the Marshall's and Weld's language – were identified as crucial elements of successful marketing. The activities of transporting and warehousing to customers, which formerly were rather peripheral concerns to management and academic research – and generally managed by the "mob in the shipping room" as reported by Drucker (1962) – became the nucleus for an emerging new discipline.

The next milestones in the creation of awareness about physical distribution and the start of the institutionalization of what became "Marketing Logistics" and "Business Logistics" were Magee's (1960) Harvard Business Review article on *The Logistics of Distribution* and **SMYKAY/BOWERSOX/MOSSMAN'S (1961)** first academic textbook on *Physical Distribution Management.* But it seems to have been Peter **DRUCKER'S (1962)** article on "The Economy's Dark Continent" and his popularity in the broad management community that began to draw wide-ranging attention to the need for systematic scientific work in this area. Soon after other authors followed suit.[10] Physical distribution and logistics were institutionalized in academic institutions and professional associations such as the National Council for Physical Distribution Management (later to become the Council of Logistics Management and recently the Council of Supply Chain Management Professionals) in the United States.

With several years delay, from the early 1970s onward, parallel developments of research, publications, and institutionalization also evolved in Europe and in other parts of the world, such as **PFOHL'S (1969)** "Everything for Replenishment!", Pfohl's (1972) "Marketing Logistics" and Buijtenen/Christopher/Wills' (1976) "Business Logistics".

By the 1980s the power of marketing and distribution issues for drawing attention to the emerging logistics field and focusing on distribution aspects seems to have

---

8 E.g. Shaw (1912), Weld (1919), and also Converse (1921).

9 Described by Keith (1960) in an article of the same title.

10 Such as Heskett/Ivie/Glaskowsky (1964), Plowman (1964), and Magee (1958).

been exhausted. Other new ideas of "just-in-time" production and "lean" industrial procurement and production became the new centre of interest and a new pillar of the field of logistics' sense of identity. These will be discussed further in the following sections about the emergence of the "flow systems" paradigm of logistics.

III. THE EVOLUTION OF LOGISTICS AS AN AUTONOMOUS FIELD OF SCIENCE

In the early phase of the development of logistics science as a subset of Marketing, the focus was on the challenges of enhancing the "serviceability", "utility" and "value" of products to customers. At a more abstract level, the first generation logisticians were concerned with the exploration and systematic science-based analysis of the "family of facts and problems" around the generic economic activities of transportation (creating utilities of "place"), of storage and warehousing (creating utilities of "time"), and of sorting, picking/packing, consolidating, changing the order and arrangement of things (creating utilities of "possession"). With this generalization, it was a small step to apply the evolving body of know-how to other fields: for example the professionalization of the transport and other service provider industries, the analysis and reorganization of industrial materials procurement systems and the internal processes of industrial production.

1. On the Evolution of Three Meanings of Logistics: "Placing/ Pacing/Patterning (PPP)", "Coordination" and "Flow"

Prior to the integrated treatment of activities creating "place", "pace", and "possession" utilities, there was a rich stream of literature and research dealing with "Transportation", "Transportation Economics" (in German "Verkehrsbetriebslehre") as an isolated function.[11] As one example of the potential effects of the science based professionalization of the transport function, there is an early "classic" contribution by the German transport economist Fritz **VOIGT (1953)**. It illustrates the impact that systematically designed and managed transportation systems can have on the industrialization developments of an economy, as well as the importance of adapting and strategically fitting those systems to the conditions of the economic environment. Comparable early research was also carried out in the area of warehousing activities and inventory management.[12]

A later effort to interpret the gradual evolution of an autonomous science of logistics is **KLAUS/SHEFFI'S (1997)** *On the Evolution of Three Meanings of Logistics*. In this paper, the science-based integrated analysis and systematization of the

---

[11] For a broad review of this line of research see Farris (1997), also Diederich (1986).

[12] Cf. Magee (1958), Jünemann (1971), and the references found in the Smykay/Bowersox/ Mossman (1961).

generic functions of "placing", "pacing", and "patterning" is referred to as the "first meaning" of logistics – "PPP-logistics". Most textbooks and older definitions of logistics of the 1960s, 1970s – and some up to today – are based on this conception of the field.

It is argued in the Klaus/Sheffi commentary, that from the 1980s onward, a second, broader meaning of logistics evolved: Focus of the community of logisticians shifted from the operational "PPP" issues to the issues of the systematic coordination, the planning, and the challenges of the cross-functional optimization of business functions. The family of facts and problems was extended beyond the "PPP"-functions. Issues between any function in the sequence of value creating activities of businesses, such as procurement, manufacturing, distribution, finance, etc. were now considered. This expansion of the scope of the field was accompanied by more focus on the uses of sophisticated decision support tools from Operations Research, from Information Technology, and Management Planning, some of which will be discussed in detail in the "instrumentation" sections of this essay below. The "second meaning" of logistics, in Klaus/Sheffi's terminology, is "coordination logistics".

Another "third" expansion and shift regarding the meaning of logistics began and gained momentum in the 1990s, as Klaus/Sheffi argue in their commentary: It is a fascination with the challenges of the dynamics of flows of objects – such as materials and goods, but also information, people, moneys – through complex chains and networks, and process sequences. At that time the dynamic phenomena and complexities related to the managed flow of objects moved to the centre of attention of the logistics community. The focus on network and flow phenomena, the development and use of instruments which are suited to their analysis and rationalization, and the exploration of new areas of applying the evolving "flow systems" knowledge – this led to the emergence of a third "flow systems" paradigm of logistics. This will be traced in more detail in the following section of the essay.

2. The Emergence of a "Flow Systems" Paradigm

The idea of using the metaphor of "flow" in the context of economic research is very old. It is obviously attractive to focus attention on the sequential interdependence of activities, and on the need to direct and control the flow of objects through complex networks (see EXHIBIT 1, on the next page).

Tracing its emergence and adaptation to the field of logistics leads to Francois **QUESNAY (1764)**, the French 18th century physician and economist and his *Tableau Économique*. This is the first extensive and often quoted use of the metaphor. He made an effort to illustrate and analyze the flow and distribution of wealth in a (then agricultural) economy.

Today, the flow metaphor is applied widely under different terms, such as "production flow" (Ohno (1988)), the "value chain" (Porter (1985)) or "value stream analysis and mapping" (McManus/Millard (2002)), a "process perspective" (e.g. Hammer (1990)), and, of course, the "supply chain" and "supply network" (e.g. Oliver/Webber (1982)).

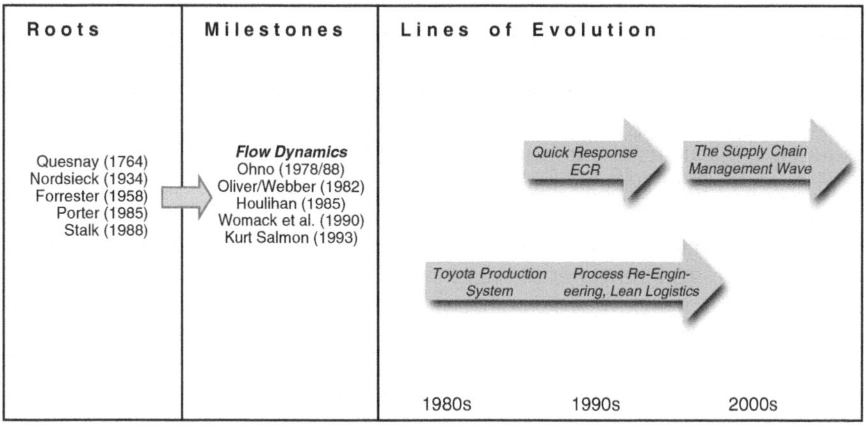

EXHIBIT 1: THE DISCOVERY AND EVOLUTION OF "FLOW DYNAMICS"

A search for earlier contributions, that may be recognized as milestones in the use of the flow metaphor, leads to – at least – two interesting discoveries.

In 1934 Fritz Nordsieck, a German professional who is often credited with having laid the foundations for the discipline of the study of formal organization in his country, produced a graphical depiction of what today would be called a complex "supply network" (**NORDSIECK (1934)**). He argued for the need to balance organizational structures with respect to both the functional similarities of activities and their interconnections in the everyday flows of materials, products, and information over half a century before "value stream" and "process" mapping, re-engineering and the systematic design of supply chain architectures became common practices in management. Approximately 25 years later, Jay Forrester of MIT presented his approach to the study of "Industrial Dynamics", which anticipated many of the problems and solutions of modern supply chain management in highly sophisticated ways (**FORRESTER (1958)**).

However it was not until the early 1980s, when some revolutionary insights about the efficient, "lean" configuration and operational dynamics of logistical systems began to receive widespread attention in the Western countries. At that time the growing competitiveness of certain Japanese companies in the automotive, optics, electronics, and other industries alarmed their Western competitors. It soon was realized that this was based on a non-traditional, new approach to managing industrial processes. The conceptual basis of this development, as Western researchers found with hindsight, had been written up in Taiichi Ohno's 1978 book – *The Toyota Production System* –, originally published in Japanese, translated only 10 years later into English (**OHNO (1988)**). At the heart of Ohno's system was the conceptualization of industrial procurement and production as a flow system. Machines and workers are strictly arranged in the sequence of the manufacturing process – in itself not a new idea, as Henry Ford had put this to use in 1913 already. Ohno's path-breaking contribution, which moved the concept "beyond large-scale production"[13] was based on a new way to control industrial "flow dynamics": Flow activities are stimulated strictly by customer demand rather than by preset, rigid clock rates. Synchronization of the flow across multi-stage sequences of activities is achieved by passing on demand signals from the end of the value chain in a "backwards", upstream direction from stage to stage, and by making sure that the demands signaled are met "just-in-time". The Toyota Production System allowed for quantum leaps in productivity, inventory reduction, and the ability to handle a significant degree of materials and product variety at the same time. It showed a way of how to replace the rigidity and complexity of centralized "synoptic" planning systems by "loosely coupled" sequences of simple loops of demand signals and just-in-time supply responses. They could be nested and changed in many ways (cf. the articles by Simon (1962) and Weick (1976) in the discussion of the instrumentation of logistics following below).

Following the discovery of the Toyota Production System in the West, the powerful idea of organizing industrial production logistics as demand-driven flow systems was continually being refined and adapted to other industries, as well as to new fields of application. Among the voices who contributed most to the diffusion of concepts of "flow" thinking and the Toyota Production System were Schonberger (1982), Goldratt (1984), Wildemann (1984),[14] Shingo (1986), and Womack/ Jones/ Roos (1990). Last but not least, it led to the recognition of the overwhelming importance of the dimension of time and of systematic management of time phenomena as a precondition to understanding and successfully managing flows. In this context a paper by George **STALK (1988)** entitled *Time – The Next Source of Competitive Advantage* is another milestone contribution.

Another closely-related development started in the 1980s as the horizon of logistics research was widened from an enterprise and "intra" logistics focus to a focus on relationships in cross-organizational chains and flow in networks between suppliers, manufacturers, distributors, and final customers. In this way the wave of supply chain management (SCM) discussions began. It expanded – and sometimes redirected – the focus of logisticians to search for new efficiencies, new sources of competitiveness, and to create more value to customers by primarily studying and re-designing flows at the inter-organizational level. First contributions marking this change were **OLIVER/WEBBER (1982)** and Houlihan (1985).

13 This is the subtitle of his 1988/1978 book.

14 Wildemann was the first to popularize the concept in German industry, helping to start the "just-in-time revolution" in European industry, too.

A trigger for the enormous momentum and popularity that the supply chain management wave gained in the mid-1990s were the sensational promises that the "ECR"-studies made after 1992.[15] In some countries and professional communities SCM even began to replace logistics as the denominator of the field and stimulated an ongoing discussion about which term should be considered the broader, overarching one, which one should be considered a sub-aspect.[16]

The fascination with SCM from the mid-1990s onwards extended and refocused research from intra-enterprise phenomena and aspects of procurement and distribution in the dyadic relationships between a focal enterprise and its immediate suppliers and customers. The SCM discussion motivated the community to address issues of the design, planning, operation, and control of value chains and networks that extend beyond individual enterprises and their immediate supplier and customer relationships. It may be critically noted, however, that often authors and practitioners are applying the new "SCM" terminology when they deal with nothing more than the long familiar narrower logistical issues of procurement in dyadic supplier-manufacturer relationships, or of producer-customer relationships in physical distribution. This leaves open the question of whether there is truly new content in their uses of the SCM terminology.

A focus on the facts and problems of flow systems and flow dynamics in the world of economic and business phenomena has not yet been systematically covered elsewhere. This is what supports the claim in this essay that the field has been evolving to become a largely autonomous and unique science of logistics. New research questions and new areas of application of logistics know-how are being discovered on the basis of the "flow systems" paradigm, opening up challenging perspectives for the future of the field.

---

[15] The ECR – Efficient Consumer Response" movement was triggered by the famous Kurt Salmon (1993) study on the potential impact of integrated supply chain Management in fast moving consumer goods supply chains.

[16] The influential American „Council of Logistics Management (CLM)" renamed themselves in 2005 to „Council of Supply Chain Management (CSCMP)", implying that "logistics" covers a subset of SCM issues only. Compare the recent discussion of this perspective in Frankel et al. (2008). In this paper "Logistics" and the "flow systems" paradigm are argued to be the more generic, broader concepts. SCM is regarded as one – if very important – field of application of logistical concepts and the flow systems paradigm.

IV. THE INSTRUMENTATION OF LOGISTICS

The story of the evolution of the science of logistics and its “flow system” paradigm, as it has been told so far, is not yet complete. As Kuhn (1970) suggested, a “normal” field of science is not only defined by focus on “certain aspects of nature”, and a “set of shared questions, facts and problems” within the chosen range of observations. It is also characterized by the set of instruments which are commonly being used by the members of the scientific community: the shared use of models, terminologies, techniques, and rules to derive hypotheses and prescriptions. Scientific progress, consequently, does not only occur by changes in paradigms, but – probably much more often – by the gradual expansion and development of the diagnostic and therapeutic instruments in use.

To complement the search for the “roots of logistics”, which is conducted in this essay, a second story needs to be told, therefore: The “instrumentation” story about the step by step adaptation of certain models, terminologies, techniques, and rules supporting the logistics scientists work in the field. A selection of contributions in the instrumentation of modern “flow system” logistics is discussed in the following sections of the essay.

1. Contributions from the Quantitative and Engineering Sciences

The need for structured and preferably quantitative scientific analysis of logistical problems was recognized very early on, and was applied to transportation, location, inventory keeping, and other questions long before integrated studies of logistical problems began (see EXHIBIT 2).

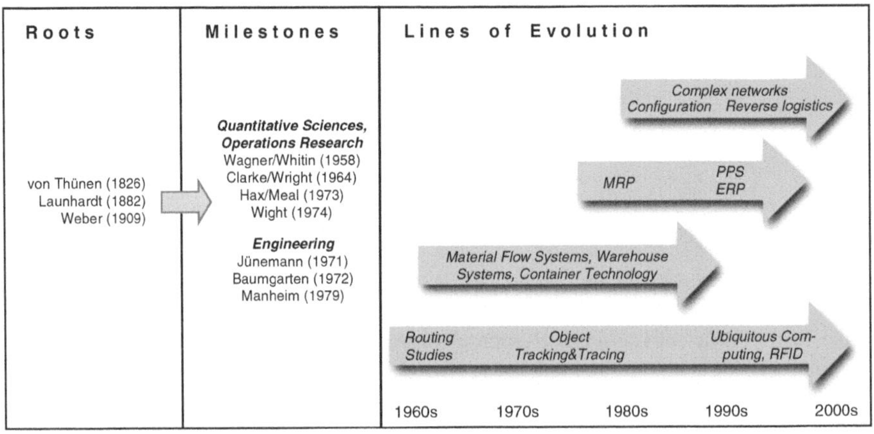

EXHIBIT 2: MODELING, OPTIMIZING AND THE CONTRIBUTIONS FROM QUANTITATIVE AND ENGINEERING SCIENCES

Some of this early work addresses geographical aspects of logistical systems design and optimization. Its academic roots date back to the 19th century. There is the famous model of the "Isolated State" by the economist von Thünen (1826) and an early approach to quantitative location analysis by **Launhardt (1882)** on *The Determination of the Most Expedient Location of an Industrial Plant.* Not to be forgotten is Alfred Weber's (1909) seminal work on the location of industries.

The management of flows through networks also requires decisions to be made about the choice of alternative routes which can be traveled to link locations, of the optimal assignment and sequencing of given orders (cargoes, objects) to vehicles, the determination of capacities, and more. There is a long, successful, still expanding line of Operations Research work about vehicle routing, scheduling and an innumerable number of variations of those problems, which has been enriching and empowering logistics. A milestone contribution and proxy for this line of research is **Clarke/Wright's (1964)** classic paper on *Scheduling of Vehicles from a Central Depot to a Number of Delivery Points.*

Still another important line of Operations Research work is concerned with time and object-quantity dimensions of the objects flowing through logistical networks: Determining optimal order sizes, lot sizes, inventory levels and related scheduling arrangements in production and distribution. Pioneering contributions in this area were Magee's (1958) and Wagner/Whitin's (1958) work on optimal ordering and inventory decisions, Wight's (1974) work about Materials Requirement Planning (MRP), and **Hax/Meal's (1973)** classic on *Hierarchical Integration of Production Planning and Scheduling.* The latter represents this line of research and the advanced computer based "production planning and control" applications which since became standard in many industries. Hax/Meal's work was followed by ever more integrated and sophisticated contributions, seeking the simultaneous quantitative optimization of geographical, time, quantity and monetary aspects of planning through concepts such as Manufacturing Resource Planning (MRP II) and Enterprise Resource Planning (ERP) systems developments (e.g. Wallace (1985)).

Recently the integration of the additional issues of "reverse logistics" and risk considerations is adding a further level of complexity to the modelling and optimization challenges in logistics, which also lead to the exploration and development of non-traditional modelling and mathematical solution approaches (e.g. the ant algorithm by Dorigo/Stützle (2004)).

Another important source for the instrumentation of logistics has been in the Engineering Sciences, i.e. the development of "hard" systems technologies to support the mechanization of logistical operations, such as intra-plant movement of materials, storage and packaging automation. Efforts to advance the integration between "hard" Mechanical Engineering and Business Logistics research appear to have been particularly active in Germany.

One of the pioneers on the Engineering side is Reinhardt Jünemann, a mechanical engineer by education, who wrote his dissertation research in the engineering of warehouse operations in the early 1970s (Jünemann (1971)) and later founded one of the largest research facilities for the engineering of logistics systems.[17] This line of work expanded to industrial materials flow equipment, picking-packing, and con-

[17] I.e. the Fraunhofer Institute for Materialfluss und Logistik (IML) at Dortmund, Germany.

tainer hardware systems, the integration of the technologies of Operations Research, computer-assisted planning, controlling, etc. (Jünemann (1989)). Parallel initiatives, also including the first creation of a dedicated academic program that jointly teaches know-how of Engineering and Management ("Wirtschaftsingenieur"), were taken by Helmut Baumgarten (1972) at the Technical University of Berlin. Baumgarten did early work about standardized container systems. He became one of the founders of the German logistics association BVL in 1978, which from its beginning followed a mission of integrating the disciplines of Business Administration and Engineering in the field of logistics.

Successful contributions to merging the research and development approaches and methods between engineers, systems scientists, and logisticians are also found in the United States and in other parts of the world.[18]

During the latest one-and-one-half decades the most important Engineering contribution was achieved through the dramatic growth – first of barcode-based tracking and tracing of objects – then of the Internet and Radio Frequency Identification (RFID) technologies and the new opportunities they are creating for distributed, low-cost communications and computing, decentralized, mobile tracing of moving objects and new options for the control of those (e.g. Fleisch/Mattern (2005)).

2. Contributions from the Management Sciences

From the 1970s and 1980s onward, when the consolidation and institutionalization of logistics in functional departments and the professionalization and concentration of "Third Party" transport and logistics service provider organizations rapidly advanced, the challenges and potentials of their perpetual rationalization became a new concern (see EXHIBIT 3).

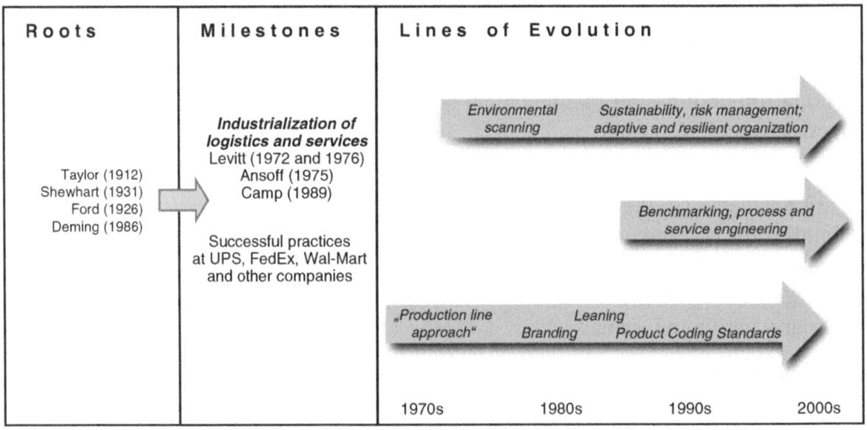

EXHIBIT 3: INDUSTRIALIZATION, RISK MANAGEMENT, AND THE CONTRIBUTION FROM MANAGEMENT SCIENCES

[18] See, e.g., Forrester (1958) and Manheim (1979).

The concepts and tools of the "industrialization" of logistics services were gradually being discovered and adopted among leading logistics organizations, marking another milestone in the instrumentation of modern logistics. Stimulation for this powerful development primarily came from "best business practices" of successful service and retail companies. The person who first drew attention to the idea and potentials of the "industrialization of services" in the academic world was Theodore **LEVITT (1972)** with his paper of the same title, and a much-quoted follow-up paper (Levitt (1976)).

The discussion about industrialization of the economic process started, however, at a time already several decades before Levitt's works. Among the academic predecessors who paved the way for the industrialization idea in the Scientific Management literature of the first half of the 20th century werc Taylor (1912), Ford (1926), Shewhart (1931), and – a few decades later – Deming (1986) with their classical contributions. Their themes were the realization of economies of scale through product standardization and production-lines, statistical quality controls, brand-name mass-marketing, and the rationalization and benchmarking (Camp (1989)) of process structures.

Industrialization concepts are making use of the full range of Operations Research instruments which were discussed in the preceding Quantitative Sciences" instrumentation section.

As a final note to this section, it is also worth mentioning one additional important early instrument and classical contribution from the field of General Management studies. As we are entering the "age of volatility", Igor **ANSOFF'S (1976)** article on *Managing Surprise and Discontinuity: Strategic Response to Weak Signals* is about to gain renewed relevance. Most current contributions to logistics are around the issues of supply chain risk managment and resilience (e.g. Sheffi (2005)).

3. Contributions from Economics and the Organizational Sciences

As the focus of logistics has been shifting towards the challenges of the integrated design of value chains, business processes, and cross-organizational supply chains and networks in turbulent, volatile environments, the issues of appropriate architectures and coordination mechanisms for very complex flow structures are becoming most critical concerns. For one, there is more recognition being given to the fact that successful flow management cannot restrict itself to the efficient planning, mobilization, and control of physical materials and information flows alone. There are intricate additional factors associated with interpersonal relationships, contractual arrangements, of the coordination and governance of the actors and activities involved in the network of supply chain relationships. Otto (2002) summarized research on the multiple levels of relationships that advanced supply chain management must consider in a model which identifies four layers of those relationships: the familiar "materials flow" and "information flow", moreover "social relationships", and finally "institutional", i.e. contractual, formal-organizational arrangements. Recent discussion suggests that a fifth network layer of money or value flows ought to be considered (e.g. Pfohl/Hofmann/Elbert (2003)).

This leads to needs for the adaptation of analytical instruments and concepts from the fields of Systems Theory, Organization Theory and the New Institutional Economics, such as:

- the "markets and hierarchies" and "clans" terminologies and concepts to choose and allocate appropriate coordination mechanisms in complex process and supply chain relationship networks (see **COASE (1937)** and **OUCHI (1980)**),
- finding the appropriate degree of complexity that can be handled within "tightly coupled", hierarchically controlled organizational units (e.g. Hagel (1989)),
- the identification of the best combinations of "loose" and "tight" coupling mechanisms between network actors and units when high degrees of uncertainty and environmental turbulence need to be accommodated (compare **SIMON'S (1962)** and **WEICK'S (1976)** pioneering works),

and also

- applying the right incentives and controls in "principal-agent" relationship networks where information is unevenly distributed between the participants in a "team" effort (Alchian/Demsetz (1972)), and
- considering the "externalities" in any complex decision about systems architectures (**AYRES/KNEESE (1969)**).

The concern with cross-organizational integration in logistical chains and networks, and the adaption of the "market and hierarchy", "transaction cost" "principal-agent" "loose-tight coupling" and related terminologies and concepts referred to have tremendously enriched the instrument set available to logisticians (see EXHIBIT 4) – and offer many more perspectives for future logistics and supply chain management research.

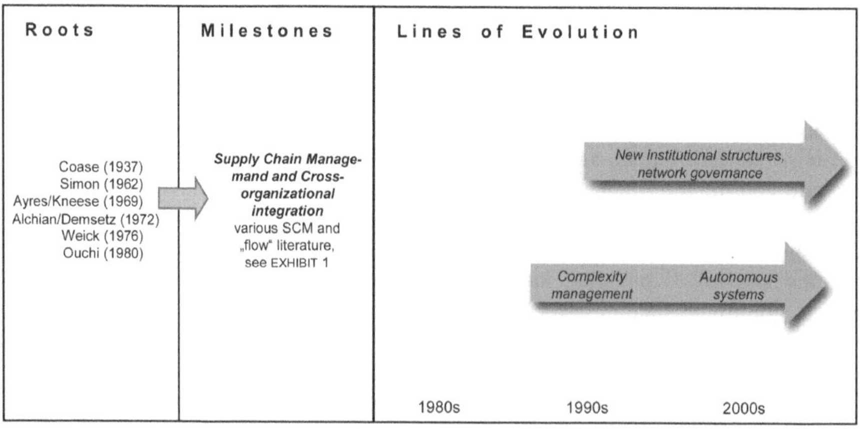

EXHIBIT 4: OPTIMAL ARCHITECTURES, GOVERNANCE, COMPLEXITY MANAGEMENT AND THE CONTRIBUTIONS FROM ECONOMICS AND ORGANIZATIONAL SCIENCES

VI. AFTERTHOUGHTS: CURRENT FRONTIERS OF LOGISTICS RESEARCH AND THE QUESTION OF WHERE TO GO NEXT?

Referring back, once again, to Kuhn's (1970) points about "what makes a science", the gradual convergence of the paths of conceptual and instrumentation developments, which was outlined in the preceding discussions and illustrated in EXHIBIT 5 seems to support the claim made in this essay: There is now a "science of logistics".

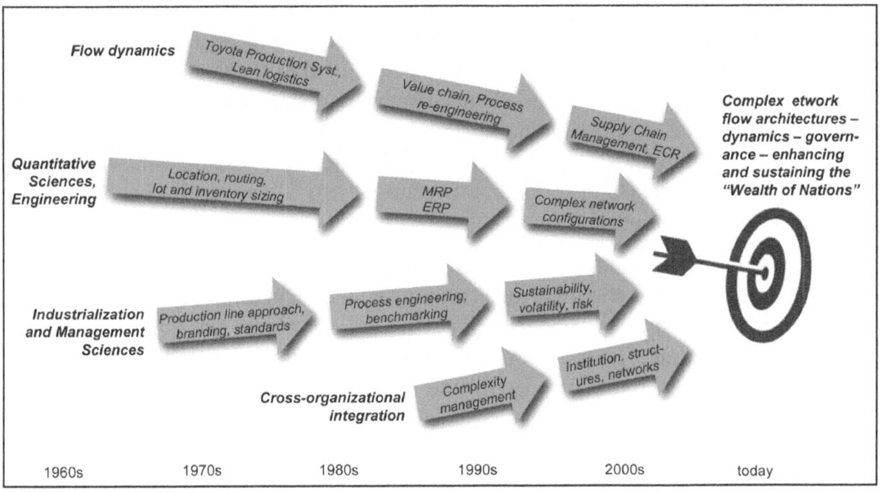

EXHIBIT 5: CONVERGING PATHS – THE SCIENCE OF LOGISTICS

There is a shared set of problems and approaches to solve these problems; there are common metaphors, models, techniques, and rules. Although some of them originate from other disciplines they today form part of the logistic "terminology" and the logistics "toolset".

This identity of logistics as a field of science is also reflected in a recent definition of the "science of logistics" that was developed jointly by members of the scientific advisory board of BVL, the German logistics association. It also seems to be a good description of the point of convergence of those decades of conceptual and instrumentation developments which were retraced in this essay: "Logistics is an applications-oriented scientific discipline. It models and analyses economic systems as networks and flows of objects through time and space (specifically goods, information, money, and people) which create value for people. It aims to supply recommendations for action on the design, implementation, and mobilization of such networks and flows through accepted scientific methods. Scientific questions of the discipline are related primarily to the configuration and organization of these networks, and to the mobilization and control of flows. The ultimate goal of logistics science is progress in the balanced achievement of economic, ecological and social objectives" (cf. Delfmann et al. (2010)).

This essay shall be concluded by reflecting on current frontiers and future research challenges in the "science of logistics". Some of these reflections may be deduced directly from the retrospective description of the paths and traces of the field's evolution.

Starting with "instrumentation", it is obvious that the amount of research that has been done in the development of models and tools to support logistics based on Mathematics, IT, and Engineering by far outweighs the research in the adaptation of "softer" concepts, models and methods from the Organization and Behavioural Sciences. The adaptation of know-how from the areas of Law, the Political Sciences has hardly begun, which are likely to have interesting potential applications to issues of appropriate contractual, institutional, international and intercultural arrangement in ever more complex networks of actors and stakeholders.

As the "flow" metaphor and flow systems paradigm will be leveraged more in the future, new areas of application for continuing conceptual refinement are opening up: The traditional focus on physical materials, goods, and on information as the primary flow objects has already started to be expanded to financial and value flows, to flows of ideas, energy, and knowledge. With the transformation of the maturing industrial economies to post-industrial, ever more service-based economies, there will be growing demands and new opportunities for the logistics of services, of know-how development and transfer, of smart energy flows, and other non-traditional areas of applications.

New promising uses of the flow metaphor, last but not least, will be offered by elevating the perspective from the traditional intra-organizational and cross-organizational supply chain focus to a still wider focus at the levels of national and international economic cooperation. The modelling of macro-flows of goods, money, and people, as certain resources such as crude oil and water become scarce and comparative advantages between nations change, will became critical to forecasting, and to the intelligent design of new types of network structures under conditions of ever more turbulent, volatile natural, economic, and political environments.

This very brief list of potential future developments of logistics is by no means meant to be exhaustive. However it may illustrate the enormous promise and potential of the science of logistics in the future.

REFERENCES

Alchian/Demsetz (1972): Alchian, Armen A. and Harold Demsetz: Production, Information Costs, and Economic Organization. In: *The American Economic Review*, Vol. 62, No. 5, 1972, pp. 777–795.

Ansoff (1975): Ansoff, H. Igor: Managing Strategic Surprise by Response to Weak Signals. In: *California Management Review*, Vol. XVIII, No. 2, 1975, pp. 21–33.

Ayres/Kneese (1969): Ayres, Robert U. and Allen V. Kneese: Production, Consumption, and Externalities. In: *The American Economic Review*, Vol. 59, No. 3, 1969, pp. 282–297.

Ballou (2007): Ballou, Ronald H.: The Evolution and Future of Logistics and Supply Chain Management. In: *European Business Review*, Vol. 19, No. 4, 2007, pp. 332–348.

Baumgarten (1972): Baumgarten, Helmut: *Über technische und organisatorische Möglichkeiten zur Anpassung der Industriebetriebe an das Container-System*. Berlin: Ladewig, 1972.

Baumgarten (2008): Baumgarten, Helmut: Das Beste in der Logistik – Auf dem Weg zu logistischer Exzellenz. In: Baumgarten, Helmut (ed.): *Das Beste in der Logistik. Innovationen, Strategien, Umsetzungen*. Berlin; Heidelberg: Springer, 2008, pp 13–19.

Buijtenen/Christopher/Wills (1976): Buijtenen, P.M. van, Martin Christopher, and Gordon Wills: *Business Logistics*. The Hague: Nijhoff, 1976.

Camp (1989): Camp, Robert C.: *Benchmarking: The Search for Industry Best Practices that Lead to Superior Performance*. Milwaukee/WI: ASQC Quality Press, 1989.

Charvet/Cooper/Gardner (2008): Charvet, Francois F., Martha C. Cooper, and John T. Gardner: The Intellectual Structure of Supply Chain Management: A Bibliographic Approach. In: *Journal of Business Logistics*, Vol. 29, No. 1, pp. 47–73.

Clarke/Wright (1964): Clarke, G. and J.W. Wright: Scheduling of Vehicles from a Central Depot to a Number of Delivery Points. In: *Operations Research*, Vol. 12, 1964, pp. 568–581.

Coase (1937): Coase, Ronald H.: The Nature of the Firm. In: *Economica*, Vol. 4, No. 16, 1937, pp. 386–405.

Converse (1921): Converse, Paul D.: *Marketing: Methods and Policies*. New York: Prentice-Hall, 1921.

Converse (1954): Converse, Paul D.: The Other Half of Marketing. *Proceedings of the 26th Annual Boston Conference on Distribution*, 1954, pp. 310–314.

Cooper/Lamberg/Pagh (1997): Cooper, Martha C., Douglas M. Lambert, and Janus D. Pagh: Supply Chain Management: More than a Name for Logistics. In: *International Journal of Logistics Management*, Vol. 8, No. 1, 1997, pp. 1–14.

Davis-Sramek/Fugate (2007): Davis-Sramek, Beth and Brian S. Fugate: State of Logistics: A Visionary Perspective. In: *Journal of Business Logistics*, Vol. 28, No. 2, pp. 1–33.

Delfmann et al. (2010): Delfmann, Werner et al.: Towards a Science of Logistics: Cornerstones of a Framework of Understanding of Logistics as an Academic Discipline. In: *Logistics Research*, Vol 2, No. 2, 2010, pp. 57–63.

Deming (1986): Deming, W. Edwards: *Out of the Crisis*. Cambridge/MA: MIT Press, 1986.

Diederich (1986): Diederich, Helmut: Entwicklung und Stand der Verkehrsbetriebslehre, *Zeitschrift für Betriebswirtschaft*, Vol. 56., No.1, 1986, pp. 51–88.

Dorigo/Stützle (2004): Dorigo, Marco and Thomas Stützle: *Ant Colony Optimization*. Cambridge/Ma.: MIT-Press, 2004.

Drucker (1962): Drucker, Peter F.: The Economy's Dark Continent. In: *Fortune*, Vol. LXV, No. 4, April 1962, pp. 265–270.

Eccles (1954): Eccles, Henry E.: Logistics – What is it? In: *Naval Research Logistics Quarterly*, Vol. 1, No. 1, 1954, pp. 5–15.

Farris (1997): Farris, Martin T.: Evolution of Academic Concerns with Transportation and Logistics. In: *Transportation Journal*, Vol. 37 No. 1, 1997, pp. 42–55.

Fleisch/Mattern (2005): Fleisch, Elgar and Friedemann Mattern: Das Internet der Dinge. Ubiquitous Computing und RFID in der Praxis. Berlin, Heidelberg: Springer, 2005.

Ford (1926): Ford, Henry: *Today and Tomorrow*. Garden City: Doubleday, 1926.

Forrester (1958): Forrester, Jay W.: Industrial Dynamics – A Major Breakthrough for Decision Makers. In: *Harvard Business Review*, Vol. 36, July-August 1958, pp. 37–66.

Frankel et al. (2008): Frankel, Robert, Yemisi A. Bolumole, Reham A. Eltantawy Antony Paulraj, and Gregory T. Gundlach: The Domain and Scope of SCM's Foundational Disciplines – Insights and Issues to Advance Research. *Journal of Business Logistics*, Vol. 29, No. 1, pp. 1–30.

Goldratt (1984): Goldratt, Eliyahu M.: *The Goal. A Process of Ongoing Improvement*. Croton-on-Hudson: North River Press, 1984.

Göpfert (1996): Göpfert, Ingrid: *Stand und Entwicklung der Logistik*, Arbeitspapier Nr. 5 des Lehrstuhls für Betriebswirtschaftslehre und Logistik, Philips Universität Marburg. Marburg, 1996.

Hagel (1989): Hagel III, John: Managing Complexity. In: *McKinsey Quarterly*, Spring 1989, pp. 2–23.

Hammer (1990): Hammer, Michael: Don't Automate, Obliterate! In: *Harvard Business Review*, July-August 1990, pp. 104–112.

Hax/Meal (1973): Hax, Arnoldo C. and Harlan C. Meal: *Hierarchical Integration of Production Planning and Scheduling*. Published as Sloan School of Management Working Paper No. 656–73, Massachusetts Institute of Technology (MIT), 1973, pp. 1–25.

Heskett/Ivie/Glaskowsky (1964): Heskett, John L., Robert M. Ivie, and Nicholas A. Glaskowsky jr.: *Business Logistics*. New York, London, Sydney: Macmillan, 1964.

Hopp (2008): Hopp, Wallace J.: *Supply Chain Science*. New York: McGraw-Hill, 2008.

Houlihan (1985): Houlihan, John B.: International Supply Chain Management. In: *International Journal of Physical Distribution and Logistics Management*, Vol. 15, No. 1, pp. 22–38.

Ihde (1972): Ihde, Gösta B.: Zur Behandlung logistischer Phänomene in der neueren Betriebswirtschaftslehre. *Betriebswirtschaftliche Forschung und Praxis*, Vol. 24, March 1972, pp. 129–145.

Ihde (1984): Ihde, Gösta B.: *Transport, Verkehr, Logistik*, Munich: Vahlen, 1984.

Jomini (1838) : Jomini, Antoine-Henri: *Précis de l'Art de la Guerre*: Des Principales Combinaisons de la Stratégie, de la Grande Tactique et de la Politique Militaire. Brussels: Meline, Cans et Copagnie, 1838. English translation as: Jomini, Antoine-Henri: *The Art of War*. Philadelphia: Lippincott, 1862 (translated from the French by Capt. G.H. Mendell and Lieut. W.P. Craighill).

Jünemann (1971): Jünemann, Reinhardt: *Systemplanung für Stückgutläger*. Mainz: Krausskopf, 1971.

Jünemann (1989): Jünemann, Reinhardt: *Materialfluss und Logistik*, Berlin; Heidelberg; New York: Springer, 1989.

Keith (1960): Keith, Robert J.: The Marketing Revolution. In: *Journal of Marketing*, Vol. 24, No. 1, pp. 35–38.

Kent/Flint (1997): Kent, John L. jr. and Daniel J. Flint: Perspectives on the Evolution of Logistics Thought. In: *Journal of Business Logistics*, Vol. 18, No. 2, pp. 5–29.

Kirsch (1971): Kirsch, Werner: Betriebswirtschaftliche Logistik. In: *Zeitschrift für Betriebswirtschaft*, Vol. 41, No. 4, April 1971, pp. 221–234.

Klaus (1993): Klaus, Peter: *Die Dritte Bedeutung der Logistik*, Logistics Working Papers No. 1, Friedrich-Alexander-University, Nuremberg 1993)

Klaus (2009): Klaus, Peter: Logistics Research: A 50 Years' March of Ideas. In: *Logistics Research*, Vol. 1, No. 1, pp. 53–65.

Klaus/Sheffi (1997): Klaus, Peter and Yossi Sheffi: Logistics at Large: Jumping the Barriers of the Logistics Function. In: *Proceeding of the Council of Logistics Management's 26th Annual Transportation and Logistics Educators Conference*, Chicago/IL Oct. 5th, 1997.

Krampe/Lucke/Lüsch (1990): Krampe, Horst, Hans J. Lucke, and Jürgen Lüsch: Ist Logistik eine Wissenschaft? In: *DDR-Verkehr*, Vol. 23, No. 2, pp. 44–46.

Kurt Salmon (1993): Kurt Salmon Associates: *Efficient Consumer Response: Enhancing Consumer Value in the Grocery Industry*. Washington D.C.: Food Marketing Institute, 1993.

Kuhn (1970): Kuhn, Thomas S.: *The Structure of Scientific Revolutions*. 2nd ed. Chicago: University of Chicago Press, 1970.

Launhardt (1882): Launhardt, Wilhelm: Die Bestimmung des zweckmäßigsten Standortes einer gewerblichen Anlage. In: *Zeitschrift des Vereines Deutscher Ingenieure*, Vol. XXVI, No. 3, March 1882, pp. 105–116.

Levitt (1972): Levitt, Theodore: Production-line Approach to Service. In: *Harvard Business Review*, Vol. 50, October-November 1972, pp. 41–52.

Levitt (1976): Levitt, Theodore: The Industrialization of Service. In: *Harvard Business Review*, September-October 1976, pp.63–71.

Magee (1958): Magee, John F.: *Production Planning and Inventory Control*. New York: McGraw-Hill, 1958.

Magee (1960): Magee, John F.: The Logistics of Distribution. In: *Harvard Business Review*, July-August 1960, pp. 89–101.

Manheim (1979): Manheim, Marvin L.: *Fundamentals of Transportation System Analysis* Cambridge/MA: MIT-Press, 1979.

Marshall (1890): Marshall, Alfred.: *Principles of Economics*, Vol. 1, London; New York: Macmillan, 1890, pp. 116f.

McManus/Millard (2002): McManus, Hugh L. and Richard L. Millard: Value Stream Analysis and Mapping for Product Development. In: *Proceedings of the International Council of the Aeronautics Sciences*, 23rd ICAS Congress, 8–13 September, Toronto, 2002.

Morgenstern (1955): Morgenstern, Oskar: Note on the Formulation of the Theory of Logistics. In: *Naval Research Logistics Quarterly*, Vol. 2, No. 3, 1955, pp. 129–136.

Nordsieck (1934): Fritz Nordsieck: *Grundlagen der Organisationslehre*, C.E. Poeschel, Stuttgart, 1934.

Novack/Rinehart/Wells (1992): Novack, Robert A.., Lloyd M. Rinehart, and Michael V. Wells: *Rethinking Concept Foundations in Logistics Management*, Journal of Business Logistics, Vol. 13, No. 2, 1992, pp. 233–267.

Ohno (1988): Ohno, Taiichi: *Toyota Production System: Beyond Large-Scale Production*. Cambridge/MA: Productivity Press, 1988 (original ed. *Toyota Seisan Hoshiki*, Tokyo, 1978).

Oliver/Webber (1982): Oliver, R. Keith and Michael D. Webber: Supply-Chain Management: Logistics Catches up with Strategy. In: *Outlook*; Booz, Allen and Hamilton, 1982.

Otto (2002): Otto, Andreas: *Management und Controlling von Supply Chains. Ein Modell auf der Basis der Netzwerktheorie*. Wiesbaden: Deutscher Universitäts-Verlag, 2002.

Ouchi (1980): Ouchi, William G.: Markets, Bureaucracies, and Clans. In: *Administrative Science Quarterly*, Vol. 25, March 1980, pp. 129–141.

Pfohl (1969): Pfohl, Hans-Christian: Alles für den Nachschub. In: *Der Volkswirt. Wirtschafts- und Finanzzeitung*, special reprod. from No. 17, 25th April, 1969, pp. 159–161.

Pfohl (1972): Pfohl, Hans-Christian: *Marketing Logistik. Gestaltung, Steuerung und Kontrolle des Warenflusses im modernem Markt*, Darmstadt: Distribution Verlag, 1972.

Pfohl/Hofmann/Elbert (2003): Pfohl, Hans-Christian, Erik Hofmann, and Ralf Elbert: Financial Supply Chain Management: Neue Herausforderungen für die Finanz- und Logistikwelt. In: *Logistik Management*, Vol. 5, No. 4, pp. 10–26.

Plowman (1964): E. Grosvenor Plowman: *Lectures on Elements of Business Logistics*, Stanford Transportation Series, private manuscript, 1964.

Porter (1985): Porter, Michael E.: *Competitive Advantage*. New York: Free Press, 1985.

Quesnay (1764): Quesnay, François: *Tableau Économique*, 3rd Edition, Ex Libris Petri Du Pont, Agriculturæ Regiæ, Societatis Suessionensia, 1764.

Rheinberger (2007): Rheinberger, Hans-Jörg: *Historische Epistomologie*, Hamburg: Junius, 2007.

Schonberger (1982): Schonberger, Richard: *Japanese Manufacturing Techniques: Nine Hidden Lessons in Simplicity*. New York: Free Press, 1982.

Shaw (1912): Shaw, Arch W.: Some Problems in Market Distribution. In: *Quarterly Journal of Economics*, August 1912, pp. 703–765.

Sheffi (2005): Sheffi, Yossi: *The Resilient Enterprise*, Cambridge/MA: MIT-Press, 2005.

Shewhart (1931): Shewhart, Walter A.: *Economic Control of Quality of Manufactured Product*. New York: Van Nostrand, 1931.

Shingo (1986): Shingo, Shigeo: *Revolution in Manufacturing: The SMED System*. Cambridge/MA: Productivity Press, 1986.

Simon (1962): Herbert A. Simon: The Architecture of Complexity. In: *Proceedings of the American Philosophical Society*, Vol. 106, 1962, pp. 62–76.

Smith (1776): Smith, Adam: *An Inquiry into the Nature and Causes of the Wealth of Nations*, London: Methuen, 1776.

Smykay/Bowersox/Mossman (1961): Smykay, Edward W., Donald L. Bowersox, and F.H. Mossman: *Physical Distribution Management*. New York: Macmillan, 1961.

Stalk (1988): Stalk, George: Time – The Next Source of Competitive Advantage. In: *Harvard Business Review*, Vol. 65, July-August 1988, pp. 41–51.

Taylor (1912): Taylor, Frederick W.: *The Principles of Scientific Management*. New York: Harper & Bros., 1912.

Thorpe (1917): Thorpe, George C.: *Pure Logistics; the Science of War Preparation.* Kansas City: Franklin Hudson, 1917.

Thünen (1826): Thünen, Heinrich von: *Der isolirte Staat.* Hamburg: Perthes, 1826.

Voigt (1953): Voigt, Fritz: Verkehr und Industrialisierung. In: *Zeitschrift für die gesamte Staatswissenschaft*, 109. Band, J.C.B. Mohr, Tübingen, 1953, pp. 193–239.

Wagner/Whitin (1958): Wagner, Harvey M. and Thomas M. Whitin: Dynamic Version of the Economic Lot Size Model. In: *Management Science*, Vol. 5, 1958, pp. 89–96.

Wallace (1985): Wallace, Thomas F.: *MRP II: Making it Happen.* Essex/NH: Oliver Wight, 1985.

Weber (1909): Weber, Alfred: *Über den Standort der Industrien. Reine Theorie des Standorts.* Tübingen: Mohr, 1909.

Weber (1992): Weber, Jürgen: Logistik als Koordinationsfunktion: Zur theoretischen Fundierung der Logistik. In: *Zeitschrift für Betriebswirtschaft*, Vol. 62, No. 8, 1992, pp. 877–895.

Weick (1976): Weick, Karl E.: Educational Organizations as Loosely Coupled Systems. In: *Administrative Science Quarterly*, Vol. 21, March 1976, pp. 1–19.

Weld (1919): Weld, L.D.H.: *The Marketing of Farm Products*, New York: Macmillan, 1919.

Wight (1974): Wight, Oliver W.: *Production and Inventory Management in the Computer Age.* Boston: CBI Publishing, 1974.

Wildemann (1984): Wildemann, Horst: Kanban-Systeme. In: *Die Betriebswirtschaft*, Vol. 44, No. 2, 1984, pp. 313–314.

Womack/Jones/Roos (1990): Womack, James P., Daniel T. Jones, and Daniel Roos: *The Machine That Changed the World.* New York: Scribner, 1990.

PART A: ON THE EARLY TERMINOLOGICAL AND CONCEPTUAL ROOTS OF LOGISTICS

Antoine-Henri Jomini (1838/1862)

THE ART OF WAR

It was the French general Antoine-Henri Jomini (1779–1869) who is credited with the first documented elaboration of an idea of logistics.

Jomini, born in Western Switzerland, served as a general in the French, later in the Russian army. Through his works on strategy and the art of war he became (along with the German general Carl von Clausewitz) one of the most celebrated military theorists in the Napoleonic era. In 1838 he published his 400+ page grand work on *Précis de l'Art de la Guerre:* Des Principales Combinaisons de la Stratégie, de la Grande Tactique et de la Politique Militaire (English translation as *The Art of War*).

Chapter VI of the book, under the title *Logistics, or the Practical Art of Moving Armies*, contains two articles about logistics, of which the first one – article 41, *A Few Remarks on Logistics in General* – is reprinted here. Jomini, in this article, first refers to earlier uses of the term logistics in the military as a "science of the details of staff duties" of army quartermasters (in French "major général des logis"). Quartermasters were "the officers whose duty it formerly was to lodge and camp the troops". With this reference the primary etymological root of the term logistics was established in the French word "loger" (standing for the accommodation of guests or soldiers) and the Latin "locare".

But then Jomini goes on to discuss the added complexity of the logistics for moving armies – war without camps. In his extended "new" definition of logistics he lists 18 principal duties, which include "the preparation of all the material necessary for setting the army in motion", references to the requirements of advance planning and two-way information processes between logisticians and the actual operations of marching and fighting troops in the field.

With ideas like these Jomini anticipated quite a few of the modern elements of logistics, such as the need to deal with timing, information exchange, and planning challenges in dynamically changing operational environments ("calculation of times and distances" under varying conditions) or the needs to set up appropriate supply networks ("becomes... necessary to have a good line of operations and depots").

Reference:

Antoine-Henri Jomini: *Précis de l'Art de la Guerre*: Des Principales Combinaisons de la Stratégie, de la Grande Tactique et de la Politique Militaire. Brussels: Meline, Cans et Copagnie, 1838. English translation as: Jomini, Antoine-Henri: *The Art of War*. Philadelphia: Lippincott, 1862 (translated from the French by Capt. G.H. Mendell and Lieut. W.P. Craighill).

P. Klaus and S. Müller (eds.), *The Roots of Logistics*, DOI 10.1007/978-3-642-27922-5_2,

Antoine-Henri Jomini: The Art of War

Article 41.

A FEW REMARKS ON LOGISTICS IN GENERAL

Is logistics simply a science of detail? Or, on the contrary, is it a general science, forming one of the most essential parts of the art of war? Or is it but a term, consecrated by long use, intended to designate collectively the different branches of staff duty, that is to say, the different means of carrying out in practice the theoretical combinations of the art?

These questions will seem singular to those persons who are firmly convinced that nothing more remains to be said about the art of war, and believe it wrong to search out new definitions where every thing seems already accurately classified. For my own part, I am persuaded that good definitions lead to clear ideas; and I acknowledge some embarrassment in answering these questions which seem so simple.

In the earlier editions of this work I followed the example of other military writers, and called by the name of **logistics** the details of staff duties, which are the subject of regulations for field-service and of special instructions relating to the corps of quartermasters. This was the result of prejudices consecrated by time. The word "logistics" is derived, as we know, from the title of the **major général des logis** (translated in English by "Quartermaster"), an officer whose duty it formerly was to lodge and camp the troops, to give direction to the marches of columns, and to locate them upon the ground. Logistics was then quite limited. But when war began to be waged without camps, movements became more complicated, and the staff officers had more extended functions. The chief of staff began to perform the duty of transmitting the conceptions of the general to the most distant points of the theater of war, and of procuring for him the necessary documents for arranging plans of operations. The chief of staff was called to the assistance of the general in arranging his plans, to give information of them to subordinates in orders and instructions, to explain them and to supervise their execution both in their **ensemble** and in their minute details: his duties were, therefore, evidently connected with all the operations of a campaign.

To be a good chief of staff, it became in this way necessary that a man should be acquainted with all the various branches of the art of war. If the term **logistics** includes all this, the two works of the Archduke Charles, the voluminous treatises of Guibert, Laroche-Aymon, Bousmard, and Ternay, all taken together, would hardly give even an incomplete sketch of what logistics is; for it would be nothing more nor less than the science of applying all possible military knowledge.

It appears from what has been said that the old term **logistics** is insufficient to designate the duties of staff officers, and that the real duties of a corps of such officers, if an attempt be made to instruct them in a proper manner for their performance, should be accurately prescribed by special regulations in accordance with the general principles of the art. Governments should take the precaution to publish well-considered regulations, which should define all the duties of staff officers and should give clear and accurate instructions as to the best methods of performing these duties.

The Austrian staff formerly had such a code of regulations for their government; but it was somewhat behind the times, and was better adapted to the old methods of carrying on war than the present. This is the only work of the kind I have seen. There are, no doubt, others, both public and secret; but I have no knowledge of their existence. Several generals – as, for instance, Grimoard and Thiebaut – have prepared manuals for staff officers, and the new royal corps of France has issued several partial sets of instructions; but there is nowhere to be found a complete manual on the subject.

If it is agreed that the old **logistics** had reference only to details of marches and camps, and, moreover, that the functions of staff officers at the present day are intimately connected with the most important strategical combinations, it must be admitted that logistics includes but a small part of the duties of staff officers; and if we retain the term we must understand it to be greatly extended and developed in signification, so as to embrace not only the duties of ordinary staff officers, but of generals-in-chief.

To convince my readers of this fact, I will mention the principal points that must be included, if we wish to embrace in one view every duty and detail relating to the movements of armies and the undertakings resulting from such movements:

1. The preparation of all the material necessary for setting the army in motion, or, in other words, for opening the campaign. Drawing up orders, instructions, and itineraries for the assemblage of the army and its subsequent launching upon its theater of operations.
2. Drawing up in a proper manner the orders of the general-in-chief for different enterprises, as well as plans of attack in expected battles.
3. Arranging with the chiefs of engineers and artillery the measures to be taken for the security of the posts which are to be used as depots, as well as those to be fortified in order to facilitate the operations of the army.
4. Ordering and directing reconnoissances of every kind, and procuring in this way, and by using spies, as exact information as possible of the positions and movements of the enemy.
5. Taking every precaution for the proper execution of movements ordered by the general. Arranging the march of the different columns, so that all may move in an orderly and connected manner. Ascertaining certainly that the means requisite for the ease and safety of marches are prepared. Regulating the manner and time of halts.
6. Giving proper composition to advanced guards, rear-guards, flankers, and all detached bodies, and preparing good instructions for their guidance. Providing all the means necessary for the performance of their duties.

7. Prescribing forms and instructions for subordinate commanders or their staff officers, relative to the different methods of drawing up the troops in columns when the enemy is at hand, as well as their formation in the most appropriate manner when the army is to engage in battle, according to the nature of the ground and the character of the enemy.[1]
8. Indicating to advanced guards and other detachments well-chosen points of assembly in case of their attack by superior numbers, and informing them what support they may hope to receive in case of need.
9. Arranging and superintending the march of trains of baggage, munitions, provisions, and ambulances, both with the columns and in their rear, in such manner that they will not interfere with the movements of the troops and will still be near at hand. Taking precautions for order and security, both on the march and when trains are halted and parked.
10. Providing for the successive arrival of convoys of supplies. Collecting all the means of transportation of the country and of the army, and regulating their use.
11. Directing the establishment of camps, and adopting regulations for their safety, good order, and police.
12. Establishing and organizing lines of operations and supplies, as well as lines of communications with these lines for detached bodies. Designating officers capable of organizing and commanding in rear of the army; looking out for the safety of detachments and convoys, furnishing them good instructions, and looking out also for preserving suitable means of communication of the army with its base.
13. Organizing depots of convalescent, wounded, and sickly men, movable hospitals, and workshops for repairs; providing for their safety.
14. Keeping accurate record of all detachments, either on the flanks or in rear; keeping an eye upon their movements, and looking out for their return to the main column as soon as their service on detachment is no longer necessary; giving them, when required, some center of action, and forming strategic reserves.
15. Organizing marching battalions or companies to gather up isolated men or small detachments moving in either direction between the army and its base of operations.
16. In case of sieges, ordering and supervising the employment of the troops in the trenches, making arrangements with the chiefs of artillery and engineers as to the labors to be performed by those troops and as to their management in sorties and assaults.

---

1 I refer here to general instructions and forms, which are not to be repeated every day: such repetition would be impracticable.

17. In retreats, taking precautionary measures for preserving order; posting fresh troops to support and relieve the rear-guard; causing intelligent officers to examine and select positions where the rear-guard may advantageously halt, engage the enemy, check his pursuit, and thus gain time; making provision in advance for the movement of trains, that nothing shall be left behind, and that they shall proceed in the most perfect order, taking all proper precautions to insure safety.
18. In cantonments, assigning positions to the different corps; indicating to each principal division of the army a place of assembly in case of alarm; taking measures to see that all orders, instructions, and regulations are implicitly observed.

An examination of this long list – which might easily be made much longer by entering into greater detail – will lead every reader to remark that these are the duties rather of the general-in-chief than of staff officers. This truth I announced some time ago; and it is for the very purpose of permitting the general-in-chief to give his whole attention to the supreme direction of the operations that he ought to be provided with staff officers competent to relieve him of details of execution. Their functions are therefore necessarily very intimately connected; and woe to an army where these authorities cease to act in concert! This want of harmony is often seen, – first, because generals are men and have faults, and secondly, because in every army there are found individual interests and pretensions, producing rivalry of the chiefs of staff and hindering them in performing their duties.[2]

It is not to be expected that this treatise shall contain rules for the guidance of staff officers in all the details of their multifarious duties; for, in the first place, every different nation has staff officers with different names and rounds of duties, – so that I should be obliged to write new rules for each army; in the second place, these details are fully entered into in special books pertaining to these subjects.

I will, therefore, content myself with enlarging a little upon some of the first articles enumerated above:

1.) The measures to be taken by the staff officers for preparing the army to enter upon active operations in the field include all those which are likely to facilitate the success of the first plan of operations. They should, as a matter of course, make sure, by frequent inspections, that the **matériel** of all the arms of the service is in good order: horses, carriages, caissons, teams, harness, shoes, etc. should be carefully examined and any deficiencies supplied. Bridge-trains, engineer-tool trains, **matériel** of artillery, siege-trains if they are to move, ambulances, – in a word, every thing which conies under the head of **matériel**, – should be carefully examined and placed in good order.

---

[2] The chiefs of artillery, of engineers, and of the administrative departments all claim to have direct connection with the general-in-chief, and not with the chief of staff. There should, of course, be no hinderance to the freest intercourse between these high officers and the commander; but he should work with them in presence of the chief of staff, and send him all their correspondence: otherwise, confusion is inevitable.

If the campaign is to be opened in the neighborhood of great rivers, gun-boats and flying bridges should be prepared, and all the small craft should be collected at the points and at the bank where they will probably be used. Intelligent officers should examine the most favourable points both for embarkations and for landings, preferring those localities which present the greatest chances of success for a primary establishment on the opposite bank.

The staff officers will prepare all the itineraries that will be necessary for the movement of the several corps of the army to the proper points of assemblage, making every effort to give such direction to the marches that the enemy shall be unable to learn from them any thing relative to the projected enterprise.

If the war is to be offensive, the staff officers arrange with the chief engineer officers what fortifications shall be erected near the base of operations, when **têtes de ponts** or intrenched camps are to be constructed there. If the war is defensive, these works will be built between the first line of defense and the second base.

2.) An essential branch of logistics is certainly that which relates to making arrangements of marches and attacks, which are fixed by the general and notice of them given to the proper persons by the chiefs of staff. The next most important qualification of a general, after that of knowing how to form good plans, is, unquestionably, that of facilitating the execution of his orders by their clearness of style. Whatever may be the real business of a chief of staff, the greatness of a commander-in-chief will be always manifested in his plans; but if the general lacks ability the chief of staff should supply it as far as he can, having a proper understanding with the responsible chief.

I have seen two very different methods employed in this branch of the service. The first, which may be styled the old school, consists in issuing daily, for the regulation of the movements of the army, general instructions filled with minute and somewhat pedantic details, so much the more out of place as they are usually addressed to chiefs of corps, who are supposed to be of sufficient experience not to require the same sort of instruction as would be given to junior subalterns just out of school.

The other method is that of the detached orders given by Napoleon to his marshals, prescribing for each one simply what concerned himself, and only informing him what corps were to operate with him, either on the right or the left, but never pointing out the connection of the operations of the whole army.[3] I have good reasons for knowing that he did this designedly, either to surround his operations with an air of mystery, or for fear that more specific orders might fall into the hands of the enemy and assist him in thwarting his plans.

It is certainly of great importance for a general to keep his plans secret; and Frederick the Great was right when he said that if his night-cap knew what was in his head he would throw it into the fire. That kind of secrecy was practicable in Frederick's

[3] I believe that at the passage of the Danube before Wagram, and at the opening of the second campaign of 1813, Napoleon deviated from his usual custom by issuing a general order.

time, when his whole army was kept closely about him; but when maneuvers of the vastness of Napoleon's are executed, and war is waged as in our day, what concert of action can be expected from generals who are utterly ignorant of what is going on around them?

Of the two systems, the last seems to me preferable. A judicious mean may be adopted between the eccentric conciseness of Napoleon and the minute verbosity which laid down for experienced generals like Barclay, Kleist, and Wittgenstein precise directions for breaking into companies and reforming again in line of battle, – a piece of nonsense all the more ridiculous because the execution of such an order in presence of the enemy is impracticable. It would be sufficient, I think, in such cases, to give the generals special orders relative to their own corps, and to add a few lines in cipher informing them briefly as to the whole plan of the operations and the part they are to take individually in executing it. When a proper cipher is wanting, the order may be transmitted verbally by an officer capable of understanding it and repeating it accurately. Indiscreet revelations need then be no longer feared, and concert of action would be secured.

3.) The army being assembled, and being in readiness to undertake some enterprise, the important thing will be to secure as much concert and precision of action as possible, whilst taking all the usual precaution's to gain accurate information of the route it is to pursue and to cover its movements thoroughly.

There are two kinds of marches, – those which are made out of sight of the enemy, and those which are made in his presence, either advancing or retiring. These marches particularly have undergone great changes in late years. Formerly, armies seldom came in collision until they had been several days in presence of each other, and the attacking party had roads opened by pioneers for the columns to move up parallel to each other. At present, the attack is made more promptly, and the existing roads usually answer all purposes. It is, however, of importance, when an army is moving, that pioneers and sappers accompany the advanced guard, to increase the number of practicable roads, to remove obstructions, throw small bridges over creeks, etc., if necessary, and secure the means of easy communication between the different corps of the army.

In the present manner of marching, the calculation of times and distances becomes more complicated: the columns having each a different distance to pass over, in determining the hour of their departure and giving them instructions the following particulars must be considered: – 1. the distances to be passed over; 2. the amount of **matériel** in each train; 3. the nature of the country; 4. the obstacles placed in the way by the enemy; 5. the fact whether or not it is important for the march to be concealed or open.

Under present circumstances, the surest and simplest method of arranging the movements of the great corps forming the wings of an army, or of all those corps not marching with the column attached to the general head-quarters, will be to trust the details to the experience of the generals commanding those corps, – being careful, however, to let them understand that the most exact punctuality is expected of them. It will then be enough to indicate to them the point to be reached and the object to be attained, the route to be pursued and the hour at which they will be expected to be in position. They should be informed what corps are marching either on the same roads with them or on side-roads to the right or left in order that they may govern themselves accordingly; they should receive whatever news there may be of the enemy, and have a line of retreat indicated to them.[4]

All those details whose object it is to prescribe each day for the chiefs of corps the method of forming their columns and placing them in position are mere pedantry, more hurtful than useful. To see that they march habitually according to regulation or custom is necessary; but they should be free to arrange their movements so as to arrive at the appointed place and time, at the risk of being removed from their command if they fail to do so without sufficient reason. In retreats, however, which are made along a single road by an army separated into divisions, the hours of departure and halts must be carefully regulated.

Each column should have its own advanced guard and flankers, that its march may be conducted with the usual precautions: it is convenient also, even when they form part of a second line, for the head of each column to be preceded by a few pioneers and sappers, provided with tools for removing obstacles or making repairs in case of accidents; a few of these workmen should also accompany each train: in like manner, a light trestle-bridge train will be found very useful.

4.) The army on the march is often preceded by a general advanced guard, or, as is more frequent in the modern system, the center and each wing may have its special advanced guard. It is customary for the reserves and the center to accompany the head-quarters; and the general advanced guard, when there is one, will usually follow the same road: so that half the army is thus assembled on the central route. Under these circumstances, the greatest care is requisite to prevent obstructing the road. It happens sometimes, however, when the important stroke is to be made in the direction of one of the wings, that the reserves, the general head-quarters, and even the general advanced guard, may be moved in that direction: in this case, all the rules usually regulating the march of the center must be applied to that wing.

Advanced guards should be accompanied by good staff officers, capable of forming correct ideas as to the enemy's movements and of giving an accurate account of them to the general, thus enabling him to make his plans understandingly. The commander of the advanced guard should assist the general in the same way. A general advanced guard should be composed of light troops of all arms, containing some of the **élite** troops of the army as a main body, a few dragoons prepared to fight on foot, some horse-artillery, pontoniers, sappers, etc., with light trestles and pontoons for passing small streams. A few good marksmen will not be out of place. A topographical officer should accompany it, to make a sketch of the country a mile or two on each side of the road. A body of irregular cavalry should always be attached, to spare the regular cavalry and to serve as scouts, because they are best suited to such service.

---

4 Napoleon never did this, because he maintained that no general should ever think seriously of the possibility of being beaten. In many marches it is certainly a useless precaution; but it is often indispensable.

5.) As the army advances and removes farther from its base, it becomes the more necessary to have a good line of operations and of depots which may keep up the connection of the army with its base. The staff officers will divide the depots into departments, the principal depot being established in the town which can lodge and supply the greatest number of men: if there is a fortress suitably situated, it should be selected as the site of the principal depot.

The secondary depots may be separated by distances from fifteen to thirty miles, usually in the towns of the country. The mean distance apart will be about twenty to twenty-five miles. This will give fifteen depots upon a line of three hundred miles, which should be divided into three or four brigades of depots. Each of these will have a commander and a detachment of troops or of convalescent soldiers, who regulate the arrangements for accommodating troops and give protection to the authorities of the country (if they remain); they furnish facilities for transmitting the mails and the necessary escorts; the commander sees that the roads and bridges are kept in good order. If possible, there should be a park of several carriages at each depot, certainly at the principal one in each brigade. The command of all the depots embraced within certain geographical limits should be intrusted to prudent and able general officers; for the security of the communications of the army often depends on their operations.[5] These commands may sometimes become strategic reserves; a few good battalions, with the assistance of movable detachments passing continually between the army and the base, will generally be able to keep open the communications.

6.) The study of the measures, partly logistical and partly tactical, to be taken by the staff officers in bringing the troops from the order of march to the different orders of battle, is very important, but requires going into such minute detail that I must pass it over nearly in silence, contenting myself with referring my readers to the numerous works specially devoted to this branch of the art of war.

Before leaving this interesting subject, I think a few examples should be given as illustrations of the great importance of a good system of logistics. One of these examples is the wonderful concentration of the French army in the plains of Gera in 1806; another is the entrance of the army upon the campaign of 1815.

In each of these cases Napoleon possessed the ability to make such arrangements that his columns, starting from points widely separated, were concentrated with wonderful precision upon the decisive point of the zone of operations; and in this way he insured the successful issue of the campaign. The choice of the decisive point was the result of a skillful application of the principles of strategy; and the arrangements for moving the troops give us an example of logistics which originated

[5] It may be objected that in some wars, as where the population is hostile, it may be very difficult, or impracticable, to organize lines of depots. In such cases they will certainly be exposed to great dangers; but these are the very cases where they are most necessary and should be most numerous. The line from Bayonne to Madrid was such a line, which resisted for four years the attacks of the guerrillas, although convoys were sometimes seized. At one time the line extended as far as Cadiz.

in his own closet. It has been long claimed that Berthier framed those instructions which were conceived with so much precision and usually transmitted with so much clearness; but I have had frequent opportunities of knowing that such was not the truth. The emperor was his own chief staff officer. Provided with a pair of dividers opened to a distance by the scale from seventeen to twenty miles in a straight line, (which made from twenty-two to twenty-five miles, taking into account the windings of the roads,) bending over and sometimes stretched at full length upon his map, where the positions of his corps and the supposed positions of the enemy were marked by pins of different colors, he was able to give orders for extensive movements with a certainty and precision which were astonishing. Turning his dividers about from point to point on the map, he decided in a moment the number of marches necessary for each of his columns to arrive at the desired point by a certain day; then, placing pins in the new positions, and bearing in mind the rate of marching that he must assign to each column, and the hour of its setting out, he dictated those instructions which are alone enough to make any man famous.

Ney coming from the shores of Lake Constance, Lannes from Upper Swabia, Soult and Davoust from Bavaria and the Palatinate, Bernadotte and Augereau from Franconia, and the Imperial Guard from Paris, were all thus arranged in line on three parallel roads, to debouch simultaneously between Saalfeld, Gera, and Plauen, few persons in the army or in Germany having any conception of the object of these movements which seemed so very complicated.

In the same manner, in 1815, when Blücher had his army quietly in cantonments between the Sambre and the Rhine, and Wellington was attending **fêtes** in Brussels, both waiting a signal for the invasion of France, Napoleon, who was supposed to be at Paris entirely engrossed with diplomatic ceremonies, at the head of his guard, which had been but recently reformed in the capital, fell like a thunderbolt upon Charleroi and Blücher's quarters, his columns arriving from all points of the compass, with rare punctuality, on the 14th of June, in the plains of Beaumont and upon the banks of the Sambre. (Napoleon did not leave Paris until the 12th.)

The combinations described above were the results of wise strategic calculations, but their execution was undoubtedly a masterpiece of logistics. In order to exhibit more clearly the merit of these measures, I will mention, by way of contrast, two cases where faults in logistics came very near leading to fatal consequences. Napoleon having been recalled from Spain in 1809 by the fact of Austria's taking up arms, and being certain that this power intended war, he sent Berthier into Bavaria upon the delicate duty of concentrating the army, which was extended from Braunau as far as Strasbourg and Erfurt. Davoust was returning from the latter city, Oudinot from Frankfort; Massena, who had been on his way to Spain, was retiring toward Ulm by the Strasbourg route; the Saxons, Bavarians, and Wurtembergers were moving from their respective countries. The corps were thus separated by great distances, and the Austrians, who had been long concentrated, might easily break through this spider's web or brush away its threads. Napoleon was justly uneasy, and ordered Berthier to assemble the army at Ratisbon if the war had not actually begun on his arrival, but, if it had, to concentrate it in a more retired position toward Ulm.

The reason for this alternative order was obvious. If the war had begun, Ratisbon was too near the Austrian frontier for a point of assembly, as the corps might thus be thrown separately into the midst of two hundred thousand enemies; but by fixing upon Ulm as the point of rendezvous the army would be concentrated sooner, or, at any rate, the enemy would have five or six marches more to make before reaching it, which was a highly-important consideration as the parties were then situated.

No great talent was needed to understand this. Hostilities having commenced, however, but a few days after Berthier's arrival at Munich, this too celebrated chief of staff was so foolish as to adhere to a literal obedience of the order he had received, without conceiving its obvious intention: he not only desired the army to assemble at Ratisbon, but even obliged Davoust to return toward that city, when that marshal had had the good sense to fall back from Amberg toward Ingolstadt.

Napoleon, having, by good fortune, been informed by telegraph of the passage of the Inn twenty-four hours after its occurrence, came with the speed of lightning to Abensberg, just as Davoust was on the point of being surrounded and his army cut in two or scattered by a mass of one hundred and eighty thousand enemies. We know how wonderfully Napoleon succeeded in rallying his army, and what victories he gained on the glorious days of Abensberg, Siegberg, Landshut, Eckmühl, and Ratisbon, that repaired the faults committed by his chief of staff with his contemptible logistics.

We shall finish these illustrations with a notice of the events which preceded and were simultaneous with the passage of the Danube before the battle of Wagram. The measures taken to bring to a specified point of the island of Lobau the corps of the Viceroy of Italy from Hungary, that of Marmont from Styria, that of Bernadotte from Linz, are less wonderful than the famous imperial decree of thirty-one articles which regulated the details of the passage and the formation of the troops in the plains of Enzersdorf, in presence of one hundred and forty thousand Austrians and five hundred cannon, as if the operation had been a military fête. These masses were all assembled upon the island on the evening of the 4th of July; three bridges were immediately thrown over an arm of the Danube one hundred and fifty yards wide, on a very dark night and amidst torrents of rain; one hundred and fifty thousand men passed over the bridges, in presence of a formidable enemy, and were drawn up before mid-day in the plain, three miles in advance of the bridges which they covered by a change of front; the whole being accomplished in less time than might have been supposed necessary had it been a simple maneuver for instruction and after being several times repeated. The enemy had, it is true, determined to offer no serious opposition to the passage; but Napoleon did not know that fact, and the merit of his dispositions is not at all diminished by it.

Singularly enough, however, the chief of staff, although he made ten copies of the famous decree, did not observe that by mistake the bridge of the center had been assigned to Davoust, who had the right wing, whilst the bridge on the right was assigned to Oudinot, who was in the center. These two corps passed each other in the night, and, had it not been for the good sense of the men and their officers, a dreadful scene of confusion might have been the result. Thanks to the supineness of the enemy, the army escaped all disorder, except that arising from a few detachments following corps to which they did not belong. The most remarkable feature of the whole transaction is found in the fact that after such a blunder Berthier should have received the title of Prince of Wagram.

The error doubtless originated with Napoleon while dictating his decree; but should it not have been detected by a chief of staff who made ten copies of the order and whose duty it was to supervise the formation of the troops?

Another no less extraordinary example of the importance of good logistics was afforded at the battle of Leipsic. In fighting this battle, with a defile in rear of the army as at Leipsic, and in the midst of low ground, wooded, and cut up by small streams and gardens, it was highly important to have a number of small bridges, to prepare the banks for approaching them with ease, and to stake out the roads. These precautions would not have prevented the loss of a decisive battle; but they would have saved the lives of a considerable number of men, as well as the guns and carriages that were abandoned on account of the disorder and of there being no roads of escape. The unaccountable blowing up of the bridge of Lindenau was also the result of unpardonable carelessness upon the part of the staff corps, which indeed existed only in name, owing to the manner of Berthier's management of it. We must also agree that Napoleon, who was perfectly conversant with the logistical measures of an offensive campaign, had then never seriously thought what would be proper precautions in the event of defeat, and when the emperor was present himself no one thought of making any arrangement for the future unless by his direction.

To complete what I proposed when I commenced this article, it becomes necessary for me to add some remarks with reference to reconnoissances. They are of two kinds: the first are entirely topographical and statistical, and their object is to gain a knowledge of a country, its accidents of ground, its roads, defiles, bridges, etc., and to learn its resources and means of every kind. At the present day, when the sciences of geography, topography, and statistics are in such an advanced state, these reconnoissances are less necessary than formerly; but they are still very useful, and it is not probable that the statistics of any country will ever be so accurate that they may be entirely dispensed with. There are many excellent books of instruction as to the art of making these reconnoissances, and I must direct the attention of my readers to them.

Reconnoissances of the other kind are ordered when it is necessary to gain information of the movements of the enemy. They are made by detachments of greater or less strength. If the enemy is drawn up in battle-order, the generals-in-chief or the chiefs of staff make the reconnoissance; if he is on the march, whole divisions of cavalry may be thrown out to break through his screen of posts.

Henry E. Eccles (1954)

LOGISTICS – WHAT IS IT?

Since at least the 19th century and Jomini's classical discussion the term "logistics" was used to describe a set of tasks and a pragmatic body of experiences and knowledge in the realm of the military. But only in the 20st century, with rapidly expanding uses of science and technology, attempts were made to elevate the field to the level of a science.

The first explicit effort may have been Cyrus Thorpe's 1917 book on *Pure Logistics. The Science of War Preparation*. Thorpe, an officer with the American Marine Corps, is quoted in the article following: "Pure logistics is «a scientific inquiry into the theory of logistics…» and «concerns itself, in accordance with general principles, with the detailed manner of dividing labour in the logistical field in the preparation for war and maintaining war during its duration»".

This idea was followed up and further developed in the post World War II years in publications by Henry Effingham Eccles (1898–1986). The author was a rear admiral in the U.S. Navy. In several military positions during and after World War II he had been dealing with logistics planning and with the coordination of planning, construction, and support of military bases thus acquiring profound knowledge of the characteristics and the requirements of military logistics planning. In his first book *Operational Naval Logistics*, which was published in 1950, Eccles discussed the fundamentals of his subject. This book became a standard manual for the United States Navy.

In Eccles' 1954 classical article reprinted here the progress in building a conceptual basis for an emerging science of logistics becomes apparent. He follows Thorpe in arguing that logistics is a concept which stands beside and supplements the concepts of strategy and tactics: "Strategy and Tactics provide the scheme for the conducts of military operations: Logistics provides the means therefore". He highlights the primary function of military logistics as "the supplying of men and material, and the rendering of services to the operating military forces". Logistics is about "furnishing the means of war, which are materials, men, facilities and services". To operationalize this task, he suggests a "simple and straightforward fundamental sequence for logistics planning": To support a "proposed scheme of deployment and action and an estimate of opposition", logistics planning encompasses:

- Determination of broad logistics requirements and determination as to which of these will be decisive or critical;
- Determination of availabilities of these critical items;
- Estimation of the effect of shortages upon the strategic and tactical courses of action;
- Determination of what can be done logistically or tactically to alleviate these shortages or to overcome the handicaps imposed by them.

P. Klaus and S. Müller (eds.), *The Roots of Logistics*, DOI 10.1007/978-3-642-27922-5_3,

Eccles also gives first explicit thought to the transferability of this idea to the realm of civil and business activities by quoting a Bureau of Supplies and Accounts paper: "Civil logistics is the mobilization of the civilian industrial economy to support the armed forces".

Last but not least one of Eccles' contributions to the emergent idea of a science of logistics is his statement that "we must have knowledge of the theories and principles of the underlying science". He also provides a – though very brief – explanation of what he means by a logistics principle. And finally he offers a framework of terms and concepts for organizing and describing the field of logistics, which include the distinction of strategic vs. tactical logistics, of logistics planning vs. planning for logistics, and thoughts about the division of labor within a logistics staff organization and its administration.

Reference:

Henry E. Eccles: Logistics – What is it? In: *Naval Research Logistics Quarterly*, Vol. 1, No. 1, 1954, pp. 5–15.

Further reading:

Thorpe, Cyrus: *Pure Logistics; the Science of War Preparation*, Kansas City/MO: Franklin Hudson, 1917.

Henry E. Eccles: Logistics – What is it?[1]

Since World War II the importance of logistics has been stressed at all levels of military activity, and military writers have discussed its various aspects at length. In many ways and forms it was a major matter of consideration in the unification of the armed forces. Today we understand it better than before the war, and therefore our organizations and plans are logistically superior to those of the past. Many of our most important unsolved problems are logistical, however, and there are still serious deficiencies in our logistic plans and capabilities. If we fail to correct these deficiencies, they may easily cause our defeat should we have to fight another global war.

While some of our present deficiencies are caused by our failure properly to apply what we already know, there are many areas where we are limited by our imperfect knowledge of the art and science of logistics. If we can perfect and apply this knowledge, it is probable, within our present economic limitations, that we can improve the effectiveness of our logistic support by fifty per cent.

Among the obstacles to improvement are the uncertainties that exist as to the meaning of the word itself and as to the proper place of logistics in military organizations and plans. This is because logistics has several distinct aspects and in each aspect the definitions and descriptions differ. Frequently, therefore, people may be talking from diverse points of view without realizing the effect this has on their descriptions and opinions. Each will be ascribing a different meaning to the word without realizing it.

In 1917 Lieutenant Colonel Cyrus Thorpe, U.S. Marine Corps, published an excellent little book, *Pure Logistics. The Science of War Preparation.* This initial attempt to develop theory and principle apparently attracted little or no attention until five copies were found in the Naval War College Library in 1945. Some students of war wonder how many billions could have been saved had the significance of Colonel Thorpe's ideas been fully appreciated before 1941. Unfortunately the book is out of print, the publishing house out of business, and only few copies remain in the hands of individuals.

In his preface Colonel Thorpe says:

"The terms 'pure' and 'applied' may be used with the same meaning as to logistics as to other sciences. Pure Logistics is merely a scientific inquiry into the theory of logistics – its scope and function in the Science of War, with a broad outline of its organization. Applied Logistics rests upon the pure, and concerns itself, in accordance with general principles, with the detailed manner of dividing labor in the logistical field in the preparation for war and maintaining war during its duration."

---

[1] This article appeared in the June 1953 issue of the *United States Naval Institute Proceedings*, copyright 1953 by the U.S. Naval Institute, and is being republished with permission of the Institute. It was prepared under sponsorship of ONR Contract N7onr-41904.

It is the purpose of this article further to develop this approach and to discuss what is meant by the word logistics and what is its place in the naval establishment. It is not my intent to solve problems but rather to establish a background that will be helpful to those who are trying to solve them.

PURE LOGISTICS

The word **logistics** is an abstract term which represents a very practical reality.

In its abstract sense it, like **strategy**, **tactics**, **economics**, and **politics**, is not susceptible to a single simple and permanent definition. Rather it can be described in a variety of ways in very general terms. If it is to be understood it must be approached from various points of view, and it must be discussed and described by reference to other intangibles and abstract terms. It is only through the consideration of one abstract term with relation to the other abstract terms with which it is naturally associated that the true picture can be developed and described.

Several of the most useful and enlightening descriptions of abstract logistics – or "pure logistics", as we may call it – are:

- "Logistics is all that part of war which is not included in Strategy and Tactics." Col. Thorpe, *Pure Logistics*.
- "Strategy and Tactics provide the scheme for the conduct of military operations: Logistics provides the means therefore". Derived from Col. Thorpe, *Pure Logistics*.
- "The logistic process is, at one and the same time, the military element in the nation's economy and the economic element in its military operations." From a Munitions Board Study.

If we carry on the thoughts expressed in these descriptions we can see war itself in its abstract sense as comprising three interlocked abstractions – **Strategy**, **Tactics** and **Logistics**. If we draw on our experience and on history, we find that war is a matter of infinite variation, in which no two situations are precisely the same. We see that in all war situations the actions and decisions are based upon a blend of strategical, tactical, and logistical considerations. If we consider these abstract terms as three overlapping discs, we can see where some situations involve all three considerations, others involve two, and some only one. However, as we study, we realize that the dividing lines tend to blur and in some cases to vanish.

A somewhat different approach would be to say that, in its broadest sense, war is a combination of military, political, economic, and geographic situations and considerations. Here again we find a variable blend of abstract terms, each of which is subject to a variety of meanings and descriptions as well as to a variety of subdivisions; for example, geographic situations may well include climate and weather; political considerations can well include sociological aspects. However, military and economic considerations are related by logistics as indicated before, since it is "the military element in the nation's economy and the economic element in its military operations."

In order to fight a modern war we must "mobilize" our economic and industrial resources in order to create and support the necessary combat forces and to maintain our civil economy and health. Again we are speaking in abstract terms, for "mobilize" is a word that can be applied to many affairs.

Two descriptions taken from a Bureau of Supplies and Accounts paper are helpful in understanding this problem:

- "Civil logistics is the mobilization of the civilian industrial economy to support the armed forces."
- "Military logistics is the supplying of men and material, and the rendering of services, to the operating military forces."

If we now return to our earlier description and statement that "logistics provides the means for the conduct of military operations," we find ourselves on a relatively simple and broad abstract level using simple terms – terms that can be readily related and are adequate to cover the major problem of war. We also find that we have placed logistics on a coequal level with strategy and tactics, thus forming the three major subdivisions of the art and science of war.

If we are to practice an art, we must have a knowledge of the theories and principles of the underlying science. In this sense, principles can be considered as the statements of known cause-and-effect relationships. As of today these cause-and-effect relationships in logistics are virtually undeveloped. Patient, free research is required to develop these principles, and after they are developed they must be frequently reexamined to confirm their validity.

Someone may well ask – What is a logistic principle? An illustration is the "Principle of Priorities" which states:

- Priorities and allocations are required only in case of shortages.
- When shortages exist, both allocations and priorities are required to provide equitable distribution.
- Priorities used alone are self defeating and merely serve to worsen the defect they are supposed to correct.

This is an example of an imperfectly developed principle. In World War II this principle seemed to be borne out, particularly by the confusion in national distribution of scarce raw material and by the manner in which priority cargo waited shipment for months in air transportation depots. In the present partial war the evidence, while not yet wholly conclusive, tends to confirm the truth of this principle. If it be true, its application can greatly improve our efficiency in war.

APPLIED LOGISTICS

Abstract speculations, theories, and principles have never prepared a nation to fight and have never won a war. All they have done is to enable man to understand his war problems and to assist him to solve them.

In order to prepare for war, we must define the practical tasks of the armed forces and we must assign these tasks to specific organizations and individuals. For this purpose we have organized the Department of Defense and the armed forces; for this purpose authoritative definitions have been published and specific tasks assigned.

The U.S. Joint Chiefs of Staff have defined logistics as:

"That part of the entire military activity which deals with: (1) design and development, acquisition, storage, movement, distribution, maintenance, evacuation and disposition of material; (2) induction, classification, training, assignment, welfare, movement, evacuation and separation of personnel; (3) acquisition or construction, maintenance, operation and disposition of facilities; and (4) acquisition or furnishing of services. It comprises both planning, including determination of requirements, and implementation."

This definition of applied or practical logistics is in no way out of harmony with the previous broad descriptions of pure logistics. Rather it amplifies them and reduces them to specific functional terms suitable for application throughout the armed forces.

Nowhere are there any orders as to how each service shall interpret this definition. This is wise, for conditions in each service vary. In the Navy Department logistics is rather loosely organized. The Deputy Chief of Naval Operations for Logistics has responsibility for coordination, and for the determination of requirements, while the technical bureaus have the operating function. In the Departments of the Army and the Air Force the Assistant Chief of Staff, G4 Logistics (Army) and Deputy Chief of Staff – Materiel (AF) have a much more direct authority.

The fact that a certain function is included within the definition of logistics does not necessarily mean that the function in question is carried on by an organization that is logistical in name or is wholly devoted to logistical activity. (Failure to remember this important fact frequently causes confusion.)

However, it is of vital importance to understand one thing: that regardless of how the logistics functions are assigned and divided, the functions themselves are the same and must be performed by qualified officers. Furthermore, these functions must be supervised and coordinated by other senior officers; and these senior officers must not only understand the full implications of their responsibility, but also must understand the relationships involved in these tasks.

In the practical field, the definition of applied logistics, or the "What is it?" varies in accordance with the level of the organization being considered. But always logistics is concerned with "furnishing the means of war," which are material, men, facilities and services.

On the international and national levels, applied logistics deals with the broadest economic and industrial matters, such as: the sources and availabilities of raw materials; the state of the domestic economy and finances; the availability of manufacturing plants, skilled and unskilled labor, design and production engineers, management; and other similar affairs. Some persons may prefer to consider this as a combination of economic mobilization, industrial mobilization and military planning. The precise labels attached to the process are not as important as understanding the nature and inter-relationship of the functions performed.

The international process is exemplified by the military assistance activities and by the mutual security programs which have been placed under the administration of the Commander in Chief, U.S. European Command (U.S. CINCEUR) General Ridgway. The process is further illustrated by the work of the Screening and Costing Committee under General McNarney in Paris in 1951, and their immediate superior, the Temporary Council Committee, "The Three Wise Men." The Lisbon Conference of 1952 was almost wholly a high level logistic conference.

On this highest level the international and national situations and decisions must be continuously inter-related. Therefore, our own governmental organizations must work with their opposite numbers in other nations and with the special **ad hoc** or permanent international organizations. It is vital to seek harmony among the national and the international policies, strategic plans, and military programs. While it is naive to expect to achieve complete harmony, it is very important that we avoid contradiction. Policies and plans are made by both international and national agencies, but action is almost always by national agencies. There is an extraordinarily complex mixture of political, economic, strategic, and logistic factors.

At this top national-international level, the activities of the Joint Chiefs of Staff, the Munitions Board, the three military departments, and many other agencies and specially appointed individuals are so fluid and intertwined that they literally defy description. This condition emphasizes the importance of understanding the basic principles that operate in these fields.

Strictly national U.S. action follows a somewhat simpler pattern. The projects authorized and the funds appropriated by the Congress actually determine the logistic capabilities of the armed forces. Within the limits of these practical capabilities, the Secretary of Defense, the Joint Chiefs of Staff, and the military departments decide as to the specific forces to be built up and they formulate the broad strategic plans. The Munitions Board makes recommendations as to policies governing the allocation of both raw materials and finished products among the three services. This board also makes recommendations as to how the productive capacity of certain industrial plants should be allocated. The three military departments, through their technical bureaus and services, then actually procure and distribute to the operating forces the ships, planes, guns, men, equipment, supplies, and services which are the means of war.

The above are the major processes which constitute, or are associated with, the international and national level of logistics. It is sometimes convenient to call this the mobilization level or to consider it as **part** of the strategic phase of logistics.

The next level of activity may be described as the operating level or field level. It includes what the Army calls the Zone of the Interior and what the Navy calls the Continental Shore Establishment, and it takes in the theaters of operation, the fleets, the armies, and the air forces. It may be called "Operational Logistics" and may be said to include both strategical logistics and tactical logistics.

The logistics of the naval operating forces are based on the Sea Frontier, Naval Base, and Naval Depot systems, which channel the flow of logistic support and services to the fleets.

Planning for the actual conduct of this "Operational Logistics" is based upon the strategic plans and the broad logistics plans and policies of the theater and fleet commanders and upon their estimates of requirements. All of these furnish the necessary guidance to the type commanders who actually submit the requisitions and operate the basic logistic services afloat.

As we move from the theaters through the fleets to the task forces, we move from strategical logistics to tactical logistics; from the realm of long range plans and forecasts, to the actual repair and replenishment of combat forces. The techniques of **tactical** logistics are under constant scrutiny and improvement in actual practice. However, the techniques and procedures of so-called theater and fleet **strategical** logistics are frequently imperfect and sometimes neglected in peacetime.

Any one can understand the effect of a ship at sea running out of fuel and ammunition. But few officers understand the importance and nature of the long range concurrent strategic-operational-logistic planning on theater and fleet level: the planning that will insure the readiness of task forces for sustained combat operations in time of sudden emergency.

In considering this division of logistics into various levels, we should always remember that each level overlaps with the other, both above and below. There can never be a sharp "chop line" of interest, although there are various "chop lines" of specific action responsibility. In all of the above stages and relationships we find changing general characteristics and emphasis.

On the highest level we deal in the broadest terms, and as we go down the chain we find ourselves being more specific. At first the emphasis is on civilian control, with important military participation. But as we go down to the operating level, the civilian interest tends to diminish and the military control increases. Civilian control tends to be strongest in the "producer" or business end of the logistic process; military control is strongest in "consumer logistics."

LOGISTICS IN PLANNING

There are two general types of planning in applied logistics: **Logistics Planning** and **Planning for Logistics**. Again, as in so many other fields, these overlap.

The first term, logistics planning, can be used to indicate the incorporation of logistics considerations in the development of strategic and tactical plans. It includes the determination of the basic logistic requirements and the general dispositions necessary to support these plans. All of this can be effectively done only by a constant and intimate relationship between the strategic planners and the logistic planners. In this relationship the former are constantly aware of the logistic capabilities and limitations, and the latter are constantly aware of the dispositions and employment being proposed for the forces, and of the estimated nature and strength of enemy opposition.

While the details of this Logistics Planning can be complex, the whole system rests on the following simple and straightforward fundamental sequence:

- A proposed scheme of deployment and action and an estimate of opposition;
- Determination of broad logistic requirements and determination as to which of these will be decisive or critical;
- Determination of availabilities of these critical items;
- Estimate of the effect of shortages upon the strategic and tactical courses of action;
- Determination of what can be done logistically or tactically to alleviate these shortages or to overcome the handicaps imposed by them.

Since this process exemplifies the meaning of the expression "Logistics is a Command responsibility," it requires the complete understanding and active personal interest of the Commander himself, as well as the full participation of his strategic and logistic assistants.

The terms **feasibility** and **calculated risk** are frequently used in discussions of the problems of Command in War. These words have probably been clouded with more ignorance and superstition than any other terms in our war and postwar vocabulary. Each should be used with caution. They are closely related; yet each requires careful study and proper qualification in use. Final decision in either is a matter of the personal professional judgment of the Commander. In each case it is a question of how much risk and how much hardship the Commander is willing to impose on his subordinate forces and personnel in order to gain his objective. There are no absolute or arbitrary limits. The decision involves a process of selection of courses of action and of the development of plans that will make the most effective use of the combat forces and logistic resources that are available. This process is the highest test of military judgment and it requires close personal relationships among the commander and his responsible assistants.

The second term, planning for logistics, can be applied to the more detailed planning for the logistic support of the combat forces. Here is where the degree of flexibility of the combat forces is almost wholly determined; it is the most "practical" type of functional logistics. Among other things it involves the details of the build up, distribution, and allocation of supplies, transportation lift, and repair facilities.

These two types of planning take place at all levels. The techniques may vary and in some instances they may telescope and become concurrent. Normally, however, the **planning for logistics** follows the basic decisions in the **logistic planning** stage. Since all of this planning requires close coordination between staffs and within staffs, we will now discuss staff organization as it applies to logistics.

LOGISTICS ASPECTS OF STAFF ORGANIZATION

Since no standard staff organizations are prescribed, there is a wide variation in the manner in which various Commanders choose to organize their staffs. The organization is derived from the wartime tasks and responsibilities of the Commander. From the logistic point of view two general types of organization can be recognized.

The first type is found where the principal function of the Commander is logistics. The Service Force of a fleet is an illustration. The second is found where logistics is only one of several major tasks of the commander. Theater commanders and sometimes fleet commanders fall into this category.

In the first type, the Service Force, individual major functions – such as personnel, supply, maintenance and repair, medical, base development, plans, and operations – are usually assigned to separate staff divisions. Fuel, electronics and ammunition usually require special arrangements which vary according to circumstances. Since the whole task of the commander is a logistics task, the whole staff is a logistics staff. Therefore the Chief of Staff has the fleet logistic situation as his major concern. He, assisted by the plans officer, coordinates the entire logistics task of the command. This coordination is based on the strategic and the broad logistics guidance provided by the Fleet or Theater Commander. Under these circumstances there is no need for a separate logistics division on the Service Force Staff.

In the case of a theater or of a fleet staff, the situation may be quite different. Here it is important to group the major logistics functions under a single logistics officer who is the principal staff adviser to the Commander in logistics matters. During World War II this was done with notable success by Admiral Nimitz. However, this sound practice has not always been followed in our post war organizations. Instead, in peacetime, there has been a tendency to diffuse the logistic function throughout the staff. In a Theater or Fleet Staff, the Chief of Staff has many urgent concerns other than logistics. Therefore he cannot act as an effective logistics coordinator. If no single officer is given specific responsibility and authority for logistic planning and coordination within the staff, many loose ends or "holidays" develop. Under such conditions, it is probable that both the strategic and the logistics plans will be fatally defective. This can lead to disaster in war.

This need for centralizing logistics staff work in a single staff division does not imply that type and force commanders with major logistic responsibilities do not also act as advisers to the theater or fleet commander. There are two legitimate sources and channels, one the personal staff, the other the chain of command. Both are necessary.

Uncertainty and diffusion of logistic responsibility within naval staffs are illustrated in certain postwar staff organizations. In one instance the logistics section was established as a subsection of the material division. This is simply a case of reversing the cart and the horse: material is an essential part, but only one essential part, of the overall logistic problem.

In another case, planning for logistics support of naval aviation was assigned as a minor task of the operations division; the logistics division being aware, only by hearsay of aviation logistic requirements.

In staff organization, a common fault is to give the communications division cognizance over all matters dealing with electronics. This will inevitably break down in war because communications is only one aspect of the huge electronics problem.

Another fault is to attempt to maintain in the plans division a small logistics plans section to formulate the logistics plans. This results in waste of effort and poor plans, for if the staff planning picture is to be kept clear, the logistics planning must be charged to the individual who is directly responsible to the Commander for advising and assisting him in logistics matters.

And finally there is the very human and understandable desire of each technical bureau to have its staff representative report directly to the Commander, without dealing through a "Logistics Officer" or through the Commander of a Logistic Force.

Many of the above variations in staff organization spring from the desire to operate with a minimum staff. Except for war planning, Theater and Fleet staffs in peacetime have relatively little to do, and what they do is relatively simple. Therefore, if logistic responsibility is diffused and scattered among several major staff divisions, such as plans, or operations, or communications, these divisions can probably handle certain current logistic matters with no undue effort and with no **apparent harm**. However, war instantly transforms this situation. The above divisions become swamped with their own primary duties and have no time for anything else. In fact they urgently require additional personnel to handle their extra load.

When war breaks out, fleet and theater logistics problems literally explode to huge size and great urgency. However, if the logistics responsibility has been diffused rather than concentrated, the logistics division will not be prepared to handle the emergency. Therefore confusion, serious trouble, and major waste ensue. These are the minimum results of the hasty improvisation which inevitably occurs. At the worst, a major military disaster can take place.

On the mobilization level as pointed out previously, the three services, the Joint Chiefs of Staff, and the Munitions Board are striving to relate war, mobilization, and budget plans to the national economy and to political factors. Here the relatively simple planning process and staff organizations previously described do not directly apply. The basic principles are the same, but the magnitude of the task and the realities of legal regulation and the budgetary and service competition require unwieldy organizations and a long drawn out planning process.

In this lengthy **Planning Cycle**, practically every major activity in the Department of Defense is involved, either directly or indirectly. For example, some people are working on budget and finance, some on ship design and construction, some on industrial plant capacity and allocation, some on manpower, some on raw material, some on transportation. Cut the cake any way you choose, call it by any name you want – Administration, Economic Mobilization, Industrial Mobilization, General Planning, or what you will – it is still the same reality. These men are working to "provide the Means of War in order to support the National Strategy." Regardless of the cut of the cake or the nomenclature used, it still must follow the basic logistic procedure. First determine your requirements; next figure out how and where you will procure what you want; and, finally, distribute it in accordance with the military needs of the situation. The consumers are the military; the producers are essentially civilian.

This process is **logistics**, logistics in a very practical sense. If those individuals doing this applied logistics have an understanding of the purpose, the relationships, and the principles of pure logistics, their work will have the coherence that is so essential. They can develop an efficiency and effectiveness impossible in the cornpartmentized operations of individual groups who see only their own day to day crises without relating them to the over-all problems and purpose.

LOGISTICS AND ADMINISTRATION

There is some difference of opinion as to the relationship between the terms **logistics** and **administration**, since these terms have always been linked in military organization and planning. Where major staffs have used deputies to relieve the Chief of Staff of detail, it is usual to find a Deputy Chief of Staff for Plans and Operations and a Deputy Chief of Staff for Logistics and Administration. The Navy and the Air Force use a "Logistic Annex" to their operation plans, while the Army and Marine Corps use an "Administrative Order" to contain the logistic and administrative portions of their operation plan. This latter usage implies that pure logistics is a subcategory of the broader abstraction, **Administration**.

In one instance the term **Administration** has been described as:

- "The management of all phases of military operations not directly involved in tactics, strategy and logistics."

This has been coupled with a secondary meaning:

- "The interior management of units."

However, other authorities give administration an even wider scope and consider that it "denotes the endeavor of each member of an organization." In this sense it includes all organized planning, execution, and supervision of the planned action. Here it would apply equally to strategy, tactics, and logistics, and to all operations from all levels from the President of the United States down to the individual sailor performing his duties in peace or war.

Still another manner of treating the subject of administration is found in some staffs that group personnel and administration in one staff division. In some instances this staff division will have cognizance of legal matters, welfare, recreation, and postal affairs and military government, in addition to the distribution of personnel.

The term "staff administration" has a well recognized special meaning which usually includes the operation of the clerical and routing system of the staff, custody of registered publications and security, etc. It sometimes includes the supervision or operation of staff housekeeping and transportation facilities.

From the above we can conclude that **Administration** has such a broad general use and such a variety of special uses that, unless the context permits no misunderstanding, the term should always be clearly and appropriately qualified. It supplements but in no way replaces the term **Logistics** in military usage.

BROAD OBJECTIVES

Sir Richard Livingstone once pointedly wrote:[2]

"... by a mere technician I mean a man who understands everything about his job except its ultimate purpose and its place in the order of the universe."

One of the most important reasons for understanding the abstract relations between strategy, logistics, and tactics is to emphasize the purpose and the objectives of our effort.

The objective of logistic and administrative work is the creation and continued support of adequate combat forces in order that these forces may effectively support our national strategy.

The nature of modern war is such that, in order to fight effectively, there must be the greatest economy in the provision of these forces and their support. But economy is not the objective; it is merely one of the essential factors in the attainment of the objective. If the objective is jeopardized by over-emphasis of economy, fatal damage may be done to our national security.

If, on the other hand, the economic limitations of our national defense effort be not recognized, our national objectives and national security may be lost. Since the Russian development of Marxist doctrine has emphasized the employment of economic warfare and inflation to bankrupt countries they wish to take over or defeat, this latter danger must be recognized.

CONCLUSION

In analyzing any subject it is desirable to reduce it to simple terms and to show them in logical relationship. In attempting to do this it is inevitable that statements be made which will appear obvious or trite and even repetitious to those who are experienced in the art. On the other hand, in attempting analysis in any complex field, it is likely that the pioneer must also use some relatively abstruse ideas and expressions. It is only after the exchange and interplay of the ideas of many enthusiasts that adequate clarification and simple statements of principles can evolve. With this in

---

2 In *Some Tasks for Education*, Oxford University Press, 1946.

mind, the following thoughts are offered as to some of the most important facts and relationships which lead to an understanding of the question "What is logistics?"

In order to understand logistics one must understand both the word and the subject.

The word **logistics** is a symbol, an abstraction, and as such can be truly described in a variety of ways. It can be strictly defined only for a particular case at a particular time.

The subject of Logistics itself is a living, changing thing, it is not dead or static. Full understanding of the subject can come only by **seeing it as a whole**, by pondering upon its many elements and functions, and by practice in the art.

Pure logistics is an abstract term used to indicate the whole complex process whereby the means of war necessary to support a national strategy are determined, procured, and finally distributed to the combat commanders. **Applied logistics** represents the everyday practical application of this abstract process.

The objective of all logistic effort is the creation and sustained support of adequate combat forces. Economy is an essential factor in the attainment of this objective.

The **applied logistics** process of providing men, materials, facilities and services, comprises the performance of many specialized and technical functions. These functions include ship and aircraft design, construction, maintenance and repair; air and naval base design, construction, and operation; the operation of an intricate supply system; the provision of fuel, ordnance, and ammunition; transportation of all sorts; and personnel and medical services. Some of these practical functions are performed by staff corps officers and line specialists, some by unrestricted line officers, and some by civilians.

Both **pure** and **applied logistics** can be roughly divided into two general classifications: mobilization logistics and operational logistics. Before operational logistics can function, there must be the prior performance of the **mobilization logistics** function: **Mobilization logistics** and **operational logistics** have a large area in common.

The nature of war and its component parts is such that sharp dividing lines cannot be drawn between strategy, tactics, and logistics. Instead, they blend and overlap in many continually varying ways. This is equally true whether one is thinking in abstract terms or in practical functional terms.

An understanding of both **pure logistics** and the broad aspects of **applied logistics** is essential to the exercise of high command. Control of over-all **applied logistics** requires a knowledge of the problems of high command – particularly a knowledge of the relationships between the functional elements of applied logistics.

And finally, in any field of human activity, in the elementary and low level stage, individuals may have to be guided by specific rules and fixed procedures in rather narrow fields. As one gains in rank and in responsibility, the fields broaden and the rules and fixed procedures are gradually replaced by mature professional judgment based on experience and understanding of principles.

In this respect **logistics** in no way differs from any other art and science that challenges man.

Oskar Morgenstern (1955)

NOTE ON THE FORMULATION OF THE THEORY OF LOGISTICS

The gradual elevation of military logistics from a pragmatic description of an army's quartermaster's tasks – as in Jomini's work – towards a more abstract, theory based and universally applicable body of knowledge is most evident in Oskar Morgenstern's *Note on the Formulation of the Theory of Logistics*.

Oskar Morgenstern (1902–1977) was not a military person. He was a German-born economist who worked in Austria and emigrated to the United States in 1938. For the major time of his academic life he was a member of Princeton University where he together with John von Neumann founded the mathematical field of game theory. Only few of Morgenstern's academic works were dedicated to the logistics discipline, like his 1936 publication *Logistics and the Social Science.* His general interest was the application of economics to subjects which did not obviously lend themselves to economic analysis.

In Morgenstern's "Theory of Logistics" paper he first tries to develop a generalized description of a logistical situation and problem. "A logistic operation consists in the supply of definite quantities of physical means and services for **activities** that according to their **missions** consume these means and services in order that the activities be maintained… The supplies come from a source and must be moved, i.e. transformed in space and time, by means of transportation to the activity". "The **logistical problem** is to identify in detail the expected requirements of the activity […]. After the identification has taken place, the quantities […] have to be computed […]".

He then goes on to specify some to the complexities of logistics decisions, such as the potential substitutability and complementarity of supplies on the one hand, the "compressibility" of requirements on the other hand. His perspective as an economist is evident in his treatment of the logistics problem as one which can be approached both from the supply-side and from the demand-side, and by his appeal for the use of markets in managing required supplies for dynamically changing, volatile "missions".

Morgenstern calls his contribution a "preliminary step to the erection of a «theory» of logistics". As such the paper reinforces the challenge to logisticians inside and beyond the military realm, which will be traced through more contributions in this volume.

P. Klaus and S. Müller (eds.), *The Roots of Logistics*, DOI 10.1007/978-3-642-27922-5_4,

Although today's logisticians will probably not agree with all of his arguments – e.g. his statements regarding the simplicity of business logistics even in large firms –, many of the logistic specifics he describes are still valid and constitute a universal characterization of logistics and logistical situations and challenges. Very appropriate and nearly predictive are the words at the end of his paper when he states:

> "The preceding remarks make it clear that a theory of [...] logistics will be difficult to establish and complicated in nature. It will have to incorporate elements from other fields, such as economics, management, mathematics, etc. It must be firmly empirical; it must be operational."

Reference:

Oskar Morgenstern: Note on the Formulation of the Theory of Logistics. In: *Naval Research Logistics Quarterly*, Vol. 2, No. 3, 1955, pp. 129–136.

Oskar Morgenstern:
Note on the Formulation of the Theory of Logistics

As a preliminary step to the erection of a "theory" of logistics, the author proposes definitions of some important concepts and identifies several types of logistical situations and problems.[1]

The lack of a rigorous and generally accepted theory of logistics at the present time is a serious impediment to making actual military logistical operations as effective as desired, although immensely complicated situations have been met, sometimes with surprisingly minor noticeable deficiencies. But, in the absence of the aid theory would render, it cannot be determined whether the solutions were really good or wasteful, and where the optimum might have been. Clearly, the future theory will have to rest on a firm empirical basis which will be won by a careful scrutiny of typical logistical operations and their descriptions in modern terms and concepts. Furthermore, it will be necessary to make as many comparative studies as possible so that the existence of perhaps several distinct types of logistical situations and problems may be discovered. Also the comparative studies – which are always called for as a quasi-experimental device when there is as yet no firm basis for a theory – must be related to the similar fields of economics and organization.

Logistical operations are primarily carried on **within** centrally directed social organizations and therefore fall under that part of the theory of organization that is supposed to cover that field. This is certainly true for military logistics which are, for the present, our main concern. There are also operations that carry outside the organization, e.g., when, instead of only distributing available materials and services, the organization has to go outside and, for example, must buy commodities that, after possible processing and packaging, are to be shipped in proper quantities to particular consuming units at specified dates. We may neglect those instances for the present. The logistical operations not only occur within an organization but usually require a special suborganization within the former. It will be formally identified as a special service if the logistic operations are sufficiently voluminous and complicated. Whether such a suborganization actually exists is not material for this discussion of the few basic principles.

In the following I shall try to give a brief description of what appears to me a possible formulation of the logistical situation and problem. This is not, by a long shot, the perfected kind of formulation which is the necessary first step towards a theory, but, more modestly, merely an outline which might at least lead towards a consistent language and henceforth to more adequate empirical descriptions in terms of this language. These, taken together, must precede any attempt towards a real theory.

---

[1] This paper is a revised and somewhat extended version of a paper contributed to the Logistics Conference held at Santa Monica, California, in 1950. It was distributed under the above title as RAND Memorandum RM-614, dated 28 May 1951.

A **logistic operation** consists in the supply of definite quantities of physical means and services for **activities** that according to their **missions** consume these means and services in order that the activities be maintained at particular present or expected future rates. The supplies come from a source and must be moved, i.e., transformed in space and time, by means of transportation to the activity. A source may itself be viewed as an activity by another source that is placed higher up in the line and so on until the final source is reached. A given activity may be connected simultaneously with several sources and any one source will, as a rule, have to supply several different activities. The flow of supplies is therefore always directed **towards** the activity, never away from it, except for disposal which may then be viewed as a separate activity. The activity can make demands, however, and makes these effective by being recognized in the general command setup and also with regard to the rate at which it is allowed to function. A flow towards the activity can also be originated by authorities other than the activity; these are then higher operating authorities not in the chain: source – transportation – activity, but comprising it.

The **logistical problem** is to identify in detail the expected requirements of the activity, if the latter are described in a form that does not already contain the complete specification and enumerations of the needs. After the identification has taken place, the quantities, corresponding to the rates of the various component phases of the activity, have to be computed in such a manner as to make the entire supply plan an "optimum" (e.g. minimizing time, or costs, or tonnage used, etc., or combinations of these), under the constraint that the desired rates should be met – within stated time limits – with some specified probability. The computation will show whether the activity can actually operate, i.e., whether its operations are feasible, at the desired rates. If the operation is found feasible, the question of the optimum will arise.

Supplies of goods can be stored at various points between the source and the activity, including the latter. Services cannot be stored, except in the form of personnel which, while held idle, may be used for other purposes. Goods and services have a **place** and a **time value**; i.e., the same good usually has a different value at one spot that at another, the difference being due to the difficulty and trouble of transporting it elsewhere and to the time that the transportation consumes. These value notions are familiar in economics. In the military setups it will be necessary to consider also **strategic value**. In economic relations place and time values are fairly stable (as a rule oranges are more desirable in the consumers' market than in the hands of the grower). But quick and frequent changes of strategic plans and situations may interfere profoundly with any established pattern of place and time values.

The values, in terms of which maximization of results or minimization of costs is to be accomplished, lead to the establishment of **priorities** for actions and sequences of actions (operations) within the logistical organization. (In business this is accomplished by profit calculation which is inapplicable to the military setup.) Since priorities can be set and sometimes have to be set simultaneously at different levels of the hierarchy, contradictions and inconsistency can easily develop. At present, there exists no theory of priorities, especially for the case when these have to be combined with monetary values possessed by items in the system to which no priorities are assigned.

Demands made by an activity shall be distinguished as falling into (a) "**vanishing supplies**," – goods which once delivered and out of the competence of the source (or of any intermediary storage point) actually disappear at the activity and do not involve the source any further; and (b) "**draining supplies**," – goods which, when having gotten into the possession of the activity, consume perceptibly of the activity's stock of goods and services and will, therefore, require further supplies to be shipped to the activity. They constitute a current of future drain on the source in addition to the direct, initial shipment. Depending on circumstances, the drain will be directly on some source or first on the activity and then be indirectly transmitted back to the source.

Clearly, all services (i.e., men) belong in the second category only. Other examples are machines requiring fuel, oil, spare parts, mechanics, etc. Examples for (a) are building equipment, food, ammunition, clothing, etc. Examples for (b) are guns (which require ammunition), men (who, having to be fed, are themselves of type (b)), etc. In composite shipments care should be taken to distinguish between these two categories if one wishes to calculate the true and full impact of such transactions upon the activity as well as upon the shipping center or source. In these estimated further allowance has to be made for the different assumptions of a steady state versus changes due to war activity, the possibility that additional local supplies may be available (transforming (b) into (a)), i.e. "living off the land," etc.

Demands made by activities can in part be based on **usage data** which are derived from past rates of consumption, as in business. They are, however, less stable, especially where combat is involved. Then new weapons are introduced there are no usage data at all so that such logistic problems as, for example, how much of different types of ammunition to store on a combat ship cannot be solved in a conventional way. The answer has to be obtained from strategical considerations that lead far beyond the narrower logistical domain.

Another profoundly important aspect of logistical operations is (1) the **substitutability** of some and (2) the **complementarity** of other goods and services. (It appears to me that these are currently too much neglected in logistics planning.) In the first case some goods can take the place of others. If the activity demands one kind of food, another might do as well; different kinds of fuel are sometimes substitutable for each other, and so on. Therefore, the source may still have a certain amount of freedom in satisfying a particular order. On the other hand, there are specific goods where no replacement is possible. Among these are specific shells, spare parts for engines, etc., etc. An order of this kind either can be filled or not, limiting the freedom of the source to possible changes in quantity. In the second case, the demand for one good is tied to the availability (somewhere along the line and at particular times) of one or more other goods. To send only guns but no ammunition (when none is at the activity) is meaningless or wasteful. Hence, it is often enough to specify one good and imply that the complementary goods will be viewed almost as component parts. The complementarity case is akin to, but not identical with, the above-mentioned of the "draining supplies." "Draining supplies" are necessarily complementary to others (namely, to those which they require over time in order to fulfill their function), but the converse is not always true, i.e., a demand for a complementary good does not necessarily involve a "draining supply." When complementarity exists, an order from the activity may restrict the freedom of the source even beyond the order itself. For example, orders may be staggered in time, but the order for good A clearly implies a simultaneous or later order for B, C, ... Of course, an entire group of complementary goods, making up a whole, may be substitutable for another, but within each the complementarity relation may hold. E.g., a stated quantity of one type of gun and ammunition may be substituted for some other quantity of another type of gun with its different ammunition.

It would be desirable that typical substitutability and complementarity relationships be worked out in advance and catalogued in far greater detail than appears now to be the case. This would increase the accuracy of foreseeing needs and the freedom for planning on the part of the source. To some extent this has been achieved by the Navy in designing hypothetical packages, such as entire bases, etc. (Acorn, for example). But in those schemes there may be too much rigidity because substitutability within the schemes probably is not sufficiently considered.

In view of the several millions of items with which the military have to deal, it is both difficult and important to secure advance information about these properties of orders. Because of the distance of the sources from the activity it will usually be impossible to work out substitutions on an intuitive basis; therefore, special devices will have to be designed that can help to yield the desired information. This work would largely fall under the cognizance of the Munitions Board Cataloging Agency.

The smaller the activity is and the shorter the distance in terms of communications to the nearest source, the easier will it be to discover such relationships, but only at the level of local information. Wider substitutions may be possible that are entirely unknown at the lower levels but should be known to the ultimate source. Here is a field that ought to be thoroughly explored because a much greater freedom of operations may exist than is now deemed possible. These considerations show that there is a need to arrive at a comprehensive system of logistical operations in which smaller operations are embedded as special units.

Next, I mention the **"compressibility" of requirements**. This is not strictly part of a logistical operation, if the latter is viewed as passive in the sense that it has to execute the demands made by the activities or to compute the requirements of a strategic plan according to rigorously specified standards. Now these standards are in fact not inviolate. They are set for a level of comfort of personnel, tactical maneuverability, state of repair of engines and machines, etc., which are defined on the basis of past experience or expected needs in the case of new situations. In those estimates a minimum is seldom chosen but rather lavish provisions are often made, depending on the economic conditions of the military establishment and of the country, as well as of the fortunes of the combat operations. The knowledge of these minima would be of greatest importance if properly distributed throughout the logistical system. The source should always know the minima of the dependent activities. The minima need not be fixed, rock-bottom values to be determined once and for all; they are more likely relative minima. They can be approached by substitutions; e.g., troops may get along on fewer calories, or get them from starches instead of from the demanded proteins. The state of engine repair can be lowered, although only temporarily, so that the minimum is attainable merely for a restricted period of time. All these phenomena appear in economic life and have to some extent been described in various forms by economic theory. However, the difficulties these descriptions encounter are considerable so that much basic work needs to be done here.

It is also known that demands made by activities are frequently overstated in the hope that at least part of the demand will be fulfilled. Such experiences are encountered within any centrally controlled organization. This is additional reason why the "true" demand is always imperfectly known. The entire logistical setup is also on these grounds of a probabilistic nature which differs from one type of activity to another and shows significant variations among the various military services. Any treatment of logistics, whether purely descriptive or aiming to exact models, must take this role of probability into account.

The question of compressibility of logistic demands is of great practical significance and is closely related to the other previously mentioned characteristics of logistical operations. The determination of optima obviously depends upon whether the set standards have to be accepted. Since compressibility does not operate proportionately across the board, the interdependency between goods due to their complementarity, to "draining", etc., must be observed, thus introducing considerable complications. If they can be resolved, it might lead to substantial economics in logistical operations but throw a heavier burden on the computational side. But this would be a desirable development.

Decisions concerning whether demands originating from an activity are true demands, whether they are compressible and ought to be compressed, what substitutions to perform, etc., make the source more than a passive organ in the whole logistic setup. It is not merely a warehouse under the control of the demands made by the activity, although such warehouses also exist. The source has a certain power of decision subject to controls exercised by higher authorities. It may therefore be viewed as a combination production and storage facility and decision point. It must make these decisions on the basis of information about its own state (including its own state of information!), the rate of the activity and its supply, including the goods in transit and about the expected future states of all these elements. Such information can only be had if there exists an information system properly embedded in the chain from source to activity. It is very likely that the greatest improvements in current logistical operations would be derived from progress in information techniques. In designing such techniques special attention will have to be paid to continuity of information patterns and storage of information in the experience of those responsible for logistical operations. The frequent changes especially of military personnel and the inadequate, and, indeed, often haphazard, record-keeping of vital components works against these requirements. Furthermore, the information system should be investigated regarding vulnerability from enemy action. Some technological perfections of information may, for example, be the exact opposite of protection of the logistical system as a whole.

There is an immediate **similarity** between **military logistics** and the logistical problems that have to be solved daily in **business**, notably in precise form in line production. There are also **differences**, the most significant of which is the size of the military problems which far exceeds anything encountered in business. Nevertheless, one may learn from a comparison, because in both fields one lacks a theory and frequently even a systematic approach. Economics has apparently neglected the study of these business phenomena altogether, but industrial engineers have made interesting contributions. In such early stages any comparative, morphological description can be of utmost importance. There is not room here to carry this out; therefore, only the following observations are made.

The simplicity of business logistics derives not only from the much smaller numbers of items even for the largest firms compared with, say, those encountered at a single Naval supply depot, but also from the fact that there is, as a rule, an instantly accessible source available in the market. Instead of having to provide a vast closed logistical system, it is possible to restrict oneself to a much smaller individual one and to supplement it as needed from the market. The overall **inclusive** individual military plan is replaced by a composite organization of many firms communicating with each other rapidly and accurately through the open market and price system. Mistakes made in supplying a producing unit can be easily corrected, so that the main problem for the single firm is that of inventory control and the positioning of men and material along a production line in quantities corresponding to the controlled flow of production. Errors are, as a rule, of minor significance and seldom of the gravity of, say, supplying a battleship with wrong caliber ammunition. Also the errors made by one firm are of only minor consequence for the economy as a whole. It is doubtful that the effective form of market organization – which means that the burden for the individual firm is minimized by shifting it to the composite organization – can be used for solving military problems. Also it should be noted that if the military setup gets out of order the disturbances are very widespread, often involving the remotest branches of the organization. Furthermore, economies with free markets have a great power of **recovery from shock**, an advantage that is apparently denied the current logistical setups of the military. It may be possible to design others where recovery is greater; here is an important field for investigation.

One of the chief advantages in solving problems of business logistics lies in the high degree of general **information** prevailing on open markets. Under most conditions a firm can easily determine to what changes its supply situation can be subjected. A price change is a signal for a firm to adjust input, output, and/or its own prices. All this information is practically offered free by the market. A military activity, on the other hand, making its requirements known to a source is generally not in a good position to know well in advance how quickly and to what extent its demands will be met. The firm operates only under the constraints of price and its own funds. I believe that deeper analyses of the problems of military logistics will show that the most difficult and most important aspects lie in the field of information and in the flow of messages and papers. It would require a separate treatment to enter upon this matter. At any rate, for the economy as a whole there exists a wide, universal information flow similar to that provided by the blood stream which carries general information to all parts of the human body, while the closed military system essentially provides separate channels and signals, akin to the nervous system. The latter, if it works well, is highly efficient, but complicated and vulnerable. It is probable that military systems in general, and military logistics systems in particular, do not provide for themselves enough general information for efficient operation. There is, in any large organization, a great need for a flow of information that is **not** specifically directed to any part of it in order to secure spontaneous adaptive response to changing conditions.

Another distinguishing feature lies in the **speed** with which **transitions** from a low state of activity to a high state occur. In business the consecutive differences in states are small or, when large, may be of a seasonal nature and are therefore well foreseen. Business production sometimes does undergo big changes in volume, but always well spread out over time compared with the sudden, almost instantaneous changes expected frequently of military logistical operations. In the latter case advance information is often totally lacking, e.g., because of the dependence on changes in strategic plans or on the fate of battles. Furthermore, the amount of change involved even within days may exceed anything known in the random changes of business production. This difference between military and business logistical situations is of major significance. It is not sufficiently appreciated by many critics of military logistical operations who, for example, do not see that a battle may transform within hours a condition of shortage of supplies into wasteful surplus or **vice versa**. This virtually never happens in business.

The best illustration of a fairly well investigated problem of business logistics and its empirical solution – not necessarily optimal – is offered by **line production**. There, principal and subsidiary conveyors consume vast quantities of materials and services which have to be held ready at precise moments of time at specified points in definite amounts. There is often a considerable variation in the final product which, of course, complicates the logistical situation. The fact that these systems exist and function well is an indication that at least empirical solutions have been found and often for significantly different conditions. It appears likely, therefore, that some theoretical penetration of these problems is possible. Such theory would offer a kernel for the theory of the wider problems of military logistics. It can only be attempted if a large amount of descriptive work has been done.

The immediate task appears to me to lie in providing adequate descriptions of typical military logistical situations and a formulation of the problems they actually contain. In this the military must help, especially officers and other operating personnel in the field itself. Any kind of functioning or malfunctioning of the various organizations must be brought out into the open. While such descriptions may exist, they are either inaccessible or are not being used in setting up models that would permit modern analysis. A study of a Naval supply depot conducted at Princeton has shown how difficult it is to discover the localization of the limitational factors, to describe the interaction of the various component parts, and to define and measure the capacities at which the Base can operate. Only after this side of the problem is straightened out can one collect the proper statistics and then formulate a model that may produce numerical results in the attempt to find an optimum solution. The difficulties in giving descriptions are formidable but they must be faced. One of the principal obstacles is that the operating personnel are seldom able (nor have the time) to do the describing of their own activities. This corresponds to the inability of an ordinary human being to say what goes on in his own body, although his experience with living is considerable. It corresponds also to the inability, or at least difficulty, of the businessman to describe and analyze his actions and to state the why and how of his decisions. Next, the military personnel in charge has its own terminology which differs from that of the economist, making complicated translations necessary in order to establish a consistent language, which is now lacking. Descriptions should be made of bases, the transport system, the decision points for inventory control, the consumptive activities, such as air fields and their units, etc. Although much is known about all this, the knowledge is not in the form in which it is required for purposes of logistical analysis which ultimately will have to be in mathematical form.

Substantial efforts have been made to establish computational procedures for determining military requirements in complete detail; for example, when a large fixed amount of money is available that has to be translated into specific orders spread over a lengthy future period. Consider an appropriation of X billion dollars to be spent over some years on aircraft. After the strategic decision has been made how many first line planes of what kind to buy, one has to spell out the precise number of spare parts, runways, hangars, pilots, canteens, hospital-beds, storage tanks, etc., etc. All these have to be ordered and produced in the right quantities, delivered at the proper places in the correct sequence and amounts. This is a typical logistical operation. Since the same program can normally be fulfilled – if at all – in a variety of manners, there is the problem of the selection of the optimal program. Methods have been developed, especially by the Air Force Project SCOOP, that allow the advance computation of the requirements well ahead of the time of placing the orders and the date of delivery. Since the details are of a highly technical character we merely refer to the existence of these techniques. The better acquainted other services become with the achievements of the kind of methods developed by SCOOP and others, the greater will be their readiness to cooperate and to provide that full and free flow of information that is indispensable.

The preceding remarks make it clear that a theory of (military) logistics will be difficult to establish and complicated in nature. It will have to incorporate elements from other fields, such as economics, management, mathematics, etc. It must be firmly empirical; it must be operational. Even though we do not have, at present, more than some elements of this future theory, we can – and, indeed, must – formulate these and other properties it will have to possess in order to satisfy. A theory of logistics will have to tell us how to determine equilibria in logistical situations, how to make transitions of new equilibria at vastly different levels of operations, how to compute optima in all cases, and how to design novel logistical organizations. These are rather formidable demands, but they are not unreasonable. Yet they show us how far we are still removed from the desired state of our knowledge. Clearly, it will take many good minds and much time and effort to bring us nearer to the goal.

Alfred Marshall (1890), Louis Weld (1919), Gösta B. Ihde (1984)

ON THE UTILITIES OF PLACE, TIME, AND POSSESSION

The use of logistics as a term and as a body of knowledge for the purpose of supplying quantities of physical materials and services to prepare for and support war has been well established and documented for more than 200 years.

But there is another, at least equally important and venerable "root of logistics" – a line of thought, which is uncovered when tracing the term backwards from some of the popular, recent definitions of the field: The American Council of Supply Chain Management Professionals (CSCMP), formerly the Council of Logistics Management (CLM), defines logistics management as "planning, implementing, and controlling the efficient, effective forward and reverse flow and storage of goods, services and related information between the point of origin and the point of consumption in order to meet customers' requirements". The question of interest here is how logistics can help meeting customers' requirements? A very popular, frequently quoted, if somewhat empty answer to this question was given by Grosvenor Plowman (cf. Stock/Lambert (1987), p. 11) through his statement of the "five rights" of logistics: logistics is about supplying "the right product, at the right place, at the right time, in the right condition, for the right cost" to customers.

At a more conceptual level these statements lead to insights about the specific ways by which logistics can add value – provide extra "utility" to products and services.

In historical order, it was the famous 19th century economist Alfred Marshall who first drew attention to the point, that there are several ways to satisfy customer needs. The obvious, traditional answer, which guided most economists' thinking in the past, was "producing things" (only nature can). Marshall made it clear that "man cannot create material things". He helped us to understand how value and utility are created beyond making resp. manufacturing products: by moving things to places and points in time, by arranging them in quantities and assortments that things are perfectly accessible und usable for customers. He discussed those value creating activities, which we now call logistics. This argument is presented in the first, very brief quote from Marshall's famous volume on the *Principles of Economics*.

Louis Weld, a professor of Business Administration at Yale University, and a former chief of the Division of Agricultural Economics at the University of Minnesota, took up Marshall's point by stating that utilities from a consumer's point of view may be of four kinds: Utilities of form, of time, of place, and of possession. Utilities of form are provided by the shape and functionality of a product, as they emerge as an output

P. Klaus and S. Müller (eds.), *The Roots of Logistics*, DOI 10.1007/978-3-642-27922-5_5,

from a manufacturing process. Utilities of time and place refer to availability of a product or service at the moment and location where they are of value to the customer. Utilities of possession refer to the adequacy of the quantities, arrangements and assortments of products according to the customer's needs, such that they can conveniently be handled and used. The latter three of those four relate to the functions which – in Weld's argument – are provided by marketing, respectively by that element of marketing which in later years was referred to as "physical distribution" and then "logistics".

Finally, as a retrospective conceptual summary of the interrelationships between the utilities provided through logistical activities (in time, place, and possession), an excerpt from an early German-language logistics textbook is added. The author Gösta B. Ihde was one of the first holders of a university chair in Business Logistics in Germany. In his text and graphs he succinctly shows how the structural characteristics of the production and consumption sectors of an economy determine – and at the same time are being dependent on – the pacing (i.e. time-related), placing (i.e. location-related), and patterning (i.e. quantity-, assortment-, and arrangement-related) services of the logistics sector.

References:

Alfred Marshall: *Principles of Economics*, Vol. 1, London; New York: Macmillan, 1890, pp. 116f.

L.D.H. Weld: *The Marketing of Farm Products*, New York: Macmillan, 1919, pp. 5ff.

Gösta B. Ihde: *Transport, Verkehr, Logistik*, Munich: Vahlen, 1984, pp. 1–3.

Further reading:

Stock, James and Douglas Lambert: *Strategic Logistics Management*, 2nd Ed., Homewood/IL: Irwin, 1987.

Alfred Marshall: Principles of Economics

Excerpt from:

Chapter III – Productive

Man cannot create material things. When he is said to produce material things, he really only produces utilities. In the mental and moral world indeed he may produce new ideas. But in the physical world, all that he can do is either to re-arrange matter so as to make it more useful, as when he makes a log of wood into a table; or to put it in the way of being made more useful by nature, as when he puts seed where the forces of nature will make it burst out into life.

It is sometimes said that traders do not produce: that while the cabinet maker produces furniture, the furniture dealer merely sells what is already produced. But there is no scientific foundation for this distinction. They both produce utilities, and neither of them can do more: the furniture-dealer moves and rearranges matter so as to make it more serviceable than it was before, and the carpenter does nothing more. The sailor or railway-man who carries coal above the ground produces it just as much as the miner who carries it underground; the dealer in fish helps to move on fish from where it is of comparatively little use to where it is of greater use; and the fisherman does no more. It is true that if there are more traders than necessary, there is a waste. But there is also waste if there are two men to a plough which can be well worked by one man; in both cases all those who are at work produce, though they may produce but little. Some American and other writers have revived the medieval attacks on trade on the ground that it does not produce. But they have not aimed at the right mark. They should have attacked the imperfect organization of trade, particularly of retail trade.

L.D.H. Weld: The Marketing of Farm Products

Excerpt from:

Chapter I: The Fundamentals of Marketing: Kinds of Utility

It has already been said that production may be defined as the creation of utilities. The classification of utilities into their various kinds is helpful in making clear the relation of the marketing process to other productive processes. Utilities may be divided into form, time, place, and possession utilities. Form utilities are created by changing the form of commodities, as in a manufacturing operation. Lumber is taken into a factory; it is sawed and cut into different shapes, and the shapes are fitted together into a chair or a table: Form utilities have been created, and the value of the lumber is enhanced. The raising of agricultural products also falls into this class. Time utilities are created by holding or storing commodities from times when they are plentiful to times when they are scarce. The storage of ice from winter to summer is an example. Likewise the storage of eggs and butter, from times of surplus production to times of deficient production, results in an addition of usefulness, or time utility. Place utilities are created by carrying commodities from one place to another. The wheat of the American Northwest would be of little value if it could not be transported from that area of surplus production to the great centers of consumption. Possession utilities are created by the exchange of goods from the hands of one person to another.

From this analysis of utilities, it is apparent that marketing deals primarily with the creation of time, place, and possession utilities. In fact production may be roughly divided into two great division, viz., manufacturing or crop raising on the one hand, and marketing on the other. Marketing begins where the manufacturing process ends. When goods emerge from the factory, or are harvested on the farm, the marketing process begins. Oftentimes, manufacturing processes intervene between marketing processes. Wool is a product of sheep raising and it has to be marketed to the clothing manufacturer. The latter manufactures clothing, which has to be marketed to consumers. At each step an increment of value is added by those who handle or transform the product.

Unfortunately, economists, in studying production, have had in mind primarily the creation of form utilities. Division of labor, large-scale production, and other problems of the organization and methods of production have been studied mainly in connection with the manufacturing end. Likewise, agricultural economists have concerned themselves primarily with the raising of crops, farm management, feeding of animals, etc., and not to any great extent with the marketing of products. Some beginnings have been made along this line within the past few years, but the marketing process has never been subjected to careful scientific analysis. And yet the marketing part of production is extremely important as compared with the manufacturing or crop-growing part. Comparison of factory or farm prices with final retail prices of almost any commodity will prove this. The difference, sometimes called the “spread” or “differential”, represents the cost of marketing – or more properly the cost and profits of marketing.

Gösta B. Ihde: Transport, Traffic, Logistics[1]

Excerpt from:

Part I: Developing the concept of logistics: From studies of the transportation function to the concept of logistical solutions

Chapter 1.1: Demand for transfers and transfer activities

Production and consumption are the most basic processes in an economy. They need to be connected in multiple ways. Production and consumption are processes of economic "transformation" – i.e. of production as the transformation of raw factors of production into products, and "consumption" as the transformation of products into the satisfaction of consumer wants and needs. They are linked and complemented through "transfer activities", taking place in a sector of the economy which is variously referred to as the transaction-, distribution-, or circulation-sector. Transfer activities are both a consequence of production and consumption processes, as they are a precondition for the efficiency of production and consumption.

More specifically: the need for transfer activities in the economy is a consequence of the division of labour between actors in the economy, of the fact of "dislocation" – that producers and consumers often are geographically dispersed – and of the fact that production and consumption take place under different time regimes. On the other hand: The structural, spatial, and temporal organization of the transformation sectors in an economy is influenced by the capabilities and conditions of the transfer sector as well. This idea is visualized in the FIGURE 1.

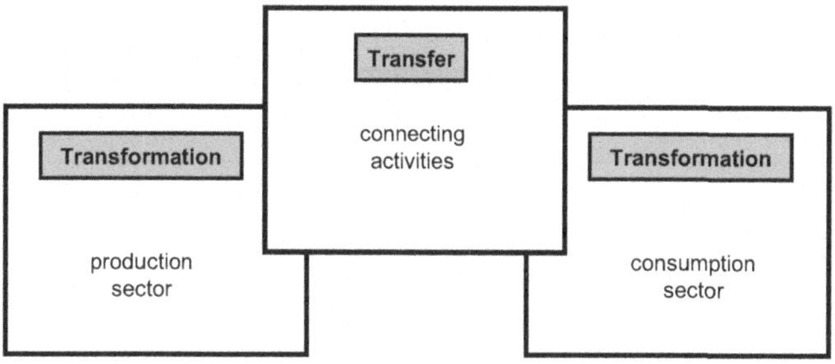

FIGURE 1: THE ELEMENTARY SECTORS OF AN ECONOMY

The sizes and characteristics of the production, consumption, and transfer sectors in an economy are mutually dependent from one another. Their interrelationship is highly variable. It depends on situational factors. The boundaries between them are not always distinct.

---

[1] Translation by the editors.

If production and consumption are seen as the primary sectors in the economy, then the transfer sector may be considered as derivative and secondary. But this is true for a snapshot moment in time, as there will be ongoing mutual adjustments between the sectors in a dynamically changing total system.

A given volume of economic activities generating a certain Gross National Product for a given population of consumers will result in a certain pattern and volume of transfer demands regarding the arrangements, quantities and types of goods to be transferred, in certain geographical patterns and distances across which products will have to be transferred, and the spans of time which will have to be bridged. The sum of those needs for transfer constitute the logistical demands in the economy.

Economic growth, and especially changes in the structures of production and consumption, will cause changes in the demand for transfer activities. If the division of labour within the production sector, or the patterns and varieties of demands from the consumption sector change, so change the demands for transfers and transaction.[2] With the reduction of the level of division of labour in an economy, for example, the need for transfers declines.

There are three fundamental types of transfer activities:

- Transfers in the order of things: arrangement and rearrangement, picking and parsing of objects;
- Tansfers in the geographical space: positioning, placing, transporting objects;
- Transfers in the dimension of time: pacing, buffering, storing.

FIGURE 2 shows the interdependencies between the fundamental structural characteristics of an economy and the types of transfer activities needed.

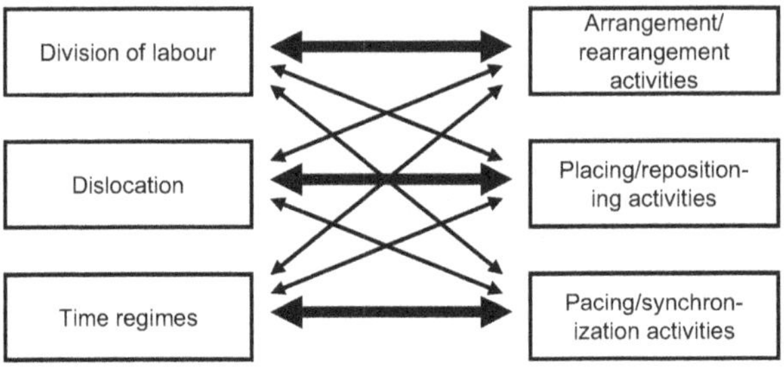

FIGURE 2: A SYSTEM OF INTERDEPENDENT INFLUENCES

The widths of the arrows indicate the relative importance of each interdependency.

---

2 Cf. the "Transaction Cost Theorem" by Ronald H. Coase: The Nature of the Firm. In: *Economica*, Vol. 4, No. 16, 1937, pp. 386–405 (comment by the editors).

The ecological and technological environment, as well as the level of socio-economic development of an economy will determine the characteristics of the transfer sector. On the other hand, the logistical capabilities and conditions of that sector will provide important stimuli for growth and structural change of that economy.

Paul D. Converse (1954)

THE OTHER HALF OF MARKETING

Paul Converse, a Professor of Marketing at the University of Illinois, was another early contributor to the understanding of logistics and its importance – without using the term yet. Logistics was seen as a part of Marketing which he – interestingly – refers to as "matter in motion".

The contribution he made through his paper on *The Other Half of Marketing* was in the recognition that Marketing really falls into at least two separate types of activity. One part of marketing, he argues, "deals with buying and selling, with creating desires and demands for goods, of transferring title to the goods". This is the immaterial aspect of marketing.

The other part "deals with the physical handling of goods, transporting them from farm and mine to mill and factory and from factory to processing plant to the consumers and storing when and where necessary. This part of marketing creates time and place utilities."

In developing his arguments Converse pointed out that a considerable part of each dollar spent in the US economy flows into marketing, as he defined it – and more than half of that into distribution activities – a fact that Peter Drucker would take up again in one of his articles a few years later. It were these considerations about the proportion of marketing in general and distribution in particular within the American Gross National Product that created awareness for logistic matters also in economic contexts.

Furthermore, Converse illustrated the enormous potentials for cost reduction in rationalizing physical distribution activities – referring to network optimization, modal choices, logistically optimized packaging, and other levers which today are part of the tool kit of professional logisticians.

Reference:

Paul D. Converse: The Other Half of Marketing. *Proceedings of the 26th Annual Boston Conference on Distribution*, 1954, pp. 310–314.

P. Klaus and S. Müller (eds.), *The Roots of Logistics*, DOI 10.1007/978-3-642-27922-5_6,

Paul D. Converse: The Other Half of Marketing

Before dividing Marketing into halves, let's look at the overall picture. Marketing accounts for approximately one-half the value created by the production and distribution of goods to the consumer. Sales of services and investment securities and military expenditures are omitted in making this statement. This division of one-half by marketing and one-half by production has held true for the period for which information is available, that is for the past forty years. All the reasonably complete estimates of the cost (or value added) by marketing since 1909 have been about the same, some a little more and some a little less than 50 per cent of that segment of the Gross National Product covering the production and sale of goods. Marketing expenses appear to be a little more "sticky" than production expenses. Hence in a Period of rising prices, marketing accounts for somewhat less than 50 per cent and in a period of falling prices for somewhat more than 50 per cent.

Some persons have reasoned that greater use of machinery has been made in production than in marketing and hence that production costs have decreased relative to marketing costs. Even some men engaged in distribution who have spoken before this convention have been more or less apologetic about the efficiency of marketing.

The trouble with the statement that production is more efficient than marketing is that it isn't so.

We have heard much about the "miracle of production". There has been a miracle of production but there has been an equally great miracle of marketing. During World War II the volume of manufactured goods doubled, the volume of minerals produced increased 30 per cent, and the volume of farm products marketed increased 25 per cent. This was done with 10,000,000 of our strongest young men under arms. This was really a miracle. However, these goods were distributed and there were probably relatively fewer military deferments for marketing institutions than for production institutions.

There have been technological advances in the production of goods. There have been matching advances in the distribution of goods. Let's cite a few examples. The cost of distribution foods through the wholesaler and retailers has been cut in half since the World War I period. At that time typical wholesaler grocers had expense of 10 per cent and margin of 12 per cent. The typical retail grocer had expense of 17 per cent and a margin of 18.5 per cent. Thus an article sold to a wholesale grocer for $1.00 would be sold to the retailer for $1.14 and the retailer would sell it to the consumer for $1.40. Today the typical chain or supermarket buys an article from the manufacturer for $1.00, takes a 16 or 17 per cent markup and sells it to the consumer for $1.20. The cost of distribution was cut from 40 cents to 20 cents. Most foods not passing through chains or supers move through voluntary or cooperative wholesalers to an independent retailer operating on a cash-carry-self-serve basis and reach the consumer at substantially the same price as charged by the super market.

Another illustration is the rapid growth of one story warehouses and the handling of goods on pallets by fork lift trucks and other mechanical devices. It has been said that the past 15 years has seen a greater improvement in the mechanical handling of goods than the previous century. One story warehouses, palletization, and fork lift trucks have reduced the cost of handling goods by some 25 per cent. A larger saving has been made by requiring the retailers to originate their own orders and so eliminate the wholesaler's salesmen. This method is almost universal in the grocery trade and several hardware wholesalers operate in this way. Another illustration of reduced costs has been in the increased use of motor trucks which save crating, and handling expenses and reduce time. A warehouse distributing to retail stores over a radius of 200 miles loads the trailers one day, the tractors pick them up that evening and deliver the goods to the retail stores the next morning. L.C.L. railroad freight would often take from 2 to 10 days. The speeding up deliveries has been so important that some middle west stockyards receive little or no livestock by rail. The truck may by-pass one, two or three middlemen and save two to four handlings of goods in the marketing of farm products.

Improvements in accounting procedures and machines have in many instances been relatively as important as the improved procedures and machines used in handling the goods through warehouses.

We people engaged in the practice or study of marketing have been too apologetic and have lacked confidence in our own work. We should be like a merchant in Illinois in the wildcat banking days when many of the local merchants were also the bankers. One store keeper decided to open a bank and hung out a sign announcing himself a banker. Very much to his surprise the very first day a customer made a deposit. The second day two more people made a deposit and still more on the third day. His confidence in his bank went up so much that by the end of the week he put some of his own money in it. We need more confidence in our work. We should realize that the overall record of the marketing segment of our economy is just as efficient as that of any other segment of our economy. There are inefficient procedures and there are inefficient distributors. If my record as a student of marketing or your record as a distributor is not good then we should be ashamed of our work and improve it. But let us realize that the overall record of market distribution is fully as good as the overall record of the production.

THE TWO HALVES OF MARKETING

Now let's look at the two halves of marketing. Marketing has been defined as matter in motion. There are two kinds of motion in moving goods from the farm, the mine, and the factory into the hands of the ultimate consumers. Goods must be moved through the trade channel and gotten into the possessions of the consumers – that is they must be bought and sold. In the study of economics you learned that a utility is something that satisfies a human want and that there are four kinds of Utility – form, place, time and possession. The creation of form utility is called production. Marketing is that part or our economy that creates possessions, place, and time utility. Goods must be moved physically they must be sorted, graded, packed, crated, loaded, unloaded, stored, and transported.

One part of marketing deals with buying and selling, with creating desires and demands for goods, of transferring title to the goods. This part of marketing creates possession utility. The other part deals with the physical handling of goods, transporting them from farm and mine to mill and factory and from factory to processing plant to the consumers and storing when and where necessary. This part of marketing creates time and place utilities.

It appears that these two parts contribute approximately the same amount to our national economy; or if you prefer involve about the same cost. This is a rough approximation owing to the fact there can be considerable difference of opinion as to just how the time of retail workers should be divided.

I want to emphasize that the two parts of Marketing involve about the same expense, or if you prefer, contribute the same amount to the national income. Yet in the study of marketing and the operation of marketing departments and businesses a great deal more attention is paid to buying and selling than to physical handling. In fact the physical handling of goods seems to be pretty much overlooked by sales executives advertising men, and market researchers. When a question is raised about physical distribution, how often have you heard: "let's refer that to the traffics manager", or "we'll let the warehouse manager worry about it", or "that's a problem for the package engineer". In other words problems of physical distribution are too often brushed aside as matters of little importance. I have many years been reading business and economic magazines. Such publications over the years have devoted relatively little space to physical distribution.

The sales manager of a medium sized company that distributes its products nationally said to me recently: "One idea that I picked up at a recent meeting of our local chapter of the American Marketing Association is saving our company $3,500 a month and we have not yet made full use of it. This is the use of in-transit rates." A student of marketing in one of our large industrial cities tells me that in-transit privileges is near the top subject of interest among the marketing executives in his city. In-transit and mixed car rates often save several hundred dollars a car in freight costs. The railroads seem willing to establish in-transit rates as this device is particularly useful in meeting truck competition since few trucking lines grant the privilege.

In-transit and mixed car rates are important but there are other devices of equal or greater importance in the physical distribution of goods. I was talking the other day to a man who only a few years ago started a wholesale business. In a very few years he has built up the business until it has sales of several million dollars annually. He preferred to start and grow in rented quarters as he could expand his space as the business grew. Now that he has the business well established he is thinking of a building of his own and estimates that he can reduce his expenses 12 per cent with a modern one-story warehouse.

Everyone needing a warehouse seems to want a one-story building on the fringe of a city. It is possible that the falling off in demand for space in multi-story warehouses will decrease until the rent of such building will decline to a point where the overall cost of operation will be as low as in a one-story warehouse.

I pointed out above that the cost of distributing groceries from manufacturer to consumer had been cut in half in past generation. This has been done by giving attention to both halves of the marketing process – buying and selling and physical distribution. The chains first and more recently the supers and wholesaler-retailer independents have very largely eliminated the selling operation between wholesale warehouse and retail store. The retail store manager or operator, as the case may be, requisitions his goods from the check lists and order blanks supplied by the wholesale warehouse. This saves time and leads to better stock keeping on the part of the retailer and saves most of the expense of salesmen on the part of the wholesaler.

Food distributors also appear to be leaders in reducing the costs of physical distribution. The chain organizations and especially the younger ones which include most of the supermarkets have been leaders in moving into one story warehouses and handling goods on pallets by fork lift trucks and other mechanical devices. Some of the wholesalers operating on the cost plus basis and sponsoring voluntary groups of retailers have very fine, modern warehouses. Receiving goods in carload rail shipments and supplying the retail stores by trucks seems to be the preferred method at present. The retail stores are usually operated on a self-serve basis in which the consumer assembles her own order, takes the goods to the check-out counter, and transports them home. Much attention is given to handling goods in the store, the selection of shelving and display racks, the most economical method of getting goods on the shelves, the fastest kind of checkout stands. It has been said that the retail food store is today primarily engaged in physical distribution, the goods having been presold by either the manufacturer or the distributing organization. This is an overstatement, since the display of goods in the store is very, very important.

The point is that the record of food distributors has been made by giving attention to both demand creation and physical distribution.

One may ask why the food distributors gave more attention to reducing expenses than other trades. The answer is that food distribution is more competitive than some other industries and they had to do so to stay in business.

SOME OTHER PROBLEMS IN PHYSICAL DISTRIBUTION

Should all goods be shipped from the factory or should regional stocks be carried? If regional stocks are desirable should wholesalers be used? Should stocks be carried in public warehouses or should the company provide its own warehouses? If regional stocks are needed where should they be located? What cities? In what locations in these cities? What are the freight rates to the towns and from the towns to the territories to be served? The use of trucks may change the preferred location of warehouses. For example most goods are manufactured in the east. In going west it was customary to locate the warehouses in the eastern part of the territory. Goods would be shipped in carloads to such towns as St. Louis, Kansas City, Omaha, or Minneapolis and be distributed in L.C.L. shipments to territories located for the most part west of the distributing centers. With truck distribution it may be better to locate the warehouse in the center of the territory. For example, the Iowa warehouse may have been located in Davenport. Today Des Moines may be a better location.

The Nebraska warehouse may have been in Omaha. Today, perhaps an interior town as Grand Island may be better.

Are transit privileges and mixed-car rates available? If not can they be secured? Is more than one railroad available? Even if the rates are the same, services may be better where there is competition. What truck services are available? Is the warehouse well located as to railroads and highways with quick access for incoming and outgoing goods? How frequent and how reliable are the carriers? What are the insurance rates? What banking facilities are available to make loans on the goods? What about the labor situation? Is the police protection adequate or must you provide watchmen?

Should goods be shipped by rail, by trucks, by water or by air? If trucks are to be used should the company operate them or use contract carriers? If operated, should they be owned or leased?

What kind of containers should be used? How should the goods be packed? Can the product be shipped in bulk or is it better to pack it? If it must be packed how and in what type and size of containers? What materials should be used for containers? Should the goods be shipped on pallets?

What types of warehouse are needed? What equipment should they have? How should the warehouse be designed? Is it to be one story or multistory? Should it be of fireproof construction? What materials? What refrigeration is needed? How many rooms are needed? What else? What temperature? Should the warehouse be heated? Is there room for expansion if the business grows?

What material handling should be used: Gasoline, or electric, or hand trucks? Overhead or subfloor belts to move the trucks or dollies? The width of the building and passageways are important. What size docks for loading and unloading, incoming and outgoing trucks? How much railroad siding is needed? Is it economical to have goods palletized in rail cars or in trucks?

What layout or arrangement of stock? Should goods be placed in warehouse in same sequence as on order blanks or requisition sheets? What kind of labor should be used? What use can be made of part-time workers such as students? How should workers be paid – straight time, or piece work basis?

Retail stores have special problems. How much self-service or self-help can be used? How should stock be replenished? What kind of fixtures? Should reserve stock be carried? If so, where? Should fixtures be made or bought? Can the customers be educated to assemble their orders as in a super market? Should goods be delivered to customers? If so, should this be free? In self-service stores what kind of check-out equipment is best? Can customers be educated to unload the carts and put their goods on the check-out counters?

VARIATION WITH TYPE OF GOODS

I said above that there were two kinds of movement involved in the marketing of goods and that the two kinds were of approximately equal importance. This is the overall picture. With some products buying, selling and demand stimulation are relatively more important and with other products the physical handling of goods is relatively more important. In general we say that demand stimulation, advertising, selling advice and personal adjustment to consumer needs is of first importance in the marketing of fashion and specialty goods. On the other hand we say that physical distribution is of major importance in the distribution of staples and convenience goods.

With a new product or a specialty the good must be called to the consumer's attention, its service to the customer must be explained and demonstrated. In other words the product must win consumer acceptance and this takes time and involves heavy expense. Often the product must be tried on and altered if a garment, or delivered to the home, installed, demonstrated, and guaranteed if a home appliance.

On the other hand a staple of convenience goods has won consumers acceptance. Little demonstration and persuasion is needed to sell it. Some 40 years ago Parlin pointed out that a woman usually shops in three stores before buying a fashion good. On the other hand once a customer enters a grocery or drug store she in nearly all cases makes a purchase. Thus the marketing of staples and convenience goods usually involves less expense in the buying and selling operation than the marketing of specialties and fashion goods.

CLASSIFICATION OF GOODS CHANCE

A fact too frequently overlooked is that a good does not always remain in the same classification. When new it may be a specialty requiring active promotion. When it wins consumer acceptance it may be a staple. It is said that when Gillette perfected his razor, it cost less than 50 cents to make it. He proposed to sell it for $1.00. However, before placing it in the market he consulted an "expert" – (We didn't have marketing researchers in those days.) This expert told him to sell the razor for $5.00 – that he could not pay the promotional and selling expenses of getting it to the consumer for 50 cents. He priced the razor at $5.00 and sold it successfully. However, once safety razors were accepted the manufacturers were glad to give them away in order to sell the blades.

Many products have gone through similar changes. Many of the leading sales managers of the 1920s won their spurs selling dry breakfast cereals in the 1900s. It took a good deal of selling to persuade us to eat the various and assorted kinds of cowfeeds that we now eat and enjoy for breakfast. But eat them we did found they soon became about as staple as anything we try. The housewife picks them up on her trip through the supermarket. We used to sell washing machines by carting to the homes of prospects and doing the family wash to prove to the housewife that the machine would get the clothes clean, would not tear off the buttons, and that its operation was safe. Today the great majority of consumers accept the washing machine and will buy without demonstrations. Silk stockings not so many years ago were a high priced luxury good. Today nylons are staples or almost so and can be bought in many super markets.

The good may not change its classification in all places at the same time. Today air conditioning units appear to staples in Texas while in Illinois they still need considerable promotion.

When a good changes in classification, the type of salesmanship and advertising changes. Instead of high-pressure salesmen, order-takers may do a satisfactory job. Instead of reason-why copy, reminder copy may be indicated. Thus the type of marketing expenses changes. Physical distribution becomes relatively more important and demand creating relatively less important.

Does not much of the present turmoil over discount houses and discount selling come from the failure to realize that many home appliances and other goods are no longer specialties and that they do not require as much advertising and salesmanship as they once did? It appears that most manufacturers have set up discounts to retailers and wholesale distributors determined and adapted to the promotional stage when appliances really had to be "sold" by the use of demonstrations and high performance advertising and salesmanship. Today many of these appliances are "presold" by the manufacturers, or by the retailer (as in the case of Sears Roebuck). If the manufacturer sells the product it seems no longer to be necessary for the retailer to sell it all over again. The problem becomes more and more one of the physical distribution – getting the appliance delivered to the home of the consumer. It seems that many discount houses are willing and anxious to do this for considerably less than the margin allowed by the manufacturers to the wholesale and retail distributors.

Some department stores are talking of discontinuing appliances and others of getting their own brands and doing their own promoting. Many of us can remember when the department stores sold groceries. When the chains reduced the expenses of selling groceries the department stores dropped their grocery departments. The department store is organized and operates primarily to sell goods needing promotion. The department store may have to stick to the promotional type of goods or else through changed methods of operation get its expenses down where it can handle the more staple goods competitively.

Edward W. Smykay/Donald L. Bowersox/F.H. Mossman (1961) and Hans-Christian Pfohl (1969)

PHYSICAL DISTRIBUTION AND MARKETING LOGISTICS

Smykay/Bowersox/Mossman's first textbook on physical distribution and logistics management, published in 1961, was the forerunner of a series of similar publications in the United States during the following years. Among the most often quoted ones were Heskett/Ivie/Glaskowsky's *Business Logistics* (1964) and Bowersox/ Lalonde/Smykay's *Readings in Physical Distribution Management* (1969).

But it was only at the end of the 1960s decade that attention was paid to this important development in other countries. In Germany it was Hans-Christian Pfohl, then a doctoral student, who – for the first time in this country and probably in all of Europe – published a short paper on the topic, which is translated and reprinted here: *Everyhing for Replenishment!*

The paper was intended as a "wakeup-call" to managers and academics in the "old world". Despite its brevity and the apparent effort by the author to keep it understandable to practicioners, it touches all the important issues and opportunities of professional logistics management which were seen at the time, such as the potentials of logistics not only to open up new ways for cost reduction, but also to increase competitiveness and value to customers, and the potential of the uses of computers.

Just two years later Pfohl published the first German language textbook on the basis of his doctoral thesis under the title *Marketing Logistics* (Pfohl (1972)). It was followed by many editions and became the most widely used logistics textbook in the German language logistics community. About the same time the first university chair in logistics in Germany was established at the University of Mannheim.

References:

Edward W. Smykay, Donald L. Bowersox, and F.H. Mossman: *Physical Distribution Management*. New York: Macmillan, 1961.

Hans-Christian Pfohl: Alles für den Nachschub. In: *Der Volkswirt. Wirtschafts- und Finanzzeitung*, special reprod. from No. 17, 25th April, 1969, pp. 159–161.

P. Klaus and S. Müller (eds.), *The Roots of Logistics*, DOI 10.1007/978-3-642-27922-5_7,

Further reading:

Bowersox, Donald L., Bernard J. Lalonde, and Edward W. Smykay: *Readings in Physical Distribution Management: The Logistics of Marketing*, New York: Macmillan, 1969.

Heskett, John L., Robert M. Ivie, and Nicholas A. Glaskowsky jr.: *Business Logistics*. New York, London, Sydney: Macmillan, 1964.

Pfohl, Hans-Christian: *Marketing Logistik. Gestaltung, Steuerung und Kontrolle des Warenflusses im modernem Markt*, Mainz: Distribution Verlag, 1972.

Edward W. Smykay/D.L. Bowersox/F.H. Mossman: Physical Distribution Management

Preface

This book is the result of investigations in the practical arena of business decision making. The long period of cooperative research and consultation between the academic and business communities stands as its underlying bedrock. The executives of large and small firms, railroads, airlines, pipelines, motor carriers, water carriers, and a long list of distinguished traffic managers of industrial firms all contributed to the conception of principles and their applications as expressed in the following pages.

The intent of the authors, in this initial work, was to present the broad outlines of distribution system analysis. Therefore, certain areas related to distribution administration and practices were purposely omitted. The whole area of communication and certain advanced mathematical models were avoided so as to concentrate more adequately on the main themes of distribution analysis.

In a sense, the authors consider this to be a pioneer work. However, many of the techniques employed have been well known in widely separated disciplines. This book brings together mathematics, statistics, transportation, marketing, and economics for the purpose of analyzing the physical flow of goods through a network.

[…]

E.W. Smykay
D.J. Bowersox
F.H. Mossman

CHAPTER I
Physical Distribution

Physical distribution can be broadly defined as that area of business management responsible for the movement of raw materials and finished products and the development of movement systems. In order to understand the need for physical distribution planning in the modern firm, it is important to realize that American industrial development has passed through a number of definable stages each characterized by a general type of business orientation. These orientations have, to a large extent, determined what factors historically received primary emphasis in business planning.

In the initial stage of American industrial development, the most significant problem revolved around the scarcity of goods and services.[1] This resulted in the application of organizational principles and technological development to the problems of production.[2] In these early stages, the dynamics of economic development encouraged market expansion and technological improvement. This tended to place the engineer and innovator at the forefront of business organization. The Carnegies, Whitneys, and Astors were characteristic of this period. Continued technological improvement resulted in the need for larger aggregates of capital. The total increase in productivity, combined with the closing of the frontier, increased the vulnerability of industry to business fluctuations.

The entrance of the American economic system into its next stage became apparent with the need for rationalization and control of the production orientation. This resulted in the financier and the lawyer gaining dominant positions in the organization structure. It was at this point that the major amalgamations of the early twentieth Century developed. Such economic amalgamations were aimed at market stability and the long run health of the individual enterprise.

Finally, when the market could no longer regularly absorb increased output, it became necessary for the business enterprise to cultivate and expand sales within geographically imposed limits. It was at this stage that business orientation became introspective and began to utilize the specialized knowledge of the marketing expert. In order to maintain or expand its competitive position, it became increasingly nec-

---

1 Some parts of this historical development originally appeared in: Physical Distribution: A New Concept, *Managerial Marketing: Perspectives and Viewpoints* (E.J. Kelley and William Lazer, ed.) pp. 359-63, Homewood, Illinois: Richard D. Irwin, Inc., 1958.

2 For one of the earliest published works in this field, see Frederick W. Taylor: *Scientific Management.* New York: Harper and Brothers, 1957. This volume contains two of Taylor's major works, *Shop Management* (1902) and *Principles of Scientific Management* (1911).

essary for the individual firm to obtain reliable and complete information on alternative courses of marketing action. These alternatives involved the relationship of supply and demand. From the marketing viewpoint, demand embodied opportunity and supply required organizational effort.[3]

Marketing Orientation and Physical Distribution

During the early stages of marketing awareness much progress was made in the application of scientific methods to the field of demand analysis. For example, the scientific techniques of market research were first applied immediately following World War I.[4] The application of statistical techniques generally began during the 1930s. In later years substantial progress was made in applying the contributions of sociology and psychology to the problem of identifying and understanding qualitative market information.[5] Each of these advances, to varying degrees, was aimed at replacing the basic assumption of "economic rationality" which was necessary in the initial development of economic models. In place of the "economic man" modern marketing became concerned with the consumer as a purchase unit having a variety of wants and normally having imperfect information. The executive became concerned with molding total corporate offerings in an effort to capitalize upon market opportunity. This embodied demand analysis in order to achieve the goal of maximum consumer satisfaction. The customer was assumed to represent the focal point of business activity, and the market was viewed as holding veto power over the business enterprise.[6]

With a full realization that sales result from organized effort applied to market opportunity, many firms created a marketing department as part of the formal organization structure. The effective marketing department became charged with the responsibility of coordinating the total activities of the firm in order to capitalize upon areas of market opportunity. A new organizational position referred to as the Marketing Manager evolved to guide this coordination process.[7]

---

3 Wroe Alderson: *Marketing Behavior and Executive Action,* p. 355. Homewood, Illinois: Richard D. Irwin, 1957.

4 C.S. Duncan: *Commercial Research,* New York: Macmillan, 1924.

5 For an example of the type and quality of information see: Lincoln H. Clark (ed.): *Consumer Behavior: Research on Consumer Reactions,* New York: Harper and Brothers, 1958. For an example of research techniques which can be employed in qualitative market research see: W. Lloyd Warner and William E. Henry: The Radio Day Time Serial: A Symbolic Analysis, *Genetic Psychology Monographs,* 1948, Vol. XXXVII, pp. 3–71.

6 Thomas A. Staudt: Business Management as a Total System of Action and: The Role of Marketing, Kelley and Lazer, op. cit., p. 6.

7 For a good discussion of the role of the marketing manager see: John A. Howard: *Marketing Management.* Homewood, Illinois: Richard D. Irwin, 1957.

Within a framework of basic Company objectives[8] the marketing manager has the responsibility of selecting among alternative courses of action. His function is to select that one course which, on the basis of market knowledge, appears consistent with the criterion of profit maximization.[9] This embodies the coordination of market research, product development, pricing, personal selling, advertising, channel selection, and location decisions to best meet changing market and competitive conditions.

The activities of integrated marketing have centered on the areas of product variation, price strategies and sales promotion. Lesser attention was directed to marketing channels, transportation and location. In part, this can be attributed to newness of integrated marketing and the lack of outstanding improvements in techniques of physical supply. However, this was a temporary situation. Innovation was soon forthcoming in the physical movement system. Improvements in transport productivity resulted from efforts on the part of transportation agencies to improve their own competitive position.[10] These improvements redounded to the benefit of industrial firms who, when applying systems analysis, could so arrange plant capacity and warehouse facilities, with methods of intercity movement, to hold costs at a minimum while meeting market requirements. Since marketing efforts are largely carried on by the individual firm, each shipper must so arrange his product movement that it complements his total marketing effort. Because of the important opportunities this approach offers for increasing movement productivity, this book is devoted to a discussion of the theoretical and practical possibilities involved.

The Field of Physical Distribution

Marketing in a broad sense is defined as "that phase or aspect of any economy that has to do with and results in the changes in custody of, responsibility for, and authority over goods, to the end that goods produced by many agencies are made available for the convenience and satisfaction of different users."[11] Marketing defined in this manner is not limited to points where production and consumption occur. Rather, it becomes an element which penetrates throughout the entire econ-

---

8 For a discussion of basic company objectives, their formulation and the dynamics of corporate change see: Peter F. Drucker: *The Practice of Management.* New York: Harper and Brothers, 1954 and Herbert A. Simon: *Administrative Behavior.* New York: Macmillan, 1959.

9 Howard describes the marketing manager as operating under a framework of previously made decisions. Within this framework, the marketing manager selects between alternatives on the criterion of contribution to profits. See: Howard, op. cit., p. 22.

10 These improvements have largely accounted for the shifts in volume which have occurred in recent years among the various forms of inter-city transport.

11 E.D. McGarry: Some Functions of Marketing Reconsidered, *Theory in Marketing* (Reavis Cox and Wroe Alderson, ed.), p. 267. Chicago: Richard D. Irwin, 1950.

omy. It is within this context that physical distribution should be included as an integral part of the marketing effort of the firm. As Professor Converse so ably pointed out, physical distribution embodies "the other half of marketing".[12]

Physical distribution is concerned with the spatial arrangement of plant capacity and warehouse facilities so that movement costs are cut to a minimum while marketing requirements of the firm are satisfied. The connecting link between such locations is transportation. The field of physical distribution, therefore, revolves around the selection of the number of plants and their locations, the determination of the number, size and geographic arrangement of warehouse facilities and the choice of transportation methods employed. The object is to maximize both short and long run profit opportunities of the firm.

The major subject areas included in physical distribution have been matters of intellectual pursuit for a number of years. The need for adequate physical supply, as a basic part of marketing, was clearly noted in the 1920s.[13] Modern literature continues to emphasize the role of transportation and storage as one element of marketing.[14] The newer managerial approach to marketing includes location decisions, a major part of marketing strategy.[15] The study of transportation, per se, has been a major part of business and economic education for a number of years. At least since 1800, transportation has been a vital part of internal American economic development.[16] The study of general economic theory, from which many techniques employed in physical distribution are borrowed, has been a major area of academic interest for years. The problems of economic location were first considered in Ger-

---

12 Although Professor Converse's definitions of physical distribution are slightly different than the ones presently employed he excellently highlights the area of physical distribution. It is interesting to note the comments concerning costs of physical supply. Professor Converse points out such costs represent one-half of the total costs of marketing. See: Paul D. Converse: The Other Half of Marketing, *Marketing in Transition* (Alfred L. Seelye, ed.). New York: Harper and Brothers, 1958.

13 Two early works noting the importance of physical supply were completed by Percival White and Fred E. Clark. White's book, probably the forerunner to modern managerial marketing, contains an interesting discussion of the forces and functions of concentration and dispersion. See: Percival White: *Scientific Marketing Management.* New York: Harper and Brothers, 1927. In 1924, Clark completed one of the first collections of marketing readings in which he saw fit to include a chapter entitled Physical Distribution. See: Fred E. Clark, *Readings in Marketing,* chapter XV. New York: Macmillan, 1924.

14 For example, see Chapter XXVI of the latest edition of Phillips and Duncan entitled, Policies and Practices Concerning Physical Supply, Charles F. Phillips and Delbert J. Duncan, *Marketing,* 4th ed. Homewood, Illinois: Richard D. Irwin, 1960.

15 Howard, op. cit., chapter XV.

16 For a historical review of transportation as an element in economic growth, see: Emery Troxel, *Economics of Transport,* chapter IV. New York: Rinehart, 1955.

many around the end of the nineteenth Century.[17] Although historically it has not been a major concern in economics, several excellent plant location contributions have appeared since World War II.[18] Warehouse literature has been centered primarily on internal operations; consequently, very little has been written purely on warehouse location.[19] However, warehousing represents a major part of physical distribution.

It may appear illogical that physical distribution is referred to as a new area of managerial concern since all integral parts are so firmly embedded in historical literature. The uniqueness of physical distribution is found in the integration of these several bodies of knowledge into a framework for marketing action. The perspectives of physical distribution allow plant location, transportation and warehousing to become a basic part of modern marketing strategy.

A common misconception is that physical distribution is a modern treatment of traditional marketing "channels of distribution". It is important to realize at the outset that the theoretical and analytical insights developed in physical distribution are independent but inclusive of traditional marketing channels. In many cases when alternative channels are available, distribution analysis will help in selecting the one which offers product movement at lowest total cost. However, there are many factors involved in channel selection other than cost. The types of channel compatibility, conflict, control and communication have received considerable attention in marketing literature and therefore are not major areas of concern in this book.[20] The marketing manager must arrive at a selection of channels which will maximize his market effort. The channel which he selects to a large extent will determine what specifications the distribution structure must satisfy. Throughout this book, the implicit assumption is that the firm employs an integrated distribution structure owning all plant and warehouse facilities necessary for product movement to the point of ownership transfer. This point may be the consumer, retailer, agent middleman, wholesaler, etc. Physical distribution analysis is applicable regardless of the channel decisions of the firm.

---

17 The contributions of the German theorists, namely von Thünen, Weber, Predohl, Palander and Lösch are reviewed by Philip Newman among others. A partial review is contained in Chapter VI of this book. See also: Philip Newman: *The Development of Economic Thought.* Englewood Cliffs, N.J.: Prentice-Hall, 1952.

18 Probably the three outstanding American contributors are: Melvin L. Greenhut: *Plant Location in Theory and Practice.* Chapel Hill: University of North Carolina Press, 1956; Walter Isard: *Location and Space Economy.* New York: John Wiley and Sons, 1956; and Edgar M. Hoover: *The Location of Economic Activity.* New York: McGraw-Hill, 1948.

19 Frederick offers one of the few theoretical insights into warehouse location theory. See: John H. Frederick: *Public Warehousing.* New York: Ronald Press, 1940.

20 Richard M. Clewett (ed.): *Marketing Channels.* Homewood, Illinois: Richard D. Irwin, 1954.

[...]

The Study of Physical Distribution

The purpose of this introduction is to develop a general understanding of the field of physical distribution including the reasons for including this area as part of the marketing endeavor. The remainder of the book is devoted to a detailed discussion of the many components of distribution systems. The general organization is from the broad to the specific. In Chapters II and III, the structure within which product flow takes place is examined in detail. Chapter II is directed to a discussion of marketing forces influencing the selection among distribution alternatives. This chapter highlights the fact that many forces shape the market effort of the individual firm. To a large extent these same forces influence the distribution system which must complement the total market effort at each stage of the marketing process. The structure of the intercity transportation system is described in Chapter III. The discussion includes the advantages and disadvantages of each form of movement, shipment size limitations and the characteristics of transport rate structure. In total, these two chapters present the framework of physical distribution, setting the stage for more analytical considerations.

The total cost technique of selecting a distribution method to serve a given market at lowest cost is introduced in Chapter IV. Included are the various methods of dividing fixed and variable expense which can be used in total cost analysis. Since the control of inventory is an important element in physical distribution, a discussion of sales forecasting and inventory control is presented in Chapter V.

The next three chapters are devoted to location theory and practice. In Chapter VI, the major contributions to spatial economics of both German and American writers are reviewed. Attention is then directed to the role of location analysis in physical distribution. The chapter is concluded with a discussion of scale economies and their impact on location decision. Chapter VII contains an elaboration of the many factors influencing plant location decisions. A plant location procedure and checklist are developed to guide the process of site selection. In Chapter VIII, the specific element of transfer cost is given detailed treatment. A mathematical model for isolating the least transfer cost location for a plant is developed. The final part of Chapter VIII traces the historical growth pattern from single plant and single warehouse to multi-plant and multi-warehouse systems.

In the next three chapters, the warehouse, as a component in physical distribution, is examined in detail. The change from a concept of the warehouse as simply a place of semi-permanent storage to that of distribution unit is described in Chapter IX. In Chapter X, a technique for determining when a warehouse is economically justified is developed and illustrated by a case study. The marketing manager is introduced to the specific problems of establishing and controlling warehouse operations in Chapter XI. The orientation is for the executive who has warehouses under his control rather than for the specialized warehouse supervisor.

There are some areas of engineering normally included in physical distribution which have been purposely omitted. Specifically, the areas of materials handling, layout and warehouse design have only been briefly discussed in Chapter XI. The complete areas of packaging and containerization have been omitted because they represent a phase of engineering only indirectly related to physical distribution. These areas of specialized concentration have been neglected in an effort to concentrate on the managerial aspects of physical distribution.

SELECTED BIBLIOGRAPHY

Alderson (1957): Alderson, Wroe: *Marketing Behavior and Executive Action.* Homewood, Illinois: Irwin, 1957.

Converse/Huegy/Mitchell (1958): Converse, Paul D., Harvey W. Huegy, and Robert V. Mitchell: *Elements of Marketing.* Englewood Cliffs, N.J.: Prentice-Hall, 1958.

Howard (1957): Howard, John: *Marketing Management.* Homewood, Illinois: Irwin, 1957.

Kelley/Lazer (1958): Kelley, E.J. and William Lazer (eds.). *Managerial Marketing: Perspectives and Viewpoints.* Homewood, Illinois: Irwin, 1958.

Maynard/Beckman/Davidson (1957): Maynard, H.H., T.N. Beckman, and W.R. Davidson: *Principles of Marketing.* New York: Ronald Press, 1957.

Phillips/Duncan (1956): Phillips, C.F. and Delbert J. Duncan: *Marketing.* Homewood, Illinois: Irwin, 1956.

Seelye (1958): Seelye, Alfred L. (ed.): *Marketing in Transition.* New York: Harper and Brothers, 1958.

Shubik (1959): Shubik, Martin: *Strategy and Market Structure.* New York: Wiley, 1959.

Hans-Christian Pfohl: Everything for Replenishment![21]

Keeping the Distribution Pipeline Optimally Filled through Marketing Logistics – Utilizing Computer Support

MARKETING LOGISTICS

Even the best efforts on the sales front will have little effect if the supply of goods comes to a halt. The ability to deliver in a buyer's market is one of the "essentials" of a successful marketing policy. Monitoring this ability has become the task of a new type of specialists: the experts of marketing logistics.

Marketing logistics has become an important topic in the USA since the early 1960s. A significant literature has developed there since, and an expanding range of teaching programs is offered by universities. Corresponding developments in Germany, as in other fields of science, are still lagging behind the American development so far. Those very few papers that were published on the topic so far in German were done by business consultants. Nevertheless, this means that the practioners in the management of German companies begin to be confronted with marketing logistics now.

The aim of marketing logistics is to make sure that the right product is supplied to customers at the right time and the right place. Marketing logistics is becoming the link between the production of goods and the consumers and user. If a certain product desired by a customer is not available, he might choose something else. All selling endeavours were then for nothing. It will depend on marketing logistics, whether a sale will be made successfully, or not (cf. Cordes (1968), p. 93).

INTERDEPENDENCIES

The logistical system is interacting closely with the other instruments of marketing (see EXHIBIT 1, on the next page).

---

[21] Translation by the editors.

EXHIBIT 1: LOGISTICS WITHIN THE MARKETING MIX

Hence, for example, product design influences the type of transportation and storage needed. If the range of products offered is changed or expanding, changes in the logistical system will have to be made. Price reductions may be justified on logistical considerations. Promotions or sales campaigns usually also require particular logistical measures in order to meet the needs of a surges in demand. And marketing logistics comprises for its part a whole complex of miscellaneous but interdependent decisions (See EXHIBIT 2).

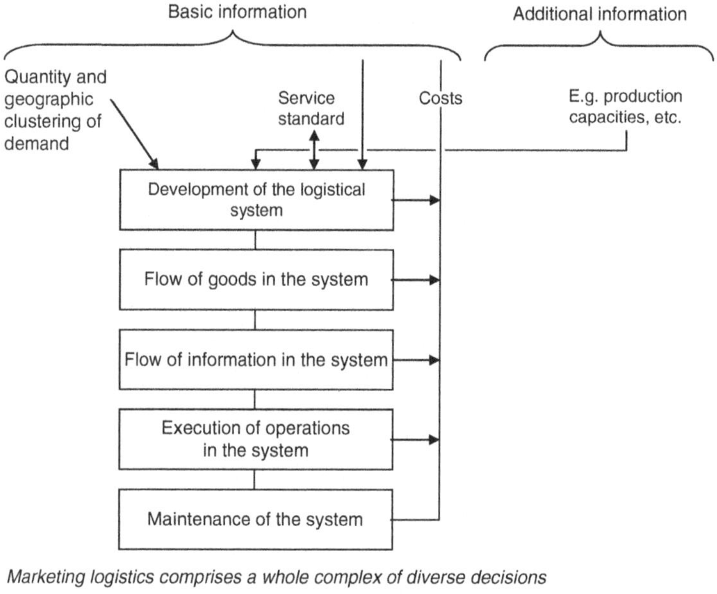

EXHIBIT 2: DECISION AREAS OF MARKETING LOGISTICS

MARKETING LOGISTICS' SCOPE OF DECISIONS

The decisions that have to be made when developing a logistical system are about the number, the type and the location of delivery points. Is it sufficient to keep inventory of product at the manufacturing plant, or is there a need to contract for additional field warehouses through third party providers?

It is of strategic importance how the warehouses in a distribution network are interconnected with each other, and how they are linked with customers. The structure of network is determined by transport routes, and the types of carriers utililized (internal or external). Once the logistical network is established, the flow of goods through the network needs to be specified: Should the full range of products be available at any field warehouse, or should certain items be held only at central locations? The determination of optimal stock keeping levels, of reordering quantities, of reordering schedules and safety inventory levels are other issues which have to be resolved. Additional questions at even deeper levels of detail are: Where and how should goods be phyisically stored inside a warehouse? What types of handling and conveyor equipment are required? The choice of protective packaging for the purposes of efficient logistical handling has to be decided. Since every system requires maintenance, further considerations are to be given to repair and replacement of warehouse facilities as well as of vehicle fleets as part of the set of marketing logistics decisions.

How those decisions will be made depends on the following basic types of information:

1. Quantity and geographic distribution of demand: market research will help to secure this information.
2. Cost: In most firms there is only insufficient information about the costs of marketing logistics activities.
3. Service standards: Marketing has to define the required levels of delivery service in cooperation with the logistics experts.

Beside the optimal supply of customers with products, marketing logistics is to develop a favourable cost-benefit-ratio between the delivery service provided and the affiliated costs. Generally this requires to find trade-off solutions between the two extremes of "instant delivery capabilities" and the "absolute minimization of cost".

The claim that the marketing conception is only a scientific paraphrase of the saying "the customer is king" or "the customer is always right" is at least not valid for industrial firms. Even though, every sales person would like to guarantee the demanded logistics service to his or her customer, the logistician needs to intervene as soon as a promised delivery service is faced with economic concerns.

Often companies are inclined to offer an excessive delivery service. They pay little attention to the service costs and are not aware of costly consequences of changing service level.

INDEPENDENTLY OR IN PROJECT TEAMS?

The question about a logistician's responsibilities is related to his place in the organization. Today, the various activities of marketing logistics are usually dispersed in different departments and are subject to the respective goals of those departments. As a consequence, is is likely that goal conflicts arise. This often requires reorganization and the alignment of goals, unless the conflicts are negligible.

Three basic forms of the organizational coordination of marketing logistics activities are suggested:

1. Centralization of all logistical activities in one department.
2. Consolidation of all activities related to marketing logistics and a reporting line to the head of marketing or the general manager.
3. Assignment of all logistical routine activities in the line hierarchy. A matrix organisation regulates non-routine activities and the resolution of specific logistical problems through project teams, which may be changed according to the demand of the prevailing problems.

COST SAVING POTENTIALS UNREALIZED

American companies have moved beyond the stage in which the logistical activities were situated somewhere between production and sales (cf. Neuschel (1967), pp. 125 et seq.). Top management recognized that the previously neglected field of marketing logistics offers opportunities for more cost reductions, once rationalization potentials are exhausted in production. Furthermore, companies have been taking on the opportunity to increase the availability of their products and gain competitive advantage over their competitors by improved logistical service levels.

The relevance of marketing logistics for individual firms depends on the industry and on the competitive environment. An investigation in the United States illustrated that the cost of logistics systems fluctuates between 10 percent (engineering industry) and 30 percent (food industry) of the turnover (cf. Snyder (1963), pp. 35 et seq.). This is why especially companies in the food industry have been pioneers in this field.

The methods of operations research and automated data processing play an important role in the further development of marketing logistics. Only with the introduction of computers the analysis and development of complex logistical systems has become feasible. The computer helps to monitor systems and allows shortening the communication process.

The growing use of computers in Europe will make sure that managers on this side of the ocean will meet the challenges of marketing logistics much more dynamically than they did in the past.

REFERENCES

Constantin (1966): Constantin, James A.: *Principles of Logistics Management*, New York: Appleton, 1966.

Cordes (1968): Cordes, Helmut: *Unternehmensforschung und Absatzplanung*, Wiesbaden: Gabler, 1968.

Heskett/Ivie/Glaskowsky (1964): Heskett, J.L., Robert M. Ivle, and Nicolas A. Glaskowsky jr.: *Business Logistics*. New York: Ronald Press, 1964.

Neuschel (1967): Neuschel, Robert P.: Physical Distribution – Forgotten Frontier. In: *Harvard Business Review*, Vol. 45, March-April, 1967, pp.125–134.

Marks/Taylor (1967): Marks, Norton E. and Robert M. Taylor (Eds.): *Marketing Logistics*. New York London, Sydney: Wiley, 1967.

Schröder (1968): Schröder, Harald J.: Von der Lagerverwaltung zum Distributionsmanagement. In: *Die Absatzwirtschaft*, Vol. 11, No. 4, 1968, pp. 28 et seq. and No. 5, pp. 5 et seq.

Snyder (1963): Snyder, R.E.: Physical Distribution Cost: A Two-Year Analysis. In: *Distribution Age*, Vol. 62, No. 1, December 1963, pp. 35 et seq.

Peter F. Drucker (1962)

THE ECONOMY'S DARK CONTINENT

Peter Drucker (1909–2005), one of the most influential authors in management sciences in the 20th century, has often been called a "visionary", since he recognized and forecasted many developments and trends in management ahead of their times. This applies to the field of logistics, too.

In his 1962 article *The Economy's Dark Continent*, which was published in the widely read American management journal *Fortune*, the word "logistics" is still missing. He talks about distribution – "the other half of marketing", as Converse called it in the paper presented earlier in this volume. Most of the analysis and the arguments are comparable between Converse and Drucker. Drucker claims that "to reach a real understanding of the role and structure of distribution in the American economy we need new concepts of economic theory and economic analysis. Most of our present concepts focus on production or on the stream of money and credit, rather than on the flow of physical goods and its economic characteristics. We also need new data; our figures today, such as those of the Census of Business, obscure distribution rather than report on it".

The article presents a number of real world business cases of his time to prove his point that there is an urgent need for more professionalism and systematic application of modern management science tools and methods. Drucker argues that a "holistic" perspective is necessary in management, rather than just the optimization of activities and functions.

The important contribution that Drucker has to get credit for is in the wide-ranging attention which the issues of distribution received. With hindsight, his voice may finally have prepared the field for the emergence of a distinct science of logistics.

Reference:

Peter F. Drucker: The Economy's Dark Continent. First Publication in: *Fortune*, Vol. LXV, No. 4, April 1962, pp. 265–270.

P. Klaus and S. Müller (eds.), *The Roots of Logistics*, DOI 10.1007/978-3-642-27922-5_8,

Peter F. Drucker: The Economy's Dark Continent

Almost 50 cents of each dollar the American consumer spends for goods goes for activities that occur after the goods are made, that is after they have come in finished form off "the dry end of the machine," to use the papermaker's graphic term. This is distribution, one of the most sadly neglected, most promising areas of American business.

The activities that are encompassed by the broad term distribution include sorting and cutting, invoicing, billing, and other paper work; labeling, packaging, storing, moving, shipping – plus wholesaling, retailing, financing, and insuring. Physically, distribution contributes little; it can only mar, soil, tear, scratch, or otherwise damage or downgrade the product. Economically, however, distribution is the process in which physical properties of matter are converted into economic value; it brings the customer to the product. But how much of the distributive cost is really "value added," how much is merely "waste added"?

We know little more about distribution today than Napoleon's contemporaries knew about the interior of Africa. We know it is there, and we know it is big; and that's about all. There are plenty of experts on individual phases: on transportation and warehousing, on retailing and consumer buying habits, on labeling and packaging, on factoring and insurance. But when a major government department recently looked for two or three men to advise on distribution, none of the many people consulted in industry, government, and the universities could name even one qualified candidate.

To reach a real understanding of the role and struture of distribution in the American economy we need new concepts of economic theory and economic analysis. Most of our present concepts focus on production or on the stream of money and credit, rather than on the flow of physical goods and its economic characteristics. We also need new data; our figures today, such as those of the Census of Business, obscure distribution rather than report on it. (And only the high-speed computer can analyze anything as complex as the distributive system.)

G.M.'S GRAND DESIGN

Business need not wait until this work is done. An industry or a company can make its own distribution manageable. This was proved more than a generation ago by the automobile industry and especially by General Motors. Henry Ford, in building River Rouge, had had the premature but great vision of the manufacturing plant as the central switchboard in a flow of physical matter from mine to customer (and from there to scrap heap and back into the furnace as raw material). With this as its starting point, General Motors, in the middle and late Twenties, built its business around the economic characteristics of the distributive process – beginning with the customers' buying habits, proceeding thence to the structure and characteristics of dealer finance, dealer inventories, and dealer compensation, going back further to the design, location, size, product mix, and scheduling of assembly plants, and finally all the way back to corporate organization. For well over three decades General Motors' distribution system has made the automobile the highly engineered mass-consumption product with the lowest total distribution costs.

By now this system may be obsolescent in important parts. How much economic sense, for instance, does the "franchised dealer" still make in a world of supermarkets and discount houses? How meaningful is his fixed commission? But G.M.'s achievement still stands as a useful example of how distribution can be brought under control. The example has been only too rarely followed. Only lately have things begun to stir – mildly. The American Management Association, for instance, has run a number of conferences on physical distribution. A few large consumer-goods companies have made distribution a specific corporate responsibility; H.J. Heinz of the "57 varieties" has a vice president-distribution. The results are encouraging. It does help, for example, to have someone around who points out that a proposed new packaging design will double freight costs, instead of having this come to light only after millions of the new packages have been made – as happened a few years ago at one of the big soap companies.

WHY DISTRIBUTION LAGS

But even at their most successful, the new approaches attempt only to do a little more systematically and a little more cheaply the old jobs of transportation, inventory keeping, and warehousing. And they are still isolated attempts rather than a broad trend. A recent study by Professor Michael Schiff of New York University (to be published on May 1 by the Financial Executives Research Foundation under the title The Financial Management of the Marketing Function) found that the importance and cost of distribution are recognized in fewer than half of the twenty-eight large companies analyzed.

Getting at the distribution problem requires cutting through a hard-rock overburden of legal, managerial, and organizational concepts. Few companies think of their distributors when they speak of "our business". Their horizon is set by the legal boundaries of their corporation. Few companies, for instance, know how large their distributors' inventories are and what is in them; this ignorance is a major cause of the persistent inventory booms and inventory busts that beset our economy. Distribution is no respecter of legal corporate boundaries. In industry after industry the price the consumer pays for the merchandise delivered and installed can be double the manufacturer's factory price or more. This means that perhaps half of the dollar the consumer pays for "our products" goes to distributors who are not legally part of "our business". The competitive position of industry and company, the reputation and availability of its products, and the use the consumer makes of them, all depend more on events outside the legal business boundaries than on what the manufacturer does within his legal walls. Perhaps the single most important lesson the generation-old G.M. approach still has to teach us is that distribution policy and distribution system must take into account the entire flow of the product regardless of lines of ownership and legal responsibility.

For example, the manufactureres who make the bulk of the products sold by Sears, Roebuck are legally and financially independent, as a rule. Yet every Sears buyer is expected to know as much about the making of the product – the manufacturer, his plant, his process, his materials, his people, and his costs – as he knows about selling it in the Sears stores. He often gets into product design. He is, in other words, responsible for the entire process even though he legally controls only a small part.

As restrictive as the legal boundaries are organizational and managerial boundaries within the manufacturing business itself. These tend as a rule to keep distributive costs out of sight and distribution acitivities unmanaged. Even in the few companies that have such a title, the "manager of distribution" usually takes over only when the product is ready to be loaded and shipped away. The major distribution costs within the manufacturing business tend, however, to lie before this stage – and so do the major opportunities for significant savings.

THE MOB IN THE SHIPPING ROOM

In a well-managed manufacturing plant the machine areas are – and should be – quiet places. There aren't too many people in sight. They don't rush around. They do not seem to exert themselves unduly but move with measured calm. There is a good reason why the "drama of industry" so rarely comes through in paintings or photographs: in a well-managed plant there is no "drama".

But even in the best-managed plant things change drastically as soon as one goes through the door labeled "Finishing Room" or "Shipping Department". There is suddenly a mob of people. Everybody seems to rush and no one seems to know why and where. If the machine areas are soporific, the areas where products are sorted, cut to size, packaged, labeled, stored or shipped are pandemonium.

The board of directors tends to hit the ceiling whenever the monthly scrap figure in the plant goes up by one percentage point. But few boards, to my knowledge, have ever been told how much spoilage there is in the distributive activities under the plant roof, and they have not been very interested in finding out. Yet this area is where the costs are in a modern plant – for this ist where the people are. An aluminium rolling mill, for instance, may employ four workers in distributive plant jobs to every one actually employed at "making". To be sure, this distribution work is unskilled work, done usually with the simplest tools. But no matter how unskilled the job, there are no more "unskilled wages." Yet the costs of these distributive activities are rarely known and almost never shown. They disappear in general catchalls: "allocations", "indirect labor", "administrative expense" or "burden". The industrial engineers sweat to pare one mill off the costs of making. But they rarely pay any attention to the dollars that might be saved in finishing, cutting, packaging, and shipping. I know of one company where a small team of industrial engineers put in ten days of routine work on the areas between the machines and the loading dock.

As a result of the study, the company was able to reduce its fleet of forklift trucks from over 200 to fifty, to cut its manpower needs by half, and to speed up shipping time of the finished products from seven to four and a half days.

Typically, no one is inidividually responsible for these distributive activities in the plant. Organizationally these activitites normally come under manufacturing management. They are lucky, however, if they are treated even as stepchildren. For people with a technical orientation – the production man, the chemical engineer, the plant manager – these activities are low-grade nuisances. And because, to a technically oriented man, most of the distributive work is donkey work, he tends to put a donkey in charge – more often than not a man of proved imcompetence for "more demanding" work as a manufacturing supervisor.

"SOLUTIONS" THAT DON'T SOLVE

As we apply management science and data processing to segments of the distribution process it becomes even more important to make sure that distribution is seen and managed as something that crosses legal and organizational boundaries. I have a suspicion – strengthened by a good deal of the work I have seen – that high-powered operations-research "solutions" for inventories, warehousing, transportation, order handling, and so on often increase the costs of distribution. For many managers tend to try and fit operations-research into the existing legal and organizational boundaries as if these were physical facts. In straightening out one kink in the stream they may only create expensive, new turbulences some place upriver or downriver, in what may be somebody else's job, if not "another business", but is still part of the same distribution process and part of the cost structure of the same product.

I have seen one example of this in a steamship line, where an operations-research team overhauled the entire loading and scheduling process. The team produced what the operating vice president wanted: a schedule that gave the fullest possible utilization of ships, that is, of expensive capital equipment. But no one told the team that there are different rates for different kinds of freight, and that therefore freight-mix has even more to do with final results than turnaround time. An the new scheduling system put such a substantial premium on getting a rather unfavorable freight-mix that total earnings plummeted even though the rate of capital utilization went up by half.

To get control of distribution, therefore, requires seeing – and managing – it as a distinct dimension of business and as a property of product and process rather than as a collection of technical jobs. Whenever this has been done (with or without a separate "manager of distribution") substantial benefits and savings have resulted.

THE DISCOUNT HOUSE AND ITS LESSONS

But effective and economical distribution also requires, in many cases, that an entire business be run differently. This is evident in three big areas in the American economy where we know distribution to be ineffectual and costly: retail selling, the new multibillion-dollar scientific-equipment market, and the buying of industrial supplies.

The discount house, it is commonly said, can sell at a lower price because it concentrates on the brands with fast turnover. Plausible – but simply not true. Many brands that turn over slowly in the ordinary retail store turn over rapidly in the discount house. In different discount houses the same brands often perform quite differently. And even the retail store's best seller turns over much faster as a rule in the discount house (or at Sears, Roebuck). The truth is that concentration on staple items is what gives the discount house both low costs of distribution and fast turnover.

Ninety per cent or so of all business in a given line – for a manufacturer as well as for a retailer – is contributed by a very small percentalg of items, 5 to 8 per cent as a rule. The remaining 90-odd per cent of all items, however, account for 90-odd per cent of the costs. For while revenue is roughly proportionate to volume, costs tend to be proportionate to the number of items or transactions. A slow seller requires just as much shelf space as a fast seller and ties down just as much capital. One requires as much paper work as the other – and paper work is so expensive that a major wholesaler has concluded that orders with a net profit of less than $10 do not even cover their ordinary paper-work expenses. In transportation, as every one knows, the smaller the unit the higher the cost – and the same goes for crating, for collection, for inventory control, for insurance, and so on. Even spoilage and losses in transit, according to the meager figures available, seem to be higher proportionately, the smaller the unit of transaction or shipment.

MARKETS WITHOUT MIDDLEMEN

For the last two years I have watched the building of a fair-sized medical center. The buying of a good many million dollars worth of hospital and classroom equipment – from beds to benches – was a reasonably well-organized job. Ample technical help was available both to design facilities and to choose equipment; there were a good many distributors, well equipped with products and competent to give technical advice; and information enabling the center to weigh the consequences of decisions – say between initial costs and costs of upkeep – was reasonably easy to obtain.

But when it came to buying the scientific equipment for laboratories and testing facilities – in biology, chemistry, bacteriology, pathology, electronics, and so on – chaos took over. No one apparently had the faintest idea what would be needed, how much, and where. No one seemed to know what was available on the market and from whom. The laboratories took at most one-tenth of the total equipment budget.

But in time, they took more than all of the other euqipment together. And worse, a score of men who should have been getting on with their scientific work spent months as amateur lab designers and purchasing agents. Even so, no one is very happy with the results: there is too much stuff that will be needed only rarely, and not enough of the bread-and-butter items; the wrong equipment is centralized, creating bottlenecks; and for too many pieces are not where they are needed the most.

Scientific equipment has become a multibillion-dollar market, with thousands of producers and with a multitude of customers: government agencies and industry, hospitals and schools of all kinds. But it is still organized as if it were a small "specialty" business in which one or two manufacturers supply a few odd customers. No wonder that the cost of equipment is going up twice as fast as the number of qualified research people – to the point where equipment costs may soon force us to cut back research. No wonder that, even on some apparatus in standard use, distribution costs take 80 cents of the customer's dollar.

This is a peculiar market: each individual buyer buys once in a lifetime – but then he buys a great deal at once (though there is, of course, also a constant replacement and expansion demand). Yet there are enough buyers to provide a constant order flow. Such a market, more perhaps than any other, needs a middleman – and it has none. A middleman knows both the market in an area and the available supply. Typically, the middleman of such a market is an adviser to the buyer – either the architect type who receives his pay as a fee from the buyer, or the advertising-agency type who receives his commission from the supplier. But out-and-out wholesale distributors also can supply such a market – as they do in many other hospital and school supplies. And there may be room for a cooperative buying organization of major users, such as the design and purchasing cooperatives some of the major universities are talking about. But without effective middlemen this rapidly expanding market cannot be supplied properly, let alone cheaply.

The scientific-equipment situation gives a conclusive answer to the old charge that the middleman is a "parasite". So, by the way, does the paperback book market, our fastest-growing market in the last few years. Cheap, paper-bound books have been around a very long time – at least since the days of the Haldeman-Julius "Little Blue Books" more than forty years ago; the "serious" reading public was there all along too – though in smaller numbers. What created the industry was the discovery and development of new distribution channels, such as the campus bookstore, and the bookrack in drugstore and supermarket.

THE SEMI-LITERATE BUYER

A good deal of work has been done these last ten or fifteen years on the marketing of industrial goods. Yet marketing men all say that selling and distribution in this field still leave much to be desired. Above all, they would increasingly agree that the key to efficient and effective industrial marketing is not the supplier but the buyer.

And the buyer of industrial products and supplies is in many cases semi-literate, or even illiterate, in a business sense.

One example is paper. For many a business, paper is a major supply: the carrier of its communications, of its "image", its relations within and without; the carrier very often also its product, its product appearance, and its product message. It is also a substantial cost item. But few companies buy paper as a major supply – if indeed they even know that they buy paper at all. Thus, in some very big banks the paper bill may be the second-biggest expense item after wages. Yet not only is there no accounting for the costs of paper, there is usually no one who knows anything about paper or about the graphic arts, no one who supervises the bank's paper buying, no one who know what papers are available and what they can and should do in terms of end results, such as graphic representation, easy control, or accessible information. Typically, forty or fifty people in as many different departments buy paper as an "incidental" office supply – and through whatever printer, merchant, or stationer they happen to have come across (check paper is an exception – it is bought as a highly engineered product for a specific end use). As a result the bank is likely to end up with far more varieties of paper than it needs, and yet with little paper that is really right for the job. This is costly for the bank – if only because it pays retail prices for wholesale quantities. It is costly above all for the manufacturer, who has to maintain a meaningless product variety with its high expenses, both in making and in distribution. What the bank needs is a paper buyer who knows the needs of the bank and the end results it aims at, as well as the paper industry's technical and economic structure. The "old battle-ax" in accounting who, only too often, places the paper order for her department knows non of these; as a paper buyer she is illiterate.

Another example is the ordering of generating equipment for electric-power companies. Technically these buyers are extremely highly qualified. But economically they know little – especially about the economics of their supplier, the equipment producer. As a result man buyers tend to insist on their own special designs – rather than ordering by performance specification, most of which can be satisfied by standard designs. If the apparatus producers could build standardized models to performance specifications and on a planned even-flow basis, their costs would tumble. It is probably not entirely accidental that one major utility system which manufacturers say really understand the economics of equipment production, American Gas & Electric, also has turned in a truly superior performance in respect to growth, technical performance, and profit.

AN URGENT JOB

The industrial purchaser has to know his own business, of course – and the progress made here in the last few years is all to the good. He has to know what the product or supply he buys is supposed to contribute to his company's end results, has to buy it as cost per unit of his own output rather than just by the price tag. But he also needs to know just as much about the structure of the supplier industry and its economics.

This sketch of the distribution challenge oversimplifies and overgeneralizes, of course. There are countless exceptions to its sweeping statements. But my purpose is to point to distribution as an area where intelligence and hard work can produce substantial results for American business. Above all, there is need for a new orientation – one that gives distribution the importance in business design, business planning, and business policy its cost warrant.

At a time when American business faces great competitive pressures from abroad – especially from unified Europe whose industries can hold their own in technology, manufacturing knowledge, equipment, and salesmanship – raising the effectiveness and cutting the costs of American distributive system may be a more important and a more urgent job than most managements yet realize.

PART B: THE EVOLUTION OF BUSINESS LOGISTICS AS AN AUTONOMOUS FIELD OF SCIENCE

Fritz Voigt (1953)

TRANSPORT AND INDUSTRIALIZATION

The modern Science of Logistics has many roots. Next to its histories in the military sector and its early development as a discipline of Marketing, there is an independent tradition of research and publications about transportation and transportation systems in many countries. It is established both at a macro-level of "Transport Economics" and more micro-levels of "Transport Systems Analysis" (e.g. Manheim (1979)) as well as – in the German-speaking countries – "Verkehrsbetriebslehre" (e.g. Diederich (1986)). Transportation studies have contributed in many ways to the understanding as well as of the elementary logistical function of "transferring objects in geographical space" (cf. MARSHALL (1890) in this volume).[1]

To represent this rich historical branch of the development of logistics a 1953 paper on *Transport and Industrialization* by Fritz Voigt (1910–1993), one of the founders of the discipline of Transport Economics in Germany and professor at the University of Bonn, has been chosen for this volume. It provides an insightful, early discussion of the impact which transport and alternative technologies and structures of transportation systems have on the development of economies and vice versa – how the choice and specification of transport systems is affected by demands from the economic and physical environment. Voigt suggests a systematic approach to describe and even measure the characteristics of transportation demand and transportation supply as a way to assess and improve their degree of "fit".

Of particular interest are Voigt's considerations about the assessment of the usability of alternative transport systems and technologies. His list of seven descriptive factors determining the usability of a transportation system to meet a given demand has become a frequently used analytical device. They include network density, capacity, speed, frequency, predictability, and others.

Due to the length of the paper, only selected sections are reprinted on the following pages.

---

[1] It is remarkable that there have not been any comparably established and deep scientific considerations of the other elementary functions of logistics, i.e. of warehousing and arranging/rearranging of items, except for rather recent studies of the engineering aspects of materials handling and flows, cf. Jünemann (1989).

P. Klaus and S. Müller (eds.), *The Roots of Logistics*, DOI 10.1007/978-3-642-27922-5_9,

Reference:

Fritz Voigt: Verkehr und Industrialisierung. In: *Zeitschrift für die gesamte Staatswissenschaft*, 109. Band, J.C.B. Mohr, Tübingen, 1953, pp. 193–239.

Further reading:

Diederich, Helmut: Entwicklung und Stand der Verkehrsbetriebslehre, *Zeitschrift für Betriebswirtschaft*, Vol. 56., No.1, 1986, pp. 51–88.

Jünemann, Reinhardt: *Materialfluss und Logistik*, Berlin; Heidelberg; New York: Springer, 1989.

Manheim, Marvin L.: *Fundamentals of Transportation System Analysis*. Cambridge/ MA: MIT-Press, 1979.

Fritz Voigt: Transport and Industrialization[2]

I. THE RELATIONSHIP BETWEEN TRANSPORT AND INDUSTRIALIZATION

Transport and industrialization represent a pair of forces that influence each other and promote reciprocal development. The term "industrialization" is used here to describe the increase in the production of commercial goods beyond the rate of population growth, achieved primarily through technological progress. "Transportation" refers to the transfer of goods and people across geographical space.

1. Industrialization influences transportation. It is possible to determine a coefficient which indicates the degree to which a unit of growth of goods production leads to additional transportation activity.

2. But transportation also influences industrialization:
   a, If a new transportation technology[3] becomes available in a given situation of a balanced economic system, it will disturb that balance and cause dynamic adjustments.
   b, Every investment into new transportation systems produces additional **income** and puts in motion acceleration and multiplier effects, as they have been described by Ohlin, Kahn, and Harrod. However, an investment into transportation systems will have a effect upon the ability of the economy to produce goods different from the effects a comparable investment into industrial production systems will have.
   c, New demand is generated through new transportation options. For example, the exploration of certain natural resources may become feasible only after they become accessible by adequate transportation. The same may be true for certain kinds of industrial production.
   d, The means of transportation shape processes of industrialization in **specific** ways that are different from industrialization developments without changes in the transportation systems.

---

2 Translation by the editors.

3 The term „Verkehrsmittel" used in the original German text is translated here as "transportation technology", referring to a given mode of transport and its technological and structural specifics.

In the following discussion I will primarily explore the first point of the effects of industrialization upon transportation, and the last argument under point two, since they are of special interest from the point of view of transportation policy.

Let us consider as model a number of regions that previously have not been engaged in any form of economic exchange. In each of these regions there shall be a perfect market economy in a state of economic equilibrium.

There is no economic region in the world where the conditions and requirements for production and consumption are completely identical. Across economic regions there are smaller or larger differences regarding the availability of natural resources, the conditions of manufacturing, the quality and intensity of human labor, the level of entrepreneurial activities, people's interest in economic affairs, and the habits of the people in everything else that plays a role for production and sales.

Let us now consider the impact of the introduction of a new transportation technology, assuming that the technology will be accepted and used, and new opportunities for additional trade and economic interchange will not be obstructed by the authorities.

The consequence will be that new transport flows develop where it is advantageous to purchase supplies from another region, if they are still cheaper than domestic supplies after transportation cost have been figured in. There will also be a redistribution of incomes. In other words: The existing economic equilibrium in the region will be destroyed by the introduction of the new transportation option. Let us call the intensity of the disturbance of the equilibrium in the affected regions, which is caused by the introduction of the new transport system option – until a new equilibrium has been reached – "factor G". As long as there are different currencies in the regions which can float against each other freely, in accordance with the law of comparative advantage there will be short-term adjustments in the currency exchange rates. Given fixed currency exchange relationships or a common currency and stable cost levels of the respective factors of production, then, over time, the market economy will favor those regions and shift production there where the production is more efficient and causes lower costs after allowing for the transportation expenditures.

This means: Even a technologically capable production facility in the previously isolated area will become unprofitable and unviable. In the area with better cost structures, as a consequence of the new transport option, there will be an incentive to invest into additional production facilities until the marginal profitability of those investments stabilizes at the level of a new equilibrium.

If production would relocate to those regions within the newly created enlarged economic area where the cost advantage is highest, then the benefits of the transportation option would be relatively easy to assess.

[...]

In reality the differences between regions may be so significant that the impact of a new transportation option may be a complete depletion of one region in terms of its income and population. On the other side of the favored region there may be undesirable effects of agglomeration and concentration, which may cause secondary economic, political and environmental problems.

These shifts, often even distortions, concentration congestion and depletion effects occur to an extent dictated by the power of impact the respected transportation option will exert: the stronger that power is – depending on its cost, its geographic range, its capacities and qualitative characteristics (such as speed, reliability, safety, comfort) – the more imbalance and dynamic adaptations in the economy will result. The dynamic impact of a new transport option can be deeper than changes in competition after some technological innovation.

[…]

It is for this reason that the assessment and measurement of the impact-factor ("Maßstab" in the German original) specific to a given transportation option is important. It shall be discussed and analyzed in more detail in the following sections of this paper.

[…]

II. ASSESSING THE USABILITY[4] OF GIVEN TRANSPORTATION TECHNOLOGY

The intensity and volume with which a new transportation technology triggers dynamic effects in an economic context depend on the degree of technological perfection which a given means of transportation has achieved.

Every technological innovation in the transport industry and every change in cost create an **additional dynamic effect**, triggering new income flows, changing the marginal productivity of capital in the affected region, and possibly rearranging the entire structure of the economy. Again the impacts of such changes depend on the technical condition of the means of transportation.

The type and power of impact of a certain transportation technology can be assessed by comparing it to a state of perfect usability. A transport system is in the ideal state of perfect usability when distances no longer play a role in the location of industries in an economy.

[…]

The factors determining the actual state of usability of a given transportation technology are the following:

1. The degree of network density which the technology allows for (respectively the number of access nodes to the transport system in a given geographical area),

---

4 "Verkehrswertigkeit" in the German original.

2. the ability of the technology to easily adapt to the topography of the terrain and to the pattern of transportation demand,
3. the ability of the technology to handle and care those volumes and types of goods, passengers and messages, whose movement is relevant for advancing industrialization,
4. the speed of transport service, the reliability (in terms of the ability to predict departure and arrival times and the duration of the transport),
5. the safety of the service (in terms of the probability of the cargoes and passengers to arrive in good, undamaged conditions),
6. the convenience of access and comfort of travel provided,
7. and, last but not least, the costs of a given transport move.

Now let us introduce a measure for the "usability" of a transportation technology. Let us assign the usability value of "1" to a technology which is perfect in the sense described above. It would be able to provide transport services in unlimited volumes at infinite speed, with perfect safety and calculability and an effort of zero. This kind of a system of transportation will thus enable an "ideal" industrial development and an "ideal" economic structure without hindrances or distortions caused by imperfections of the transportation means.

A transportation technology that is not able to provide any economically relevant transport services at all will be assigned a usability value of "0".

The usability values of 0 and 1 then represent extreme values, whereby – in reality – the value 1 will never occur, because there is always a need to use energy and other resources to move physical objects.

[...]

III. ASSESSING THE TRANSPORT AFFINITY OF GOODS

It is the task of a transportation system to enable the transfer in geographic space of goods, people and messages. The value of a transportation system can thus not only be based on its own technological features and capabilities. It may be technically capable of performing the biggest, most perfect kinds of services, but may only produce negligible economic effects if there is no demand for this kind of transport service. It is necessary, therefore, to consider the relationship between the technological features and capabilities of the transportation system, and the profile of needs and characteristics of the objects which require movement, namely the transported goods and passengers. In this discussion about the interdependencies between transportation and industrialization we are primarily interested in goods transports. We shall therefore neglect the otherwise important problem of people and information movements.

Every type of good has distinct characteristics that are potentially **sensitive** against the handling and influences it is subject to as it is processed through a given transport system. The quality of cargo is never improving as it is being transported. Certain cargo is more sensitive against the influences and occurrences in transit. Certain goods only survive a transportation move in acceptable condition if the transportation system is equipped to protect those sensitivities of the cargo. Some other goods are quite unsusceptible to damages due to their rigid structure. Let us call the degree of sensitivity of a good to be moved a certain geographic distance its **transport sensitivity**.

The given transport sensitivity of cargo therefore requires certain features and capabilities of the transportation system and the technologies employed. These features may vary in their degree with different kinds of goods: perhaps it is the case that the good is perishable, susceptible to bumps, easily blown away by the wind. It may demand certain shapes of the transportation equipment, a degree susceptibility to inflammation, or certain chemical or physical characteristics.

In some cases the developments of industrialization in an economy will increase the demands for bulk transport systems and technologies, in others the need for small, urgent shipments may pose new demands upon the transportation system capabilities. Some types of industrial productions require high levels of regularity and reliability of transportation services. These factors concerning the transport sensitivity of goods must be evaluated from the perspective of the respective industry which changes in the course of industrialization. It should be possible to develop a similar list of factors determining the affinity of a certain type of cargo to a given transportation system, as we have suggested to describe and assess the usability of a transportation supply system. Among the nearly unlimited variety of types of cargo there are goods which can be transported only if the transportation move is carried out quickly during their short life time (chemicals with short life cycles, day-old chicks, flowers). Others allow for a lot of time. Similar differences in transport sensitivity degrees occur with respect to how easily goods can break, to what extent there is a need to provide for high capacity bulk transport moves, to secure definite delivery schedules, to be capable to access shipping and delivery points of difficult geographical and topological conditions. Last not least the value of the cargo is a critical property to be considered, since it affects the amount of capital and interest cost tied up in-transit.

A transportation system which would have a **transport usability coefficient** of 1 for all kinds of cargo would be capable to move all kinds equally well. But since there will be no transportation which is perfect in all respects, there will be different degrees of "fit" between a given type of cargo and a given transportation system and technology. The coefficient describing the degree of fit I suggest to call the **transport affinity coefficient**. One obtains the transport affinity coefficient by assessing the usability of the transport system under consideration with respect to the sensitivity assessed of the type of cargo in question and putting them in a relationship. This may be done by applying the descriptive factors which describe the profile of the usability of the transportation system, as suggested above. The same scale in the range between 0 and 1 should be used.

This measure can be refined if desired. One element of the coefficient describes the degree of usability of the transportation system. The other element describes the degree of sensitivity of the cargo. The level of demand directed at the transportation system is less in the case of a transportation sensitivity of 0.1, it is high in the case of a traffic sensitivity of 0.8.

Through this comparison we can see that there are transportation systems that meets the transport requirements for certain goods better than others. To put it the other way: **every good possesses a measurable specific affinity to a given type of transportation system**.

Let us imagine an economic region now with distinct chances of developing industrialization. Let us now introduce a new or improved transportation option into this region. Not all opportunities for advancing industrialization will be exploited, because each industrialization development demands specific patterns of transport flows. In the same initial situation, each transportation option considered will support a different aspect of the industrialization development. Depending on the choice made, different income flows, different changes in marginal productivity of capital employed, will be triggered.

As a consequence, a **crystallization effect** sets in which varies depending on the choice of the new transportation system option.

[…]

In concluding let us remember that these impacts do not only occur when a new transportation system and technology option is introduced, but **also when the transporation usability factors of a given system change, or when the relative cost and prices for transportation change**. A decisive factor for the intensity of these effects will be the degree of change in the usability.

Peter Klaus/Yossi Sheffi (1997)

INTERMEDIATE COMMENTARY: ON THE EVOLUTION OF THREE MEANINGS OF LOGISTICS

The following excerpt from a paper, which was presented at the Council of Logistics Management's 26th Annual Educators' Conference at Chicago, 1997, does not claim to rank as a "classic". It is inserted as an intermediate commentary explaining how since the 1990s – when apparently the development of the field of Business Logistics had reached a level of maturity – academic logisticians started raising the question again, which Henry Eccles had asked from the perspective of a military logistician in 1954: "Logistics – what is it?" What is the essence of the field today? What constitutes its identity as a field of academic research? Does it have the right to call itself a science?

Klaus and Sheffi's paper reports on the results of that kind of reflection.[1] It suggests that the – often implicit – premises in the writings and thoughts of authors on logistics may be interpreted to represent three different meanings of logistics:

Its first meaning – "PPP" logistics – refers to the field as the integrative, systematic, science-based inquiry and body of know-how about the economic functions of transportation (Placing), of storing and buffering (Pacing), and of ordering and arranging (Patterning) for the purpose of creating value to customers. This conception is based on the classical works of Marshall, Converse and the discussions about what creates value in an economy, which have been reprinted and discussed in earlier sections of this volume. (A paper intended to represent the tradition of studying the logistical functions individually – prior to studying them as an integrated set of "transfer" activities – was presented earlier in this volume: Voigt on *Traffic and Industrialization*).

The second meaning of logistics as the science "of systematic coordination of activities and processes that procure, produce, and deliver products…" is implicit in many contributions to the field, primarily from the 1970s through the 1990s, where focus was shifted to techniques and instruments of planning and coordinating complex sequences of activities, as, for example, in production planning and control systems (see the article by Hax/Meal in the "instrumentation" section of this volume).

But only with the third meaning of logistics as the "science of flows and flow systems", which is emerging and being understood more recently, logistics seems to find an identity clearly separated from older concepts of Economics and Marketing.

---

[1] Selected other papers of this kind are: Novack, Rinehart, and Wells (1992); Weber (1992); Göpfert/Wehberg (1996); Ballou (2007).

P. Klaus and S. Müller (eds.), *The Roots of Logistics*, DOI 10.1007/978-3-642-27922-5_10,

It is the recent concern of the field with the phenomena of flows, processes, supply chains and supply networks, with the management of the dynamics inherent in those flows and networks, which seems to open up to logistics its own domain of research questions and new answers to problems. Typical problems are evident in the disharmonies of bullwhip effects, in the inefficiencies and volatility of overly complex systems. The promise is in new solutions, is in the leanness, harmonious operation and easy self-adaptation of well-designed semi-autonomous on-demand systems, as well as in the transferability of flow system concepts from the narrower world of goods flows to the much wider worlds of information flows, service, people, idea and other flows.

Some early roots of flow thinking and several important contributions to the analysis, design and operation of flow systems will be presented in the following parts of the volume.

Reference:

Peter Klaus and Yossi Sheffi: Logistics at Large: Jumping the Barriers of the Logistics Function. In: *Proceeding of the Council of Logistics Management's 26th Annual Transportation and Logistics Educators Conference*, Chicago/IL Oct. 5th, 1997 (The paper is partially based on an earlier paper by Peter Klaus *Die Dritte Bedeutung der Logistik*, Logistics Working Paper No. 1 at Friedrich-Alexander-University, Nuremberg, 1993).

Further reading:

Ballou, Ronald H.: The Evolution and Future of Logistics and Supply Chain Management. In: *European Business Review*, Vol. 19, No. 4, 2007, pp. 332–348.

Göpfert, Ingrid and Götz Wehberg: Logistikentwicklung – ein pulsierendes System wechselnder Profilierung. In: *Logistik im Unternehmen*, Vol. 10, No. 3, 1996, pp. 92–93.

Novack, Robert A.., Lloyd M. Rinehart, and Michael V. Wells: *Rethinking Concept Foundations in Logistics Management*, Journal of Business Logistics, Vol. 13, No. 2, 1992, pp. 233–267.

Weber, Jürgen: Logistik als Koordinationsfunktion: Zur theoretischen Fundierung der Logistik. In: *Zeitschrift für Betriebswirtschaft*, Vol. 62, No. 8, 1992, pp. 877–895.

Peter Klaus/Yossi Sheffi:
On the Evolution of Three Meanings of Logistics

The importance of logistics has been continuously increasing and becoming evident over the last 20 years. The "foot prints" of logistics are everywhere: there is a growing community of people identifying themselves as logisticians; there is an increasing number of logistics books and publications devoted to it in the commercial and academic arenas; more business functions, corporate positions and enterprises include logistics in their titles, membership in professional organizations, such as the Council of Logistics Management (CLM) and its international counterparts is growing steadily, and there are more stories of business success and significant competitive advantages associated with logistics activities.

The reasons for the growing significance of the profession are based on several well-known challenges and opportunities. Some of the challenges include the following:

- The globalization of commerce has created the need to procure material and parts from far-flung location and sell their finished products globally. Furthermore, many global companies who truly manage operations worldwide have emerged, creating need for intra-company international movements of material and increasing the importance of logistics-related decisions such as facility location.
- Recognition of the true costs of inventory has led many enterprises to decrease inventory levels, thereby creating tightly-coupled supply chains which can react faster but require more and different skills to put together and improve continuously. Furthermore, such tightly coupled supply chains are more vulnerable to breakdowns and interruptions, thus requiring closer monitoring and higher skill levels to operate.
- Increasing customer expectations for ever better service, lower costs and more choices have created a pressure to deliver products faster and cheaper and to react more nimbly to changing customer requirements. Furthermore there is an explosion of stock keeping units (SKUs) in response to consumers' demand for variety meaning that, in addition to the complexity of size, corporations deal with more products, each of which is subject to higher random fluctuations since it targets a smaller niche.
- A new emphasis on environmental issues is leading corporations to manage the return of products after their use for recycling and/or safe disposal and the deployment of reusable containers which have to be managed.
- Corporate realignment due to mergers and acquisitions and the emergence of extended enterprises mean that inbound and outbound networks have to be redesigned and tuned with increasing frequency.

The opportunities stem from the following:

1. The ability of new information and communication technologies to overcome space and time thereby transforming transportation and inventory

management processes. Of particular importance are technologies such as global positioning, shipment/equipment identification, electronic data interchange, enterprise planning systems and decision support systems.

2. The world's movement into further deregulation of the transportation industry and harmonization of standards and rules, create opportunities to plan and move products based on analysis rather than bureaucratic dictates; the deregulation of all transportation service in the US in the late 1970s and early 1980s, NAFTA, the European Union, Mercosur, ASEAN, open skies agreements, are all examples of continuing deregulation. In addition, the privatization of national carriers and national infrastructure will continue to bring new players into the logistics arena.

The needs and the requirements on the one side and the myriad of opportunities on the other explain the growing importance and recognition of the profession.

Yet, even as logistics is leaving such large foot prints on business and commerce, there are many, in and out of the profession, who are groping with the question of what it is.

[…]

The Council of Logistics Management (CLM) defines logistics as:

> "… the process of planning, implementing, and controlling, the efficient, cost effective flow of raw material, in-process inventory, finished goods and related information from point of origin to point of consumption for the purpose of conforming to customer requirements."

While rooted in military science, modern business logistics is seen by CLM and most logisticians as a business function, requiring a set of skills and based on a developing academic discipline. As such it is developing its own set of analytical tools and a "philosophy". This paper is based on the hypothesis that the logistics "phenomenon" does not have a single identity. It argues that there are at least three distinct meanings of logistics that have been evolving over time. Based on observations what logistics practitioners and academics actually have done and lived by in the name of logistics, the paper tries to articulate those meanings. It suggests terminology and analogies that may help logistics professionals and outsiders to the field communicate more effectively about logistics. Finally, it discusses some implications of the broadest – the third – meaning of logistics, including the skills and tools it may require and how to teach it.

The First Meaning of Logistics: Optimization of the Operational "Three "Ps"

The original and most elementary notion of logistics is that of a set of three operational activities – movement, storage and rearrangement – that add value to goods. The value-added inherent in these activities is based on the following:

- **Place** – adding place value to items by moving them from locations of lower value for the customer to locations of higher value to the customer;

- **Period and Pace** – Adding time value to items by storing them and thereby (i) moving items from periods when they are available (following extraction, harvest, or manufacturing) to periods when customers require them, and (ii) making all processes more effective – inventory allows for de-coupling the processes along the value chain from each other, so they can each run at their optimal pace according to their own economics;
- **Pattern** – Adding order value to items by arranging them in desired quantities and patterns. Examples include consolidation, break-bulk, sequencing, picking/packing, etc.

The cost of adding place value is the transportation cost. The cost of adding the period value is the inventory carrying cost, and the cost of adding the pattern value is the cost of picking and packing, sorting, and sequencing. As items are moved through a distribution channel from a vendor to, say, a distributor, they go from the vendor's inventory through a picking and packing operation, loading, transportation, receiving and are put away as part of the distributor's inventory management system. The same activities are then repeated when the distributor is replenishing the retailer.[2]

Transportation and storage operations have been long recognized as the building blocks of logistics. In many cases the creation of a "pattern" or ordering items according to certain rules, (which may include direction, consolidation, line-sequence, or whatever) has been thought of as part of either the transportation or the storage function. Indeed, carriers have been sorting their freight as part of their consolidation operation and warehouse operators were placing items in storage in specific patterns and picking items and packing them into shipments as part of warehousing activities. Lately, however, logistics operators have been offering "ordering" or pattern-creation activities as stand-alone value addition activities. Examples include line sequencing, where logistics companies would sequence parts and deliver then to an assembly plant in the order the items are needed at the plant, thus allowing the plant to synchronize the incoming flow with the production sequence. Another example is packaging, where logistics companies change the packaging of incoming items so they will fit better into a downstream process.

Another aspect of pattern-creation or "arranging" is the processing of information. For example, after entering orders for certain goods, the software "rearranges" certain indices in the computer's base so orders can be bunched and organized according to the distribution centre that has to supply them, the plant responsible for manufacturing them, the carrier responsible for hauling them, the party responsible for payment, the time of shipment, or whatever. All the reporting associated with logistics can be seen as information pattern-creation.

Thus, a definition of what operational logistics encompasses, can be based on the 3Ps of logistics operations mentioned:

---

[2] In his 1890 book *Principles of Economics*, Alfred Marshall already recognizes the three value-added functions mentioned here (pp. 63–64, 8th Ed., 1920).

Logistics is the addition of place, period and pattern values to items.

This definition is independent of what the "items" are. So it may be applicable not only to inbound material, work-in-process inventory and finished goods, but also to semi-finished goods, sub-assemblies and returns. Note that this definition does not mention explicitly the customers' requirement, since value is provided only if the movement, storage, or other activity conforms to those requirements.

The notion of "PPP logistics" (or operational logistics) has been traditionally performed as a set of functions: traffic, distribution, material management, inventory management, warehousing, and as part of manufacturing. The first stage in the evolution of logistics was to recognize the leverage of this set of activities for business success (which previously was seen only in the traditional "core" business functions such as production and marketing), and the benefits in the professional management and integration of these activities.

[...]

The Second Meaning of Logistics: A Managerial Process of Coordination – Adding to more "Ps"

Operational activities have been the cornerstone of logistics for many years, well before the current emphasis on "business logistics" began to be established and recognized. It was realized, however, that a focus on the improvement of transfer activities in isolation from each other was not sufficient. It became evident that most leverage for cost reductions and operational improvements was in the interfaces and interactions between the functions. The integration and coordination of such activities became a primary concern, foreshadowing what is now often called "supply chain management".

The primary focus was no longer on operational, warehouse-floor and road-level activities, but shifted to the managerial level. This second meaning of logistics may be articulated as:

Logistics is the systematic coordination of activities and processes that procure, produce, and deliver products and/or services to customers.

The second meaning of logistics – as an art and science of coordination activities and processes – is reflected in the current popularity of "supply chain" terminology and activities in the field. The contribution of logistics to business success may be highlighted by a fourth "P": **process coordination**. The management of supply chains involves setting up many partnerships – both inside the enterprise and between enterprises – so this fourth P of logistics can also be interpreted as **partnerships' management**.

The processes to be coordinated include not only the three Ps of "placing", "pacing" and "patterning". Instead they involve manufacturing and other operational activities.

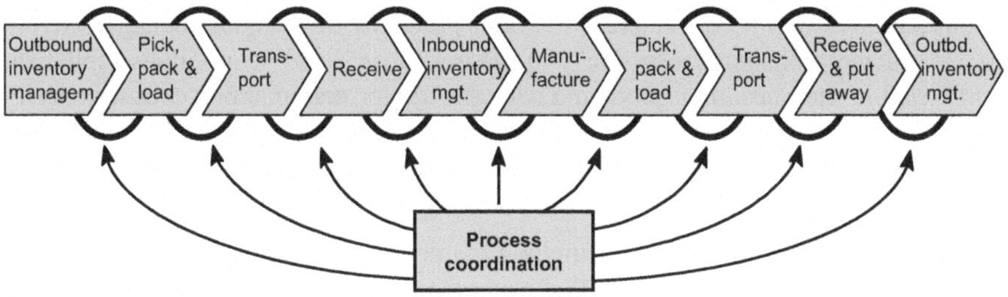

EXHIBIT: POINTS OF COORDINATION

Regardless of the type of objects being processed (parts inbound or products outbound; primary or return, etc.), operational activities are always accompanied by informational activities. These include purchase order processing, status (location and inventory) message reporting, invoicing, shipment content information notification, period reporting, etc. The management of these informational activities is a crucial part in supply chain management. Not only does the information generated help execute the operational three Ps of logistics, but it has a value in its own right. It enables planning of logistics-related activities as well as control of all other corporate activities touched by logistics.

In addition to materials and information, there is also a need for the coordination of the flow of cash. The cash moves from customers to vendors, as payment for the items, and from vendor to customer to account for the reverse flow of items.

Last, leading corporations are also including logisticians in concurrent engineering teams. According to Volkswagen AG for example, 75% of logistics costs are set by the time "Job 1" (the first car of a given model) rolls off the assembly line. Thus, issues of design for logistics, plant location, choice of vendors and manufacturing discipline are all important determinants of logistics costs. Leading corporations make sure, therefore that logistics considerations are accounted for at the earliest design stages and the process of designing the logistics supply chain is integrated with the product design process.

Supply chain management, as a coordination process is not a new concept. It has been taking root both within companies and in inter-companies operations, but has been known by different names. Within companies it has been known as "lean manufacturing", "re-engineering", "process orientation", "concurrent engineering", and even "team work". All these concepts are different manifestations of coordination of activities and processes along the internal enterprise's "operations" process, at different time frames. This process of managing the flow of items, from material and parts supplied by vendors to finished goods in the customer's hands, while coordinating with Manufacturing, Purchasing, Marketing, Finance, Legal and all other corporate functions is the essence of internal supply chain management.

Supply chain coordination activities among companies have also been described in many different contexts by various terms, all implying closer relationships and tighter coordination between the companies. These terms include the Japanese "Keiretzu"; the practice of vendor-customer "partnerships"; the movements of corporations to "core supplier", leading to a reduction in the number of suppliers so coordination activities can be executed better; the practices of "vendor managed inventory", "efficient consumer response", "quick response", and "continuous replenishment" used in the consumer goods industry for tighter manufacturer-retailer coordination; the practice of "just in time" replenishment used originally in the automobile industry and now an accepted practice of vendor-manufacturer coordination; and "pull" vs. "push" methods to move material along the supply chain.

Most of these terms and practices include elements of "coupling", and "coordination". To see the reason for it, note that aside from work-in-process, the only reason for having inventory is so that processes can be de-coupled from each other. The de-coupling allows each function to operate at its maximum efficiency in terms of its own parameters, including economic order quantities, process fluctuations, and process variability. All of the terms above are manifestations of the realization that not only is the inventory required in order to de-couple processes expensive and wasteful, but optimizing independently each function within the enterprise and each enterprise within the supply chain does not lead to over-all optimal processes. Many companies have demonstrated that quantum leaps in cost improvements and customer service can be achieved by integrating supply chain activities through coordination.

One can look at every enterprise as a collection of two fundamental processes: product development and "operations", or supply chain management. These two fundamental processes are supported by a range of support services such as Legal, Human Resources, Finance, Marketing, etc.

Embracing supply chain management is not "just another fad" for the logistics profession. Regardless of the type of process coordination activity which is relevant for particular enterprises, supply chain management is fundamentally a different activity for logistics professionals and it redefines the profession. It moves the logistician from a functional manager of one or more support activities to an owner of one of the two fundamental processes in every organization. In fact, recognizing the equal importance of activities that happen inside the factory walls and outside them, many organizations, such as Goodyear Tire and Rubber Company, Digital Equipment Corporation, and others, have an executive with the title of "Manufacturing and Logistics" in charge of that process.

An Evolving Third Meaning of Logistics: Flow Management – The fifth "P"

If one observes the activities of those businesses, which appear to be the most advanced and successful practitioners of logistics, it becomes apparent that beyond the functions of "placing", "pacing", and "patterning", and even beyond "process coordination" among adjacent activities or partners in a supply chain – there is more. Companies as diverse as Toyota, Wal-Mart, Dell, The Gap and many others, less visible, "champions" in today's dynamic and volatile markets, seem to have found ways to set themselves apart in yet another way. They view the enterprise as a fabric of flows of goods and information. They design and program their operations and processes to "mobilize" those flows for optimal continuity, speed and responsiveness to customer needs. It is a focus on the dynamics of flows, by which the evolving third meaning of logistics is characterized. Therefore, agility, or **pliancy** may be added as the fifth "**P**" of logistics. The classical example for successful application of "pliant logistics" – even though the term is not used – is Toyota (see Ohno's book about the *Toyota Production System*, Productivity Press (1988); an extract is presented in this volume).

At the core of the success of logistics leaders, hence, is the concept of pliant flow, which leads them to:

- View their businesses as networks of interrelated flows that bring value to their customers, and hence focus on network structures and corporate cultures that allow for – indeed demand – unobstructed, speedy flows of goods, information, cash, and ideas, which easily rise, ebb and change with customer demands.
- Focus on the dynamics of getting products, parts and information to move according to the customers changing requirements, having the ability to adjust operations dynamically backwards, in an optimal fashion, throughout their own supply chain.
- Ensure that all parts of the business "oscillate together" in a holistic fashion by organizing resources and orienting people along the flows of the ordering, fulfilment and payment processes that are the lifeline of each business.

The primary flow in each business is, of course, the physical supply chain, which entails the management of the flow of parts into manufacturing plants and finished products outbound to customers. It also includes the management of reverse flows rooted in returns, recycled items, exchanges, reusable Containers, trade-ins, etc. The flow perspective can also be applied to work flows at the administrative levels of each organization, to cash flows, to service flows, to decision and idea flows, and elsewhere.

Exploring the Riches of the Flow Analogy

From the notion of pliant flow, one can develop a wealth of new ideas for the conceptualization and practice of logistics management. Hydrodynamics concepts of characterizing flows, which can be transferred to the logistics context include, for example:

- Gradient – which, for any flow, is the force causing the movement. In logistics it is the service requirement. Parts/products flows are moved fast or slow depending on the customer service requirements and how well the control System pumps the flow.
- Velocity – the rate of flow, which for items in the supply chain is measured in inventory turns.

- Viscosity – is an internal characteristic of the fluid moving. When we deal with the flow of items we deal with characteristics such as bulk, standardization, homogeneity, and inventory carrying costs, etc.[3]
- Friction – measures the interaction of the fluid with its channel. With logistics flows we may be dealing with perishability, susceptibility to damage and pilferage, susceptibility to information loss en route, etc.
- Turbulent flow – characterizes the nature of the flow. Turbulent logistics flows involve rework, unscheduled returns, shortages and other unexpected phenomena. This contrasts with regular flow, where items, information, cash, and ideas flow smoothly.

These characteristics influence the speed of moving the items and the nature of the PPP logistics activities involved. Similarly, hydraulics concepts such as pressure, pumping and metering can be applied to the process of managing the flow along a supply chain.

While the analogy to fluid dynamics has its limitation, it can provide a strong guidance and a mental picture of what is required and how different items' flows can be compared and managed.

Pliant flow management requires efficient PPP (operational) logistics, and it requires that the right processes will be identified and controlled inside the firm and the right partnerships set along the supply chain. Beyond these prerequisites, however, the challenge facing logisticians is the development and management of smooth-flowing, demand-stimulated, well-structured flows of parts and products. This requires a tuning of the organization by pruning non-value-creating activities and programming the directions and responses that ensure smooth flows.

Based on the hypothesis of the evolution of logistics through three distinct meanings – from the operational functions of placing, pacing, and patterning, through the managerial functions of process coordination and partnerships management towards the design, programming and mobilizing of pliant flows – a new, cumulative definition of logistics may be suggested:

> *Logistics is the dynamic management of flows of items, information, cash and ideas, based on coordination of all the supply chain processes and the addition of place, period and pattern values.*

[…]

---

[3] Packaging is generally intended to improve those "viscosity" characteristics of the items flowing, to make tie flow smoother and faster.

François Quesnay (1764)

TABLEAU ÉCONOMIQUE

In searching for the roots of logistics and its conceptualization as a science of flows and flow systems, the earliest reference found is from the French physician and surgeon François Quesnay (1694–1774). Later in his life he turned to economics and became one of the founders of the Physiocratic school of economics.

His most famous contribution, the *Tableau Économique*, which was published in 1758, is an analysis and model of the economy as a system of revolving flows of money. He recognizes several types of economic actors and activities: "Productives" (farmers), "Propriétaires" (landowners, especially the nobility), and "Stériles" (members of the productive industry and trade). In his model, the interrelationships and flows of products and services between those actors are shown. Quesnay considered the time sequence of incomes earned in one period, being spent, reinvested, and reproduced in the following period. His model anticipates many ideas of modern macro-economic input-output analyses and of the structure of supply networks. The inspiration for his model, apparently, came from Quesnay's training as a physician. It is based on the analogy between the circulation of blood in the human body and the flows of moneys and supplies in an economy.

From today's perspective Quesnay's work is difficult to read. Many of the thoughts of his time (such as the idea that the agricultural sector is the only sector in the economy truly contributing to wealth and value creation) are no longer seen as valid. In reverence to Quesnay as the first economist who explicitly applied the concept of flows as a useful metaphor to model and analyse complex economic systems, the *Tableau Économique* and some explanatory comments are included in this volume.

On the following pages the Tableau (in the 3rd, advanced version of 1764) is reprinted – the facsimile of the French original, as well as the English counterpart. The reproductions are followed by a brief excerpt from Quesnay's explanations to his *Tableau Économique*.

Reference:

François Quesnay: *Tableau Économique*, 3rd Edition, Ex Libris Petri Du Pont, Agriculturæ Regiæ, Societatis Suessionensia, 1764.

P. Klaus and S. Müller (eds.), *The Roots of Logistics*, DOI 10.1007/978-3-642-27922-5_11,

François Quesnay: Tableau Économique

TABLEAU ECONOMIQUE.

Objets à considerer, 1.° Trois sortes de depenses; 2.° leur source; 3.° leurs avances; 4.° leur distribution; 5.° leurs effets; 6.° leur reproduction; 7.° leurs rapports entr'elles; 8.° leurs rapports avec la population; 9.° avec l'Agriculture; 10.° avec l'industrie; 11.° avec le commerce; 12.° avec la masse des richesses d'une Nation.

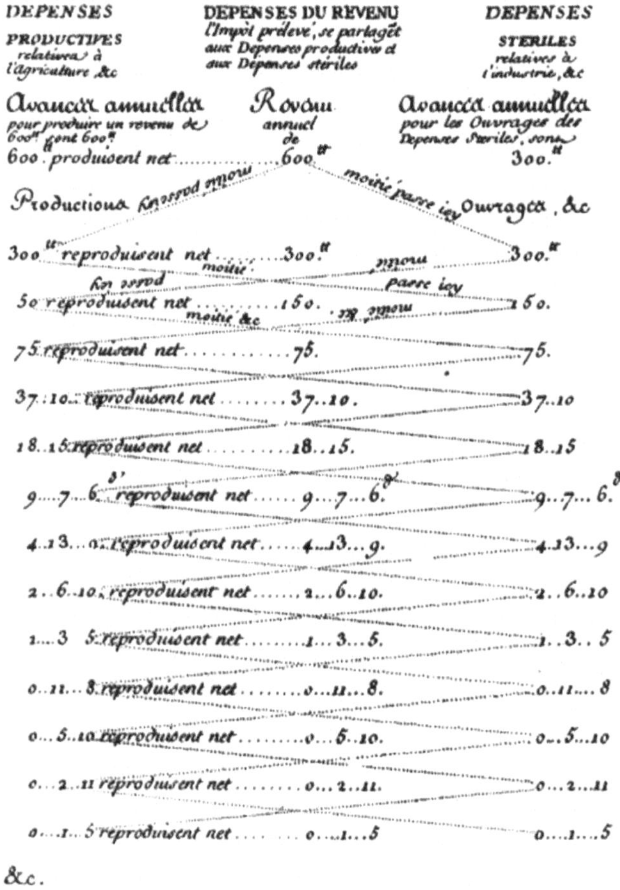

REPRODUIT TOTAL ....600.tt de revenu; de plus, les frais annuels de 600.tt et les interets des avances primitives du Laboureur, de 300.tt que la terre restitue. Ainsi la reproduction est de 1500.tt compris le revenu de 600.tt qui est la base du calcul, abstraction faite de l'impôt prélevé, et des avances qu'exige sa reproduction annuelle, &c. Voyez l'Explication à la page suivante.

TABLEAU ÉCONOMIQUE

Objects to be considered: (1) three kinds of expenditure; (2) their source; (3) their advances; (4) their distribution; (5) their effects; (6) their reproduction; (7) their relations with one another; (8) their relations with the population; (9) with agriculture; (10) with industry; (11) with trade;(12) with the total wealth of nation.

PRODUCTIVE EXPENDITURE *relative to agriculture, etc.*	**EXPENDITURE OF THE REVENUE** *after decuction of taxes, is divided between productive and sterile expenditure*	***STERILE EXPENDITURE*** *relative to industry, etc.*
Annual Advances required to produce a revenue of 600[liv.] are 600[liv.]	***Annual Revenue***	***Annual Advances*** for the works of sterile expenditure are
600[liv] produce net	*600[liv]*	*300[liv]*
PRODUCTS	*one half goes here* / *one half goes here*	**WORKS, ETC.**
300[liv]	*reproduce net ... 300[liv]* (*one half goes here* / *one half goes here*)	*300[liv]*
150	*reproduce net ... 150* (*one half etc.* / *one half*)	*150*
75	*reproduce net ... 75*	*75*
37 ... 10[s]	*reproduce net ... 37 ... 10[s]*	*37 ... 10[s]*
18 ... 15	*reproduce net ... 18 ... 15*	*18 ... 15*
9 .... 7 .... 6[d]	*reproduce net ... 9 .... 7 .... 6[d]*	*9 .... 7 .... 6[d]*
4 ... 13 ... 9	*reproduce net ... 4 ... 13 ... 9*	*4 ... 13 ... 9*
2 .... 6 ... 10	*reproduce net ... 2 .... 6 ... 10*	*2 .... 6 ... 10*
1 .... 3 5	*reproduce net ... 1 .... 3 .... 5*	*1 .... 3 .... 5*
0 ... 11 8	*reproduce net ... 0 ... 11 .... 8*	*0 ... 11 ... 8*
0 .... 5 ... 10	*reproduce net ... 0 .... 5 ... 10*	*0 .... 5 ... 10*
0 .... 2 ... 11	*reproduce net ... 0 .... 2 ... 11*	*0 .... 2 ... 11*
0 .... 1 5	*reproduce net ... 0 .... 1 .... 5*	*0 .... 1 .... 5*

etc.

***TOTAL REPRODUCED**..... 600 [liv] of revenue; in addition, the annual costs of 600 [liv] and the interest on the original advances of the husbandman amouting to 300 [liv], which the land restores. Thus the reproduction is 1,500 [liv], including the revenue of 600 [liv] which forms the base of the calculation, abstraction being made of the taxes deducted and of the advances which their annual reproduction entails, etc. See the explanation on the following page.*

EXPLANATION OF THE TABLEAU ÉCONOMIQUE

Productive expenditure is employed in agriculture, grasslands, pastures, forests, mines, fishing, etc., in order to perpetuate wealth in the form of corn, drink, wood, livestock, raw materials for manufactured goods, etc.

Sterile expenditure is on manufactured commodities, house-room, clothing, interest on money, servants, commercial costs, foreign produce, etc.

The sale of the net product which the cultivator has generated in the previous year, by means of the **annual advances** of 600 livres[1] employed in cultivation by the farmer, results in the payment to the proprietor of **a revenue** of 600 livres.

The **annual advances** of the sterile expenditure class, amounting to 300 livres, are employed for the capital and costs of trade, for the purchase of raw materials for manufactured goods, and for the subsistence and other needs of the artisan until he has completed and sold his work.

Of the **600 livres of revenue**, one-half is spent by the proprietor in purchasing bread, wine, meat, etc., from the productive expenditure class, and the other half in purchasing clothing, furnishings, utensils, etc., from the sterile expenditure class.

This expenditure may go more or less to one side or the other, according as the man who engages in it goes in more or less for luxury in the way of subsistence or for luxury in the way of ornamentation. We assume here a medium situation in which the reproductive expenditure renews the same revenue from year to year. But it is easy to estimate the changes which would take place in the annual reproduction of revenue, according as sterile expenditure or productive expenditure preponderated to a greater or lesser degree. It is easy to estimate them, I say, simply from the changes which would occur in the order of the **tableau**. Suppose, for example, that luxury in the way of ornamentation increased by one-sixth in the case of the proprietor, by one-sixth in the case of the artisan, and by one-sixth in the case of the cultivator. Then the revenue reproduced, which is now 600 livres, would be reduced to 400 livres. Suppose, on the other hand, that an increase of the same degree took place in expenditure on the consumption or export of raw produce. Then the revenue reproduced would increase from 600 to 800 livres, and so on in progression. Thus it can be seen that an opulent nation which indulges in excessive luxury in the way of ornamentation can very quickly be overwhelmed by its sumptuousness.

The 300 livres of revenue which according to the order of the **tableau** have passed into the hands of the class of productive expenditure, return to this class its **advances** in the form of money. These advances reproduce 300 livres net, which represents the reproduction of part of the proprietor's revenue and it is by means of the

[1] Livre (liv.) = French silver currency unit, in use up to the 18th century (comment by the editors).

remainder of the distribution of the sums of money which are returned to this same class that the total revenue is reproduced each year. These 300 livres, I say, which are returned at the beginning of the process to the productive expenditure class, by means of the sale of the products which the proprietor buys from it, are spent by the farmer, one-half in the consumption of products provided by this class itself, and the other half in keeping itself in clothing, utensils, implements, etc., for which it makes payment to the sterile expenditure class. And the 300 livres are regenerated with the net product.

The 300 livres of the proprietor's revenue which have passed into the hands of the sterile expenditure class are spent by the artisan, as to one-half, in the purchase of products for his subsistence, for raw materials for his work, and for foreign trade, from the productive expenditure class; and the other half is distributed among the sterile expenditure class itself for its maintenance and for the restitution of its advances. This circulation and mutual distribution are continued in the same way by means of subdivisions down to the last penny of the sums of money which mutually pass from the hands of one expenditure class into those of the other.

Circulation brings 600 livres to the sterile expenditure class, from which 300 livres have to be kept back for the **annual advances**, which leaves 300 livres for wages. These wages are equal to the 300 livres which this class receives from the productive expenditure class, and the advances are equal to the 300 livres of revenue which pass into the hands of this same sterile expenditure class.

The products of the other class amount to 1,200 livres, abstracting from taxes, tithes, and interest on the husbandman's advances, which will be considered separately in order not to complicate the order of expenditure too much. The 1,200 livres' worth of product are disposed of as follows: The proprietor of the revenue buys 300 livres' worth of them. 300 livres' worth passes into the hands of the sterile expenditure class, of which one-half, amounting to 150 livres, is consumed for subsistence within this class, and the other half, amounting to 150 livres, is taken for external trade, which is included in this same class. Finally, 300 livres' worth are consumed within the productive expenditure class by the men who cause them to be generated; and 300 livres' worth are used for the feeding and maintenance of livestock. Thus of the 1200 livres' worth of product, 600 are consumed by this class, and its advances of 600 livres are returned to it in the form of money through the sales which it makes to the proprietor and to the sterile expenditure class. One-eighth of the total of this product enters into external trade, either as exports or as raw materials and subsistence for the country's workers who sell their goods to other nations. The sales of the merchant counterbalance the purchases of the commodities and bullion which are obtained from abroad.

Such is the order of the distribution and consumption of raw produce as between the different classes of citizens; and such is the view which we ought to take of the use and extent of external trade in a flourishing agricultural nation.

Mutual sales from one expenditure class to the other distribute the revenue of 600 livres to both sides, giving 300 livres to each, in addition to the advances which are maintained intact. The proprietor subsists by means of the 600 livres which he spends. The 300 livres distributed to each expenditure class, together with the product of the taxes, the tithes, etc., which is added to them, can support one man in each: thus 600 livres of revenue together with the appurtenant sums can enable three heads of families to subsist. On this basis 600 millions of revenue can enable three million families to subsist, estimated at four persons of all ages per family.

Fritz Nordsieck (1934)

FUNDAMENTALS OF A SCIENCE OF ORGANIZATION

The idea of conceptualizing and analysing macro-economic and business activities as "value chains" and "processes", and recognizing the impact of lean, straightforward, transparent processes on business success became mainstream in the business world not before the mid-1980s. It were the widely read and widely quoted publications by Porter (1985) and Hammer (1990), followed by a huge stream of related academic and popular management literature, which started this development (e.g. Keen (1997)). A perspective of using the metaphors of "chain", "network", "stream", "process" and "flow" to draw attention to the interconnectedness of functions and activities in an economy and at the micro-level of business activity analysis to the temporal, spatial and cause-effect interrelationships between those activities proved to be highly attractive. It caused a reorientation of analyses from more traditional functional and institutional perspectives towards systematic "end-to-end" analyses and fundamentally changed the field.

But with hindsight, in the course of searching for early applications and "roots" to the ideas which are central to modern logistical thinking, it was found that a German expert on organizational analysis, Fritz Nordsieck (1906–1984), discussed and published such ideas already 50 years earlier, thus anticipating business process and process organization concepts which now are at the heart of modern logistics and supply chain management thinking.

His volume on *Fundamentals of a Science of Organization* was published in 1934. Nordsieck may have been the first academic author to explicitly distinguish between functional, hierarchical organization structures and analyses, and process-oriented analysis. The logic of process organization is appropriate to show how a business mission is being fulfilled, how a product is evolving as it moves through a series of value-enhancing activities. It is suited to show the "rhythm" of the application of activities to the object.

On the following pages an extract from Nordsieck's volume is reprinted – this is a part of his paragraph 4 where he discusses principles of relationships within the organization. He describes the nature, specifics, and relationships of organizational tasks and the requirements of a process-oriented organization, including a graphical illustration which may be considered a very early depiction of what today would be called a supply network.

P. Klaus and S. Müller (eds.), *The Roots of Logistics*, DOI 10.1007/978-3-642-27922-5_12,

Reference:

Fritz Nordsieck: *Grundlagen der Organisationslehre*, Stuttgart: C.E. Poeschel, 1934, pp. 73–79.

Further reading:

Hammer, Michael: Don't Automate, Obliterate! In: *Harvard Business Review*, July-August 1990, pp. 104–112.

Keen, Peter G.W.: *The Process Edge. Creating Value Where it Counts*. Cambridge/MA: Harvard Business School Press, 1997.

Porter, Michael E.: *Competitive Advantage*. New York: Free Press, 1985.

Fritz Nordsieck: Fundamentals of a Science of Organization[1]

§ 4 THE THEORY OF ORGANIZATIONAL RELATIONSHIPS

Section 1: Tasks; Interrelationships between Tasks; especially the Organization of Tasks

a) Origin, Significance and Transformation of the Tasks

The concept of task has been discussed already at various points in this study.

Tasks are being adopted by a person or an organization due to their self-interest. If a task is to be accepted, this will only happen if some kind of adequate compensation is offered to the "agent"[2] who is expected to execute it – even if only of non-material nature, like expecting and receiving a principal's appreciation (the "reward principle"). On the other hand, it is essential to the concept of tasks that needs and demands of the person or organization assigning the task (the "principal", in a market situation usually the "customer") are to be met. What kinds of expectation and self-interest the agent has in mind when accepting a task is crucial for the way it will be executed. In the end, only the latter is of interest here.

Customer demands and needs which are shared by many and come up frequently create permanent demands for the products or services of a business. The more uniform an underlying demand from customers and markets is, the more it lends itself to the establishment of ever more routinized and structured business processes. The rhythm of the occurrence of market demands determines the rhythm of their fulfilment through the business organization aiming to meet those demands. In some cases the maintenance of inventories may allow to reduce the immediate interdependence of the rhythms of demand and supply. The moment in which a demand becomes universal it is suitable to be satisfied through organized industrial activity.

---

1 Translation by the editors. Passages of this 1930s academic text which would have been hard to understand in a translation close to the original were freely interpreted in a more contemporary way in the effort to keep them understandable.

2 I.e. the prospective task executor. For easy understandability in this translation the now familiar terminology of "principal" for the provider and assigner of a task, and of "agent" for the adopter and executor of a task is used here.

On the other hand: the more individual and stochastic a demand is, the less it is suitable for fulfilment through organized industrial activity. This is why the first, most important question for improving organizational tasks in a business is: What is the nature of the customer and market requirements and demands which are to be satisfied by the intended organizational arrangements?

For further clarification of organizational tasks it is important to answer another question. It concerns the relevance of a particular task within a more complex set of tasks, because this determines the importance of each of the various activities which have to take place in fulfilling a task. The answer to this question is found in identifying the underlying needs and demands of the customer (respectively the "principal") by their quantities and urgency. While the quantities demanded are determined by the number of those wanting the respective need to be met and by the scope of their demands, it is more difficult to establish the respective urgency. Urgency can be clearly specified only in certain situations, such as in cases with actual time-related urgency (e.g. in fire fighting). In my opinion, the best way to assess the urgency of a task is considering the question: What would happen if the respective need or demand was not satisfied at all or not in time?

In the assessment of the objective value of a task well executed the quantity and quality of work and resources that have been spent in the process of execution should not be considered. Quite the opposite: The allowable amount of work and resources to be spent for a given task should be derived from the value of the product or service which the customer (the "principal") assigns to it. One should speak of an objective value only – for the sake of clarity this should be noted – with regard to tasks of a character which can be assessed objectively through mutual agreement of several parties. The fulfilment of other kinds of tasks only has a subjective, not an objective value.[3] The evaluation of business tasks and of individual activities undertaken in the course of their execution would have to be evaluated with regard to their contribution in meeting the needs and demands of a society. The objective value of a task fulfilled well, would be expressed best in form of a price, which reflects the needs and demands of society.

[...]

Beside the question of the relevance of a task, a further concern is: How do tasks change with regard to their type and relevance? In order to resolve this, continuous monitoring of the market's demands according to their type, scope, and urgency is necessary. Changes observed – caused by whatever reasons, such as a rearrangement of societal priorities and values, or new products and technologies – will lead to new insights about the nature of the tasks and their transformation.

---

[3] It cannot be our task to deal with the value and price problem. We will be dealing with organizational matters only.

b) Interrelationships between Tasks; the Organization of Tasks

Next to the clarification of the quantitative significance, the urgency, and the variability of tasks, the most important problem is in understanding the interrelationship between tasks. Regarding the types of relations between tasks, one may distinguish between structural relations and loose, secondary relations. The structural relations are especially important. In order to clarify these we have to define several levels of tasks, such as a primary mission ("Oberaufgabe"), intermediate tasks, subordinate tasks, etc.

A central principle of our previous argument was to derive the significance of all organizational interventions from the primary mission and objective of the business. The primary mission and objectives are usually broken down into a range of lower level tasks and sub-objectives. In other words: The interrelationship of the business tasks is given through a task structure, which organizes the various levels of tasks of objectives in a rational, hierarchical way – a system.

All considerations about organizing must be related to this important fact. Clarifying the structural context of the tasks involves clarifying the task structure, which, in turn, means answering these questions: How are sub-objectives determined and differentiated? What are the principles to be applied in the construction of the task structure?

It should be noted first that the principles of deducing a goal and task structure can only be implicit, appearing as logical derivations of the term of the main task. The content of the primary mission is considered as given a priori. Then it may be structured into intermediate and specific subordinate levels of sub-tasks. A specific task appears as the smallest logical unit to be considered in the analysis of a process of executing a mission. It is identified as the result of a sequence of differentiations of task and objectives. The formal characteristics of a well formulated specific task are:

1) that the relationship between the sub-objective in question of the higher level objectives is clearly recognizable,
2) that it is relatively independent in the sense that the task follows its own, autonomous rhythm, that is has its own contiguous range of activities and adapts to a specific set of demands and requirements with relative autonomy.

Tasks must be broken up into subtasks, step by step, until the lowest possible level of differentiation is reached which still meets the criteria stated. The ordering of subtasks should follow their sequence in time – the principle of a process structure.

The structure of a task hierarchy may differ, depending on what the objective of the primary mission is. A process means that a series of activities is performed upon an object. The characteristics of the process are described by those activities and the objects in question. We will speak of activities, when the activity as such is in focus, detached from the object to which it is applied. Activities are, for example, writing, calculating; drilling, or lathing. Every complex task is composed of several activities and often of several objects.

This means that there are alternative options to describe the fulfilment of a task – in principle – primarily by the types of activities performed, or primarily be the objects to which the activities applied.

[...]

One should always aim for clarity in the organization of the process. The logic of process organization is most appropriate to show how a mission is being fulfilled, how the object[4] is evolving as it moves through a sequence of activities. It is especially suited to show the rhythm of the application of activities to the object. If tasks are organized as processes the challenge of assigning management responsibilities, of linking the activities, and of determining the most appropriate rhythm of the process is much simplified and most likely to secure the successful execution of the mission and tasks. How important the logic of organizing is for assigning responsibilities will be discussed in more detail in a later section.

To make the considerations discussed so far still better understood, EXHIBITS 1 and 2 may be useful: Both exhibits are intended to provide a schematic representation of the processes of a business. EXHIBIT 1 provides – so to speak – a birds eye's snapshot of the fabric of the business' processes. EXHIBIT 2 provides the corresponding view, showing the vertical breakdown of the tasks by their functional relationships. The letters and numbers mark the same tasks in the process and the cross-sectional depiction. The structure primarily by functions appears in both representations. For easy identification it is labeled by special symbols.

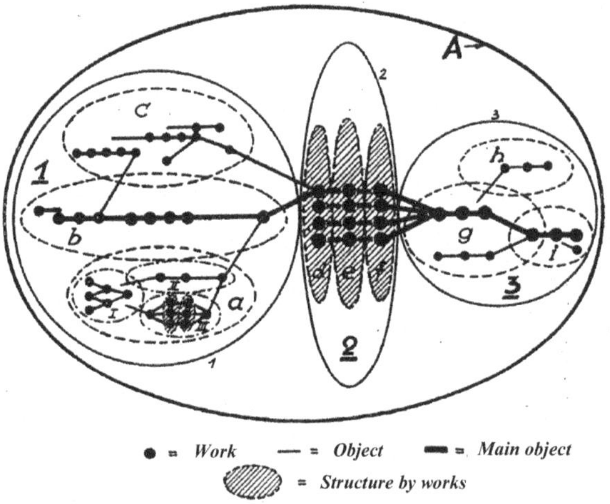

EXHIBIT 1: A PROCESS-ORIENTED REPRESENTATION OF TASK OF A BUSINESS

4 Such as a product being manufactured and distributed (comment by the editors).

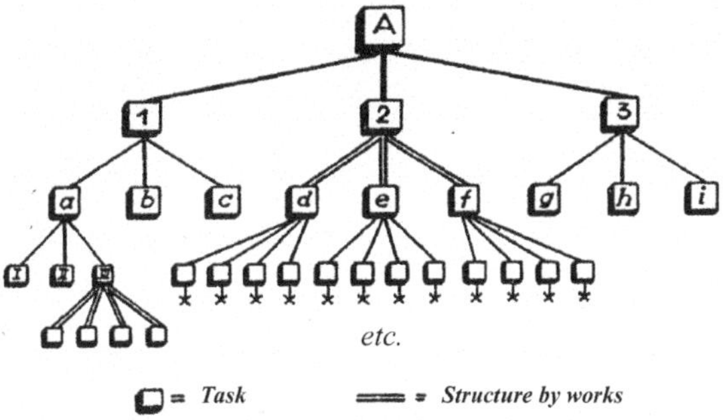

EXHIBIT 2: A REPRESENTATION OF THE TYPES OF FUNCTIONAL INTERRELATIONSHIPS BETWEEN THE TASKS OF A BUSINESS

In structuring tasks according to the process logic subtasks are being identified and arranged according to the stages of the process along the dimensions of time and space. In this kind of representation, functions may appear separated and torn apart from the perspective of their functional similarity. On the other hand, it becomes evident that a traditional hierarchical and functional organization cuts into pieces the interfaces and connections which are critical in the process.

If the process representation is preferred as the primary orientation for organizing a complex mission or task, then there should be supplementary representations used to show the other kinds of relationships, such as the functional ones.

[…]

Jay W. Forrester (1958)

INDUSTRIAL DYNAMICS: A MAJOR BREAKTHROUGH FOR DECISION MAKERS

In the introduction to his 1958 article about *Industrial Dynamics* Jay W. Forrester (born 1918), pioneer in computer engineering, systems scientist, and professor at the MIT Sloan School of Management, made a statement which anticipates one of the premises on which the importance of modern logistics rests: "company success depends on the interaction between the flows of information, materials, money, manpower, and capital equipment".

This article, which is reprinted here, and later work by Forrester and his Systems Dynamics research group founded at the MIT in the early 1960s are devoted to a better understanding – even the hope to develop a "unifying theory" – of the behaviors which complex systems display as a consequence of the interactions of those multiple interrelated flows.

Forrester discusses the new "tools of progress", which became available at the time for this kind of task – computers, simulation, and understanding of feedback control in systems – and then illustrates what "industrial dynamics" means. He uses the example of a closely coupled production and distribution process – the model of a supply chain. He shows how to simulate various types of behaviors and their consequences for inventory levels and schedules, and he derives first recommendations for the control and management of industrial dynamics. (Later his group developed the popular and widely used educational exercise, the "beer game" to allow students to actively play, try out and experience the effects of industrial dynamics.)

The paper reads like an early blueprint for research on modern logistics and supply chain management – without using the terms. Among other insights, it perfectly describes what was rediscovered nearly forty years later as the "bullwhip effect" in supply chains (Lee/Padmanbhan/Whang (1997)). Yet Forrester's ideas of studying the dynamics between the flows of information, materials, money, manpower, capital equipment as a basis for gaining new insights about underlying principles of business seem to have still much more promise for the future of the science of logistics.

Reference:

Jay W. Forrester: Industrial Dynamics – A Major Breakthrough for Decision Makers. In: *Harvard Business Review*, Vol. 36, July-August 1958, pp. 37–66.

Further reading:

Lee, Hau L., V. Padmanabhan, and Seungjin Whang: The Bullwhip Effect in Supply Chains. In: *Sloan Management Review*, Vol. 38, No. 3, 1997, pp. 93–102.

P. Klaus and S. Müller (eds.), *The Roots of Logistics*, DOI 10.1007/978-3-642-27922-5_13,

Jay W. Forrester: Industrial Dynamics: A Major Breakthrough for Decision-Makers[1]

Management is on the verge of a major breakthrough in understanding how industrial company success depends on the interaction between the flows of information, materials, money, manpower, and capital equipment. The way these five flow systems interlock to amplify one another and to cause change and fluctuation will form a basis for anticipating the effects of decisions, policies, organizational forms, and investment choices.

My aim in this article is to look ahead at the specific kinds of progress which will be achieved and the concepts which will make this progress possible. While I shall suggest certain ways of thinking about management that should be helpful to executives today in working on inventory control, production, scheduling, advertising, sales, and other problems, my primary concern here is not with techniques and prescriptions. Rather, I am interested in the development of a professional approach to management.

Business leaders, like leaders in other areas, are influenced by an "image of the future". Their ideas about where they are going may be clear or vague, but in any event they have a subtle and far-reaching impact on administrative thinking and decisions. A look at some promising new concepts of management should, I believe, convince even the skeptical executive that his job is developing into much more than an art, that conceptual skill will play an increasingly vital role in company success, and that management is fast becoming second to none as an exciting, dynamic, and intellectually demanding profession.

TOWARD A THEORY

To develop the status of a profession, management must discover the underlying principles which unify its separate aspects. It must develop a basic theory of behavior. It must learn how to convert experiences and particular case examples into a contribution to this general theory. And, finally, it must be able to employ the basic principles of the theory as a useful practical guide for explaining and solving new

---

[1] Author's note: The studies on which this article is based were made possible through the support of the Sloan Research Fund of the School of Industrial Management at Massachusetts Institute of Technology, the Ford Foundation, and the use of the IBM 704 computer at the Massachusetts Institute of Technology Computation Center.

problems as they arise. By accomplishing these aims, management will become a true profession during the next generation.

The task of management is to interrelate the flows of information, materials, manpower, money, and capital equipment so as to achieve a higher standard of living, stability of employment, profit to the owners, and rewards appropriate to the success of the managers. Looked at in this way, its goals are rooted as deeply in the public interest as the broad objectives of the legal, medical, and engineering professions.

In the past, with management considered more of an art than a profession, education and practice have been highly fragmentized. Manufacturing, finance, distribution, organization, advertising, and research have too often been viewed as separate skills and not as part of a unified system. Too often management education consists of gathering current industrial practice and presenting it to the student as a sequence of unrelated subjects. Similarly, in his work in industry the manager specializes within departments where his experience perpetuates the atmosphere of unrelated compartmentalization.

The next big step in management education will be the development of a basis for fitting together the many management functions into a meaningful whole. Around this central core specialized subjects and experience will take on more significance. Men can be developed more rapidly. They will be able to start from a point now accessible only through long training or fortuitous experience.

Such strides will far exceed in importance recent steps in using computing machines to execute clerical tasks or in applying operations research methods to isolated company problems. For we can expect to gain, during the next 25 years, a far better understanding of the dynamic, ever-changing forces which shape the destiny of a company. This understanding will lead to better usage of available information, to improved understanding of advertising effectiveness and the dynamic behavior of the consumer market, and to company policies that keep pace with technological change.

Beyond these achievements, there will be improvements in company organization resulting from a sounder basis for effective decentralization, from altering the relationships between line and staff tasks in the company, from the more effective utilization of scientific manpower, and from reducing the routine duties and enhancing the creativity of managers. And executives will gain in "clairvoyance". For example, they will be able to anticipate clearly (as I shall illustrate later in this article):

- How small changes in retail sales can lead to large swings in factory production.
- How reducing clerical delays may fail to improve management decisions significantly.
- How a factory manager may find himself unable to fill orders although at all times able to produce more goods than are being sold to customers.
- How an advertising policy can have a magnifying effect on production variations.

TOOLS OF PROGRESS

The new management concepts will rest in part on recent advances in the data-processing industry, in part on military research (which has given us an improved understanding of decision making and experience in analyzing and simulating the characteristics of complex systems), and largely on 20 years of research in information-feedback systems.

Electronic Data Processing

The performance of electronic computers has increased annually by a factor of nearly 10 per year over the last decade; in almost every year we have seen a tenfold increase in speed, memory capacity, or reliability. This represents a technological change greater than that effected in going from chemical to atomic explosives. Society cannot absorb so big a change in a mere ten years. We therefore have a tremendous untapped backlog of potential devices and applications.

Viewing data processing in another way, we can trace the shifting frontiers of the field by dividing progress into five-year periods:

- From 1945 to 1950 was the period of electronic research and the demonstration that machines having many thousand vacuum tubes would indeed operate.
- From 1950 to 1955 was the pioneering period in applying computers to the solution of scientific and engineering problems.
- During the present period, 1955 to 1960, electronic machines are being substituted for clerical effort in commercial organizations.
- From 1960 to 1965 we can expect to see the application of digital computers to physical process control. Already there are digital machine tool controls and the SAGE air defense system; we can look forward to dramatic improvement in processes, to reduction of capital investment requirements in the oil and chemical industries, and to civilian air traffic control systems, all based upon new kinds of information utilization for control purposes.
- From 1965 to 1970 we should see all these developments converging into pioneering improvements in the central management process. The routine, repetitive types of decisions will become more formalized, while management creativeness will be directed to **how** decisions and policies should be made rather than to the actual repetitive making of such decisions.

During the 1965–1970 period of widespread company testing of new management methods, electronic data processing will be an essential tool. Let me emphasize, however, that it is no more the focal point of the future management profession than a slide rule is the essence of engineering.

Decision Making

Historically, military necessity has often led not only to new devices like aircraft and digital computers but also to new organizational forms and to a new understanding of social forces. These developments have then been adapted to civilian usage.

Such new exploration is now happening in the military **command** (or **management**) function. As the pace of warfare has quickened, there has of necessity been a shift of emphasis from the tactical decision (moment-by-moment direction of the battle) to strategic planning (preparing for possible eventualities, establishing policy, and determining in advance how tactical decisions **will** be made). The battle commander can no longer plot the course of his enemy on a chart and personally calculate the aiming point. In fact, with a ballistic missile he would have no time even to select his defensive weapon.

Likewise in business: as the pace of technological change quickens, corporate management, even at the lower levels, must focus more and more on the **strategic** problems of running the business and less and less on the everyday operating problems.

In the systems development for the military, it has been amply demonstrated that carefully selected formal rules can lead to tactical decisions that excel those made by human judgment under the pressure of time and with insufficient experience and practice. Furthermore, it has been found that men are just as adaptable to the more abstract strategic planning as they are to tactical decision making, once their outlook has been lifted to the broader and longer-range picture.

Simulation

Also from military research we have available the methods of simulation for determining the behavior of complex systems. These techniques have reached the state of development where they can now be usefully applied to industrial organizations. Simulation is being used in the design of air defense systems and in engineering work. For example:

In planning the development of a river basin, numbers in a digital computer represent water volumes, flow rates, electric demand, and rainfall. A few seconds of computer time can solve a whole day of systems operation. Dams can be located and designed for an optimum compromise between power generation, irrigation, navigation, and flood control.

In business, simulation means setting up in a digital computer the conditions which describe company operations. On the basis of the descriptions and assumptions about the company, the computer then generates the resulting charts of information concerning finance, manpower, product movement, and so on. Different management policies and market assumptions can be tested to determine their effect on company success.

To use simulation studies will not require undue mathematical ability. To be sure, details of setting up a model will need to be handled by experts because there are special skills required and pitfalls to be avoided. However, the job of directing the situations to be explored, judging the assumptions, and interpreting the results will be within the ability of the type of men we now see in management schools and executive development programs.

Feedback Control

Systems of information feedback control are fundamental to all life and human endeavor, from the slow pace of biological evolution to the launching of the latest satellite. A feedback control system exists whenever the environment causes a decision which in turn affects the original environment. To illustrate

- A thermostat receives temperature information, decides to start the furnace, and changes the temperature.
- A person senses that he may fall, corrects his balance, and thereby is able to stand erect.
- In business, orders and inventory levels lead to manufacturing decisions which fill orders and correct inventories.
- A profitable industry attracts competitors until, to use the economist's terms, the profit margin is reduced to equilibrium with other economic forces.
- The competitive need for a new product leads to research and development expenditure that produces technological change.

All of these are information feedback control loops. The regenerative process is continuous, and new results lead to new decisions which keep the system in continuous motion.

The study of feedback systems deals with the way information is used for the purpose of control. It helps us to understand how the amount of corrective action and the time delays in interconnected systems can lead to unstable fluctuation. Driving an automobile provides a good example:

The information and control loop extends from steering wheel, to auto, to street, to eye, to hand, and back to steering wheel. Suppose the driver were blindfolded and drove only by instructions from his front-seat companion. The resulting information delay and distortion would cause erratic driving. If the blindfolded driver could get instructions only on where he had been from a companion who could see only through the rear window, his driving would be even more erratic.

Yet this is analogous to the situation in business. Top executives do not see the salesmen calling on customers, do not see the prospective buyers watching a TV commercial. They do not attend the board meeting of competitors. They do not have a clear view of the road ahead. The only thing they can tell with reasonable certainty (and even here there is sometimes doubt) is what happened to wages, sales, material costs, interest rates, and so on last year.

Smoother Operations

The quality of management control depends upon what information executives use and for what they use it, as well as on their skills as administrators. As a system the company has certain characteristics which are completely independent of individual functions or departments, just as an electronic computer has certain characteristics as a system of parts. For instance, there is what the engineer might call "amplification", caused by inventory accumulation, filling of supply pipelines, and inept extrapolation of trends. There are delays in decisions, shipping, communications, and accounting. These all combine to cause production fluctuations, construction of excess plant capacity, creation of company-generated rather than customer-generated seasonal sales, and detrimental advertising policies.

Without an awareness of basic information-flow principles, it is only through costly errors that managers can develop an effective intuitive judgment. For example:

In a company manufacturing consumer durables, it was discovered, after all data became available, that in a certain year retail sales had varied by 30%. In the same year the inventory and ordering practices in the distribution system at the factory caused this small retail variation to be amplified to a four-to-one, or 400%, variation in factory production. The estimated avoidable costs of increasing and decreasing production and of carrying excess inventory were equal to the normal anticipated profit margin! (We shall see later how this could happen.)

Feedback theory explains how decisions, delays, and predictions can produce either good control or dramatically unstable operation. It relates sales promotion to production swings, purchasing and pricing policies to inventory fluctuations, and typical life cycles of products to the need for research.

PRODUCTION & DISTRIBUTION

How do the concepts and approaches I have been discussing apply to specific business problems? Or, to be more precise, how do the delays, amplifications, and oscillations in the circuital flow of information in a company affect its operation?

Now, in order to illustrate, I must compromise one of the main ideas I have specified; namely, that the system (meaning not the paperwork forms and procedures, but the interrelationships between all the company operations) behaves according to the characteristics of the whole and not according to the characteristics of the individual parts. For the sake of brevity, I shall have to limit myself to the production and distribution functions, forgetting research, engineering, sales, and, until a later point in the article, promotional effort. In a sense this is a subsystem, with its own complex of interrelationships that can be studied as such, but it is not the whole. Even within this production-distribution subsystem, I shall have space to include only the flows of information and materials. Omitted altogether are considerations of money, manpower, and capital equipment.

If I can make the principles of the new approach clear, however, it should not be difficult for the reader to extend his own picture to include the areas that I must omit.

Needed Information

To begin the study of our example, we need to know three kinds of information about the system: its organizational structure, the delays in decisions and actions, and the policies governing purchases and inventories.

1. Organizational Structure. EXHIBIT 1 shows a typical organizational structure for the production and distribution functions in a hard goods industry like household appliances. If we examine the basic internal behavior of the system when customer orders bare independently specified, we will see that, even with very simple retail sales changes and with no other external disturbances affecting the company, typical manufacturing and distribution practices can generate the types of business disturbances which are often blamed on conditions outside the company. For random, meaningless sales fluctuations can be converted into annual, seasonal production cycles. Factory capacity, even though always exceeding retail sales, can seem to fall short of meeting demand with the result that production capacity is overexpanded.

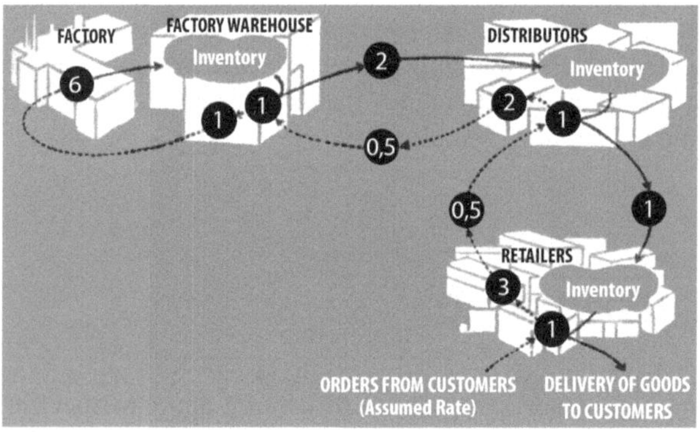

EXHIBIT 1: ORGANIZATION OF PRODUCTION-DISTRIBUTION SYSTEM

Examining EXHIBIT 1 more closely, we see that the bottom box represents the retail level. Next above are the distributors, and at the left the factory and factory warehouse. The dotted lines show the information flow, here consisting of orders for goods flowing upward. The solid lines show the shipment of goods flowing downward.

2. Delays in Decisions and Actions. To be able to determine some of the dynamic characteristics of this system we must also know the delays in the flow of information and goods. The time delays are shown on the diagram in weeks and are reasonable values for a consumer durable product line.

Delivery of goods to the consumer averages a week after the customer places an order. At the retail level, the accounting and purchasing delays average three weeks between the time of a sale and the time when that sale is reflected in an order sent out to obtain a replacement. Mailing delay for the order is half a week. The distributor takes a week to process the order, and shipment of goods to the retailer takes another week. Similar delays exist between the distributor and the factory warehouse. The factory lead time averages six weeks between a decision to change production rate and the time that factory output reaches the new level. Note that three levels of inventory exist – factory, distributor, and retailer.

3. Policy on Purchasing Orders and Inventories. To complete the initial description of our example, we need to know the policies followed in placing orders and maintaining inventory at each distribution level. Let us consider three principal types of orders: (a) orders which directly reflect sales, (b) orders to adjust inventories with changes in business volume, and (c) orders to fill the supply line with in-process orders and shipments. Let us suppose further that orders are treated in the following ways:

- After a sales analysis and purchasing delay (three, two, and one weeks for the three levels), orders to the next higher level of the system include the actual sales made by the ordering level.
- After a time for averaging out sales fluctuations (eight weeks), a gradual upward or downward adjustment is made in inventories as the rate of sales increases or decreases.
- The orders in process (orders in the mail, unfilled orders at the supplier, and goods in transit) are proportional to the level of business activity and to the length of time required to fill an order. Both an increased sales volume and an increased delivery lead time necessarily result in increased total orders in the supply pipeline.

The ordering rate will also depend on some presumptions about future sales. Prediction methods that amount to extending forward (extrapolating) the present sales trend will in general produce a more unstable and fluctuating system. For our example, however, we will use the conservative practice of basing the ordering rate on the assumption that sales are most likely to continue at their present level.

Simulation Methods

Before we can determine how our system will function over a period of time, all of the above rather general descriptions of the system must be expressed in explicit quantitative form. Illustrations of these relationships are shown in EXHIBIT 2.

To determine the behavior of a system by simulating the performance of its parts requires that one describe exactly, and in detail, the characteristics which are to be included. The validity of the outcome of the system studies depends on the judgment of what is pertinent to include in the system description. The following examples show some of the kinds of relationships that are needed in the study of the production-distribution functions of Exhibit 1:

(1) The inventory level at retail at the end of any time period is the inventory at the beginnen of the period, plus goods received, minus goods delivered during the period. The relationship can be expressed as:

$$I_{r,k} = I_{r,j} + \Delta t R_{d,jk} - \Delta t S_{r,jk}$$

which says that the inventory at retail, I_r, at time k is equal to the inventory one time period earlier at time j, plus the goods received in the time interval (the length of the time interval, Δt, multiplied by the rate at which goods are received from the distributor, R_d, over the time interval j to k) minus the goods delivered (Δt multiplied by the rate of sales to customers, S_r, in the interval j to k).

Δt must be a solution time interval that is short compared with any significant time delays which are to be represented in the system. (Δt is here 1/20 of a week, since third-order exponential delays as short as half a week – the mailing delays – are used in the system description.)

(2) Similarly, unfilled orders equal previous unfilled orders plus new orders minus shipments; that is:

$$U_{r,k} = U_{r,j} + \Delta t N_{r,jk} - \Delta t S_{r,jk}$$

where U_r represents unfilled orders at the retail level; N_r is the rate of receipt of new orders; and S_r is the rate of delivery of goods.

(3) Shipments by the retailers are dependent on various factors, here assumed to be the normal order-processing delay, d_{ur}, the level of unfilled orders, U_r, and the physical ability to fill orders based on a ratio of actual inventory level, I_r, to be a desirable level of inventory, D_r. Expressed mathe-matically:

$$S_{r,kl} = \frac{U_{r,k}}{d_{ur}} \sqrt{\frac{I_{r,k}}{D_{r,k}}}$$

This is to represent the aggregate of many retailers each selling a multiplicity of catalogue items in the product line.

(4) We need to know the rate at which the retail level purchases from the distributor level. This reflects (a) the rate of customer purchases from retail plus (b) the rate necessary to adjust the retail inventories plus (c) the adjusting of orders in process in the pipeline between retailers and distributors. The mathematical statement is:

$$P_{r,kl} = P_{e,jk} + \frac{1}{d_{ir}} [(D_{r,k} + d_{pr} A_{r,k}) - (I_{r,k} + G_{r,k})]$$

This says that the purchasing rate by retailers, P_r, in the next time interval, kl, is equal to the purchasing rate by customers, P_c, in the previous time interval, jk, as it is thought to be after data-processing delays (provided by another equation), plus an allowance for inventory and pipeline adjustments which are made gradually over the length of time, d_{ir}.

The „desired ownership“, consisting of the desired inventory, D_r, plus the orders which need to be in process to sustain the current business level (obtained by multiplying the total delay, d_{pr}, a variable time, in filling orders by the average rate of recent sales, $A_{r,k}$), is:

$$(D_{r,k} + d_{pr} A_{r,k})$$

The „actual ownership“, represented by the present inventory, I_r, plus the actual orders in process, G_r, is:

$$I_{r,k} + G_{r,k}$$

Some 40 relationships to the above type serve to relate orders, shipments, purchases, mailing delays, shipping delays, factory lead time, and so forth. It is not the purpose of this article to do more than illustrate the nature of the relationships which are required. Much study of any particular industrial system would be necessary to determine the relationships that meaningfully portray its behavior. Line executives must set the stage for such a study, and specify assumptions to be used and results to be sought, but the detailed work is for the experts.

EXHIBIT 2: FORMAL QUANTITATIVE STATEMENTS OF THE RELATIONSHIPS IN EXHIBIT 1

The next step is to determine how the system as a whole behaves. To do so, we might use some pattern of consumer purchases as an input and observe the resulting inventory and production changes. A very good test would be to see what happens after a small sudden change in retail sales. The effect on the company system can be obtained by "simulation" methods – simulation being the technique of obtaining results from a model. For the industrial system, simulation consists of tracing through, step by step, the actual flow of orders, goods, and information, and observing the series of new decisions required. For example:

Our production-distribution system might be simulated by a group of men around a table, one to represent retailers, another the postal service, another transportation, another the factory, and so on. Five minutes might represent a week, and in each time interval the proper purchase orders and deliveries would be made according to the rules illustrated in EXHIBIT 2. Alternatively, the whole exercise can be done by one person in tabular form on paper. Better still, the entire sequence can be programmed on a digital computer.

Digital computer simulation was used to obtain the results in EXHIBIT 3. A sudden 10% increase in retail sales was introduced in January. The resulting fluctuations are shown in order rates, factory output, factory warehouse inventory, and unfilled orders. (Retail orders are here given as an input which is independent of what happens in the production-distribution system itself. Retail orders are, of course, not actually independent since they are affected by availability of the product and by advertising, a point to be discussed later.)

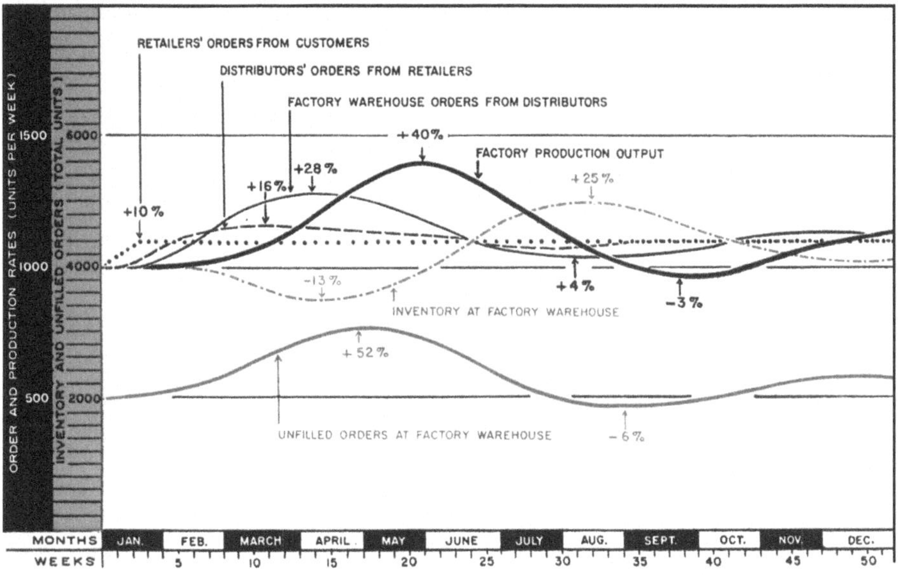

EXHIBIT 3: RESPONSE OF PRODUCTION-DISTRIBUTION SYSTEM TO A SUDDEN 10% INCREASE IN RETAIL SALES

Fluctuating Behavior

Because of accounting, purchasing, and mailing delays, the increase in distributors' orders from retailers lags about a month in reaching the 10% level. It is important to note, however, that the rise does not stop at 10%. Instead it reaches a peak of 16% in March because of the new orders that have been added at the retail level (a) to increase inventories somewhat and (b) to raise the level of orders and goods in transit in the supply pipeline by 10% to correspond to the 10% increase in sales rate. These inventory and pipeline increments occur as "transient" or "nonrepeating" additions to the order rate and, when they have been satisfied, the retailers' orders to the distributors drop back to the enduring 10% increase.

The factory warehouse orders from distributors show an even greater swing. The upturn in orders at the factory warehouse lags the upturn at the distributor level. The incoming order level for distributors is above retails sales for four months, reaching a 16% increase in March, which is readily mistaken for a corresponding true and enduring increase in business volume of the same amount. The distributors' orders to the factory, therefore, include not only the 16% increase in orders they themselves receive but also a corresponding increase for distributor inventories and for orders and goods in transit between distributor and factory. As a result, orders at the factory warehouse, reach in April a peak of 28% above the previous December.

Let us turn now to manufacturing orders. They are placed on the basis of the increasing factory warehouse orders and the falling warehouse inventory, which drops 13%. As a result, the factory output, delayed by a factory lead time of six weeks in this example and then a week of data-processing delays, reaches a peak in June which is 40% above December. While retails sales are still at 10% above December, the factory increase is four times as great.

Note that all of these effects are reversible, another very important point. As retailers satisfy their inventory requirements, they decrease their order rate. The distributors find they have built up an order rate, an inventory, and a supply-line rate in excess of needs. They take the excess out of current orders to the factory so that in August this order rate is 6% **below** retail sales and 4% above the previous December. In September and October factory output drops to 3% below December and 13% below current retail sales.

As a direct result of the typical organizational form shown in EXHIBIT 1 and of the customary inventory and ordering policies described, over a year is required before all ordering and manufacturing rates stabilize at their proper levels corresponding to the 10% retail sales increase. Ironically, this increase was minor compared to the fluctuations in company operations.

Retail Sales Variations

The problem of unexpected variations in retail sales will now be examined. Imagine first that our system has operated in the past with constant sales, and then sales rise and fall gradually over a one-year interval.

EXHIBIT 4 shows the way in which an annual retail sales variation is accentuated as orders travel toward the factory. Retail sales have been stable at 1,000 per week in the past, and in January they start rising toward a 10% increase at the end of March. They fall to 10% below "normal" by the end of September and return to normal by the end of December.

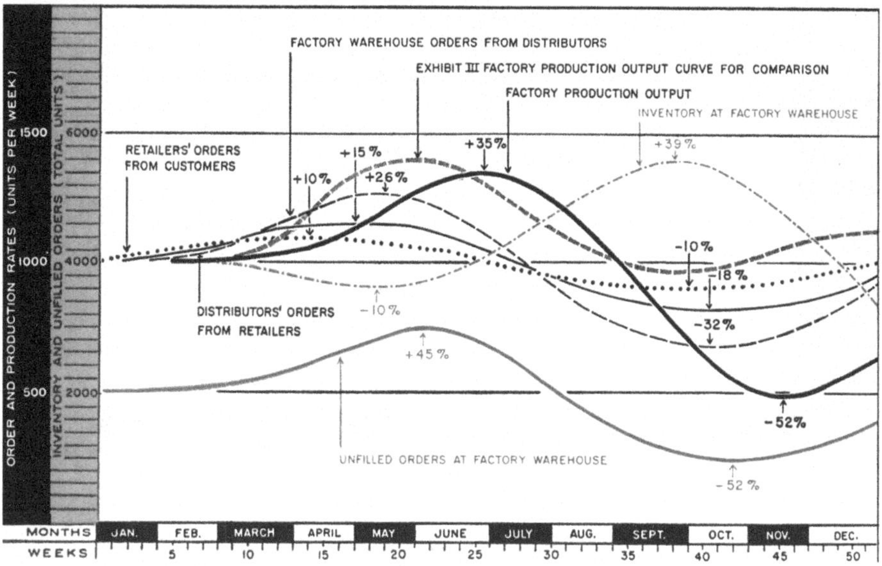

EXHIBIT 4: RESPONSE OF PRODUCTION-DISTRIBUTION SYSTEM TO A 10% ANNUAL RISE AND FALL IN RETAIL SALES

The initial upswing in orders and factory output is much like that in EXHIBIT 3 except that the initial peaks are lower and later. However, at the time the system would naturally rebound from its overproduction state, it is given an additional downward shove by the declining retail sales, which are amplified for the same reasons as previously discussed. As a combined result, factory orders from distributors reach a low of 32% below normal in October, and factory output falls to 52% below normal in November.

If the following year were shown, we would see factory production continuing to fluctuate between peaks and valleys which are about 50% above and below normal, factory inventory fluctuating from 35% below to 42% above, distributor inventory from 28% below to 25% above, and retail inventory from 9.6% below to 9.3% above. During the same year the total inventory in the system varies from 21% below normal to 20% above, which is less than the sum of the separate distribution level peaks because the peaks are reached at different times.

Random Fluctuations

Note that the curves of EXHIBITS 3 and 4 are smooth and free of the short-term random fluctuations which are seen in most charts of actual industrial orders. Uncertainty and random behavior can be inserted into systems studies of this type, as shown in EXHIBIT 5. Here the size of orders placed by distributors has been made subject to a week-by-week random uncertainty. Because of the tendency of factory scheduling to smooth out short-run order fluctuations, and because little amplification exists between distributor orders and the factory, the factory output is almost identical to that of EXHIBIT 3.

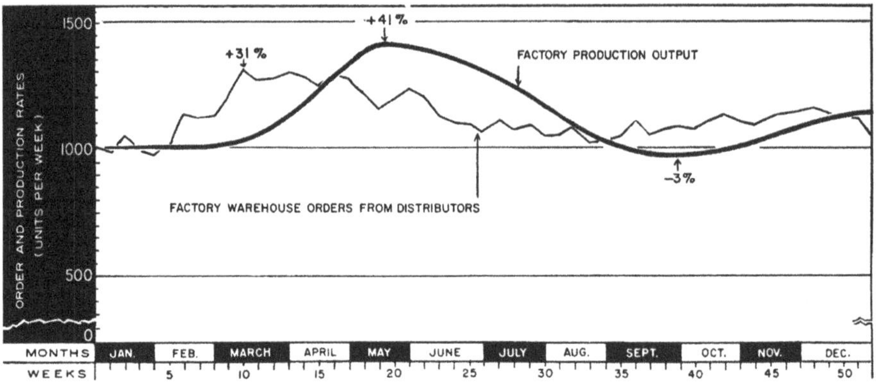

EXHIBIT 5: RANDOM SHORT-RUN UNCERTAINTY IN DISTRIBUTOR PURCHASES

However, if the random-buying fluctuation is introduced in **consumer** purchases, more amplification develops. An oscillatory system like this will respond to the random disturbances by fluctuating at a speed determined by the characteristics of the system itself, rather than in a pattern that is easily traceable to the external disturbance (in this case, a tendency toward peak-to-peak production fluctuations some seven to ten months apart).

Even if there is no average change in retail sales level (as in EXHIBIT 3) and no regular periodic change (as in EXHIBIT 4), the system will, if it is naturally oscillatory, convert random events into upswings and downswings in orders and production:

FIGURE B (at the end of the article) shows four years of week-by-week consumer purchases containing random deviations from the average, but no meaningful changes in sales rate. Also shown is the resulting factory production output for the production-distribution system described earlier in the article.

The system, by virtue of its policies, organization, and delays, tends to amplify those retail sales changes to which the system is sensitive. Retail sales fluctuate from week to week over a range of 10% or 15% with one week falling 31%. Factory output rises and falls over periods of several months with amplitudes of 20%, 30%, and 40% away from the average.

Just as waves in a bathtub build up slowly when disturbed by a small movement of the hand, so it takes most of the first year to build up serious amplitudes in the factory production swings. These waves then feed on the sales fluctuations. By the time of the third or fourth year, examination of the factory production curve could easily lead the casual observer who lacked the retail sales information to conclude that a seasonal sales pattern was present.

I know of company situations in which such an erroneous conclusion about seasonal sales has led to the establishment of employment, inventory, and advertising policies which in succeeding years **caused** a seasonal manufacturing pattern and thereby confirmed the original error.

The possibility of this happening in any company should be carefully considered in the design of management policies to give optimum stability to operations.

Production Limitation

So far we have considered simple situations that lack many important realities. Additional characteristics of a company can be added as we become ready. In the preceding exhibits we assumed the factory was able to produce at whatever level was desired. In a more realistic situation, with limited factory capacity, some surprising new effects develop:

FIGURE A (at the end of the article) shows the effect of a maximum factory capacity 10% above the average sales level. As before, the system is completely stabilized at the start of the first year; then it is assumed that retail sales rise and fall 10% during each year.

Retail sales never exceed production capacity. Yet, because of the inventory and pipeline effects, distributor orders to the factory do exceed its capacity. Furthermore, as factory deliveries become slower, distributors begin to order further in advance of needs and still more orders are put into the system. As a result, the factory operates at full capacity for four and one-half months during the first year.

Then, inventory demands are filled at a time that coincides with falling retail sales. In the last half of the year, falling retail sales and inventory liquidation combine with the reduction of in-process orders which results from improving delivery. In September and October we see a combination of rapidly falling factory orders, a swiftly declining backlog of unfilled orders, and a suddenly rising factory inventory. The natural result is to curtail production from its maximum level to 52% **below** normal.

Unlike the first year, we now enter the second year in a state of imbalance with production curtailed, orders rising, and inventory falling. The result in the second year is an accentuation of the fist-year performance. Factory orders from distributors rise to 49% above normal and show a broadened peak which is sustained by the distributors' tendency to order ahead when deliveries become slow. Factory production runs at full capacity for eight months to meet demand from the distributors. Unfilled orders at the factory rise to 337% above normal and represent a backlog which is six weeks of production higher than normal. In the meantime factory inventory has been depleted from the normal four weeks' productions to less than a week.

In spite of low factory inventory and slow deliveries, note that retail inventories deviate only 10% and that actually there is little likelihood of losing retail sales.

Note that at point A in the 84th week shipments equal incoming orders; inventory is constant. This is therefore the point at which unfilled orders reach a peak and start down. Again, since the factory has produced in excess of sales for eight months and because the long-run average production must equal retail sales, manufacturing must be sharply curtailed in the remaining four months. It drops to a minimum of 54% below normal and produces at a level below average retail sales for a period of 14 weeks.

In succeeding years the system would act much as it did in the second year, though unfilled orders at the factory would go to somewhat higher peaks.

We see in FIGURE A an occurrence which may lead a company to overexpand production capacity. For 12 months in the second year, inventory was below the desired level; for 9 months from the first of November until July the order backlog was rising. Viewed from the circumstances at the factory, this condition might easily lead management to undertake expansion plans – even though retail sales never exceed production capacity.

IMPROVING CONTROL

To explain the behavior of an industrial organization is only the first step. After adequately representing the current operations of a particular company or industry, the next step is to determine ways to improve management control for company success.

Faster Order Handling

To improve industrial stability and to make the company less vulnerable to external influences, executives may consider various alternatives. For instance, they might wish to explore the effect of reduced clerical and data-processing delays, which is often suggested as a quick and easy step toward better management control. However, companies often find the results disappointing, a point illustrated in EXHIBIT 6. Here, the clerical delays assumed in EXHIBIT 1 (the three, two, and one weeks of sales analysis and purchasing time) have been reduced to one-third their previous values. But the effect is only a slight improvement: the production peaks are reduced from five times to four times the retail sales changes. Because clerical delays are such a small factor in the operation of the system as a whole, no amount of speed-up can radically change over-all performance.

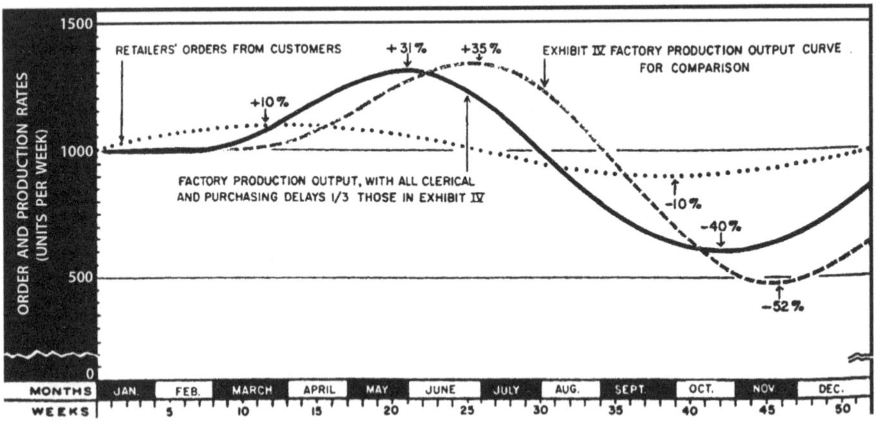

EXHIBIT 6: EFFECT OF REDUCING CLERICAL DELAYS

What about changes in ordering procedures? One possibility would be to place retails orders directly with the factory. The result is shown in EXHIBIT 7. When the inventory accumulation, fluctuation of in-process orders, and delays of one of the three stages are removed, production overshoot is 27% instead of the 40% in EXHIBIT 3.

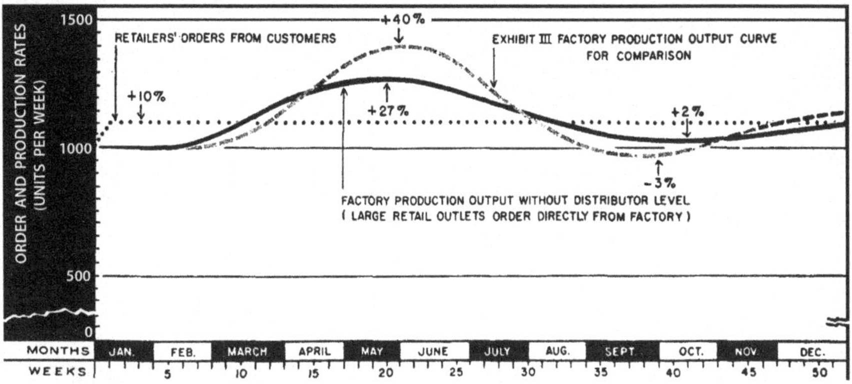

EXHIBIT 7: DISTRIBUTOR LEVEL ELIMINATED

The foregoing raises an interesting question about industries having more than three distribution levels. For example, in the textile industry, which shows marked instability, there are often four or five distribution levels from yarn manufacture to the final customer. May not a good deal of instability be caused by the existence of so many levels?

Better Sales Data

Another way to improve factory production stability is to have reliable retail sales information on which to base factory production schedules. EXHIBIT 8 (on the next page) assumes that we have "reasonably" reliable retail sales data that are eight weeks late in becoming available to management. (This lag allows for time to collect the data and average out short-run random fluctuations.) By reliable data we mean here that the retail sales information is good enough so that we are willing to give it six times as much weight in scheduling production as we do to the actual orders received at the factory. Here are the results:

The first production peak is 16% above the start, instead of 35% as in EXHIBIT 4. Later peaks are two times the retail sales changes, instead of five times as in EXHIBIT 4. The supply pipeline and inventory changes at the retailer and distributor levels are being met more from factory inventory and less from peaks in production, which necessitates larger swings in factory inventory, in the unfilled order backlog, or in both.

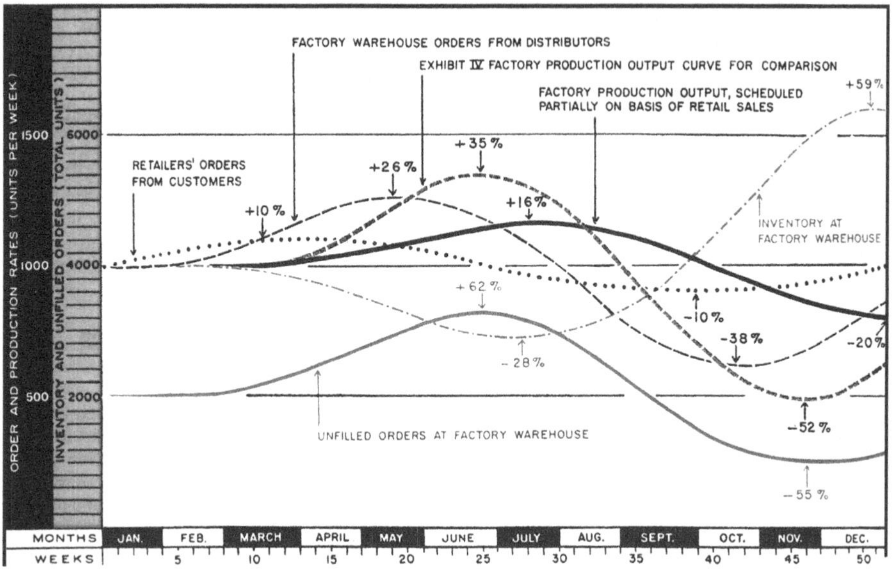

Exhibit 8: Effect on Factory Schedules of Using Reasonably Reliable Sales Data

The need for confidence in the retail sales information is apparent in October, when the factory output is being reduced only gradually in spite of incoming orders being down nearly 40% from normal, and in spite of a rapidly falling backlog of orders and a rapidly rising inventory. To achieve this degree of stabilization of production, one must be able to let factory inventory fluctuate from 35% below normal to 60% above.

Inventory Adjustment

The behavior of a simple production-distribution system like that shown in the exhibits is probably more affected by the practices followed in adjusting inventories and in-process orders than by any other single characteristic. The amount of inventory change and its timing are both important. These points deserve special emphasis, and should be considered with especial care by decision-making executives. To illustrate what can be accomplished:

In Exhibit 9 we take the same production-distribution system as before but assume that management has changed the rapidity with which the inventory corrections are made (but not the total amount of adjustments made). Management has specified that an inventory correction should be based on sales smoothed over the previous eight weeks (exponential smoothing with eight weeks time constant) so that in no case can the adjustment follow a sales change more rapidly than the delay introduced by the smoothing. The question is what time interval to specify for making inventory correction. When inventories are lower than desired for the current level of business activity, how much of the difference or imbalance should be added to the orders sent to the factory in the next week?

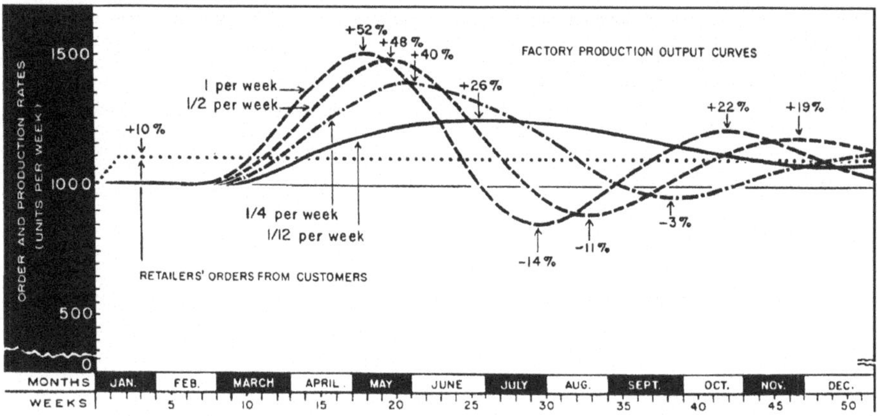

EXHIBIT 9: CHANGING THE TIME TO MAKE INVENTORY AND IN-PROCESS ORDER CORRECTIONS
(Fractions refer to amount of remaining imbalance corrected in following week)

The factory output curves portray the data management needs for a decision. Each curve is based on a different correction rate – a different fraction per week of any inventory discrepancy which is to be corrected through adjustment of the ordering rate to the next higher supply level. (This is the d_{ir} term defined in the fourth point in EXHIBIT 2.) This correction rate varies from the top curve, in which orders for any imbalance in inventory and in-process orders are fully placed in the following week, to the lowest curve, in which only 1/12 of any **remaining** imbalance is corrected in the following week. The lowest curve leads to about 60% of an initial imbalance being corrected in 12 weeks and 85% corrected in 24 weeks.

We see that the production fluctuations for the 1-week correction (the top curve) reach 52% above, 14% below, and 22% above the initial value as a result of a sudden 10% rise in retail sales. There is an interval of 24 weeks between successive peaks. At the other extreme, the 12-week correction time (the lowest curve) leads to peaks of only 25% above, 3% below, and 14% above the initial value as the production rate approaches the 10% higher level. The interval between peaks is longer – 34 weeks. As for the curve for the 4-week-correction, it is the same as that used in EXHIBIT 3 and is the basic system used in the other exhibits.

We see from EXHIBIT 9 that the more gradual inventory correction to business level changes leads to improved stability. Further, the reduction of manufacturing fluctuations is achieved without increasing the inventory extremes.

EXHIBIT 10 (on the next page) presents, for various correction times, peak percentage changes in inventory at each level in the system and the total change for the sum of all inventories in the system. The total inventory changes are less than the sum of the separate levels because the extremes do not occur at the same time.

Adjustment time	*Retail*	*Distributor*	*Factory*	*Total*
1 week	– 3.8% to + 5.6%	– 10.9% to + 12.9%	– 14.3% to + 32.3%	– 7.1% to + 13.6%
2 weeks	– 4.0 to + 5.4	– 11.3 to + 13.5	– 13.9 to + 30.2	– 7.3 to + 13.6
4 weeks	– 4.3 to + 5.5	– 11.5 to + 12.9	– 12.7 to + 24.6	– 7.4 to + 12.2
12 weeks	– 4.8 to + 5.3	– 11.3 to + 8.9	– 10.2 to + 13.2	– 7.4 to + 8.2

EXHIBIT 10: EFFECT OF CORRECTION TIME ON INVENTORIES

Under many forecasting methods, the kind of gradual adjustment of inventories I have described is impossible. The unstabilizing effects are accentuated by superimposing changes in sales, in-process orders, and inventories.

EFFECT OF ADVERTISING

To show how we would begin introducing into our model additional realism which would be a part of actual company operations, we will now look at **one** of the many effects of advertising. Incorporating this aspect illustrated the step-by-step approach of adding new features to our analysis. An adequate representation of a complete company would, of course, be a very large task. In a single article one can only illustrate the possibilities.

Role in System

First we need a schematic diagram of the advertising and marketing behavior which is to be considered. EXHIBIT 11 is oversimplified but will serve our present purposes. It includes the production-distribution organization set forth in EXHIBIT 1 and adds (1) an advertising function and (2) a special segment of the market, prospective purchasers. We assume that advertising expenditure decisions are based on (a) sales, particularly as reflected in the factory production schedule, and (b) excess inventories.

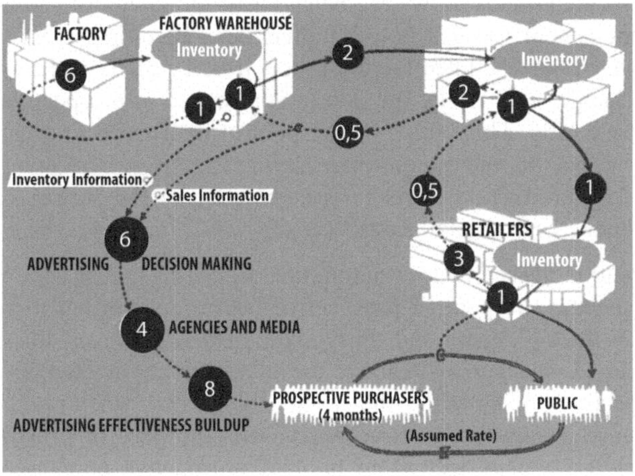

EXHIBIT 11: ADVERTISING AND CONSUMER MARKET

Now let us look more closely at the relationships shown in EXHIBIT 11:

- A time interval of six weeks is allowed for analysis of sales conditions, waiting to be sure of business trends, and the reaching of a decision to change the advertising expenditure rate.
- Acting on advertising decisions by agencies and media is assumed to require four weeks.
- Full buyer awareness of the advertising is assumed to develop over a period of eight weeks after the appearance of a new level of advertising.
- The rate at which consumers become prospective buyers is determined by such factors as the rate of wear-out of their existing household appliances and new housing construction.
- The average time that prospective buyers contemplate purchase and during which they are subject to advertising influence is assumed to be four months in the presence of an "average" level of advertising.
- Each 10% change in the advertising level is taken to shorten or lengthen by one week the four-month period during which the average prospective buyer considers his purchase.

Obviously, this underrates the impact of advertising; in reality, it would normally affect not only the timing of purchases but also the size of the potential market. But even so, as we shall see, it plays a most important role in the working of the system.

Aggravated Production Cycle

Now we can compare the effect of different advertising policies. What are the implications of gearing the advertising budget to the incoming sales dollar, for example, rather than to a preconceived plan?

FIGURE C (at the end of the article) shows the behavior of the system illustrated in EXHIBIT 11, when management's policy is to let advertising be proportional to expected sales, as represented by scheduled factory production. To see how the system responds to a market change, we here let the rate of appearance of prospective purchasers suddenly increase 10%. This builds up the pool of **prospective** purchasers, and gradually over 20 weeks the rate of consumer buying increases. Beginning with the twelfth week, factory production starts to rise; because of the factors earlier described, it exceeds the increase in retail sales.

The decision that leads to increased factory output also leads to a higher advertising commitment. This influences the pool of prospective buyers and further increases the retail sales between the eighteenth and thirty-sixth weeks. However, this retail sales rate, which rises in 43 weeks to 16% above the start, is 6% above the new rate of generation of prospective buyers. After the twenty-third week, when retail sales surpass the 10% increase in new prospective purchasers, the pool of prospective purchasers starts to decrease. The decreasing pool of prospective purchasers becomes the controlling factor after 43 weeks, and retail sales start to fall despite a still rising advertising effectiveness.

We then see falling retail sales and a more rapidly falling production rate. The consequent fall in advertising commitments and effectiveness lets retail sales sink to a still lower level. Retail sales fall for nearly a full year, from October of the first year to September of the second year. In the meantime, retail sales have been below average demand since March of the second year because buyers are reverting to a slightly longer average wait before purchasing (since advertising has been cut back).

The pool of prospective buyers starts to rise in March of the second year; by September it overpowers the falling advertising effectiveness and retail sales start to rise again.

Because of the sequence of production swings that are greater than retails sales swings, advertising that is proportional to production plans, and a slight advertising influence on how soon the average prospective buyer makes his purchase, there is created a two-year production cycle which rises and falls some 30% above and below the new 10% increase in average demand. Furthermore, the amplitude of the production swings is gradually increasing and will do so until the swings are curbed by manufacturing limitations, lost sales due to delays in shipments, and other factors.

This pattern, with nearly two years elapsing between production peaks, reminds us of recent occurrences in household appliances and some other industries. It indicates that advertising is a **possible** cause of ling cyclic disturbances. It would be premature to assume that advertising is **the** cause, however, until many other factors (e.g., consumer credit, sales forecasting methods, price discounting, and growth and saturation characteristics of the market) are studied.

Longer-Run Effects

To make our study still more realistic, let us introduce a few more factors and project the behavior of the system over a longer period.

FIGURE D (at the end of the article) portrays the results obtained from a more elaborate model of the company. It is based on the same kind of production-distribution-advertising system as that used in FIGURE C, and on the added assumptions that:

1) Consumer purchases follow a random pattern which, while not affecting the long-run average of sales, does affect week-by-week sales.

2) Factory production capacity is limited to 40% above the value of the initial sales level.

As in FIGURE C, a sudden, permanent 10% increase in the rate of appearance of prospective customers is followed by long, approximately two-year, fluctuations. Superimposed on this broad pattern are shorter disturbances which are caused by the system's response to random fluctuations in retail sales.

During the third year, production just reaches the maximum production capacity; not until the fourth and fifth years does the cyclic behavior of the industry reach an amplitude where the capacity limit is strongly evident. Periods of 8 months appear when the industry is producing at full capacity and much in excess of retail demand. These periods are likely to lead to production-capacity expansion, with consequent heightening of the peaks and deepening of the recessions in the following periods.

The preceding discussion has shown how deferrability of purchase can join with other influences (advertising in the above) to produce fluctuating conditions in an industry. Economic conditions and price discounts can also interact with purchase deferrability. For example, the farm equipment industry sells to a market which can defer purchases for several years. Sales are then concentrated, depending not only on years of farm prosperity, but also on the history of machine usage and the existing inventory in the hands of the customer.

How can this "roller-coaster" effect be reduced? What can flatten out the cycles? Management policies themselves are an important influence. As it is, the production and inventory policies in this example are helping to accentuate the detrimental effect from the advertising policy. Yet advertising policy might have been most effective in smoothing out the peaks and valleys in FIGURES C and D if the effort had been consciously planned to stabilize sales. Another management tool for stabilizing production and sales is pricing policy.

COMPLETE ANALYSIS

I have indicated only in a crude way what can be learned from the study of a company's system of operations. The examples used have included simplified flows of information and materials but not capital equipment, manpower, or money. In actual practice, management would want to have all the latter elements included in the model.

For instance, by adding money flow, including factors determining the cost of carrying inventory, of changing production levels, and of losses from special inventory clearance sales, executives would be able to get direct profit comparisons between various methods of operation. The model could include the restrictions imposed by available labor supply, maximum hiring and training rate, overtime, and policies regarding work-force reduction.

To set up a dynamic model for simulating company or industry behavior, one must adequately describe the real system which it represents. Getting the data will often be difficult. However, the kinds of information needed involve the basic characteristics of the company. We need factory lead time, inventory policies, construction time for new plants, union contract provisions, and advertising policy. These are the same items that one considers and tries to interrelate in the everyday management of a business.

In a complete company model more intangible items must also be specified, such as the likelihood of results from research expenditures, consumer response to advertising, and market behavior. Many of these quantities will be unknown at first; a few can be measured; some can be closely estimated; others will be guessed at. Today these factors are combined intuitively, and sometimes quite effectively, but the system-analysis approach should help reduce the margin of error.

When the exact value of an operating characteristic cannot be established, one can test the system with a range of possible values. Sometimes the system will be found insensitive to most differences in value or a factor; in one of the cases earlier described, for example, a speed-up in data processing was found to have little effect on the behavior of the total operation. In such cases it does not make too much difference what assumption management directs the researchers to make about the activity so long as the quantity or relationship assigned is "in the right ball park". By contrast, if it is found that the system is highly sensitive, additional effort can be put into determining the correct value. At the very least, the risks of not knowing it become apparent.

SHAPE OF THE FUTURE

Industrial dynamics is in an early stage of development. At a few isolated places, in both companies and universities, research is under way as an outgrowth of military or operations research projects, but the full impact of the new approach will not be felt for one or two decades.

What kinds of advances are in prospect? When are they likely to be made? Obviously, any predictions that we make are subject to the usual "if's" (business conditions, military pressures, and so on), but we can now foresee what is **probable**.

Two-Stage Progress

Progress will come in two stages. It is likely that the next five years will be devoted to exploratory research, development of basic analytical techniques, handling of enough specific industrial situations to demonstrate success, and establishment of new academic programs for training future managers.

After this period there will come general recognition of the advantage enjoyed by the pioneering managements who have been the first to improve their understanding of the interrelationships between separate company functions and between the company and its markets, its industry, and the national economy. Competitive pressures will then lead other managements to seek the same advantage.

In other words, the company will come to be recognized not as a collection of separate functions but as a system in which the **flows** of information, materials, manpower, capital equipment, and money set up forces that determine the basic tendencies toward growth, fluctuation, and decline. I want to emphasize the idea of movement here because it is not just the simple three-dimensional relationships of functions that count, but the constant ebb and flow of change in these functions – their relationships as **dynamic** activities. Such a concept is complicated and not easy to grasp. And, as with any major social change, a new approach to the management process will take time – time to develop ideas, time to test them, time to train a new generation of managers, and time for these men to reach positions of responsibility.

Many companies are already developing the background skills that will be needed. Computers that are now being acquired for other purposes can be used as systems simulators. Operations research departments can expand into broader activity of studying management systems. Simulation techniques, feedback-control system theory, and analysis of decision processes are being developed in the engineering and research departments of many firms.

It is already clear that some companies will lead, rather than be overtaken by, these changes. For them, research on the management process is a part of the long-range planning of company improvement. But such companies are still the exception. In case after case we have the anomaly of thorough systems studies on products and almost non on the behavior and future prospects of the company itself. Research on devices is fashionable; research toward better management has yet to be fully appreciated. At the same time more companies fail from poor management than from poor engineering.

Areas for Study

What is the nature of the job to be done in the immediate future? I believe that business and educational leaders should concentrate on six high-priority areas of research:

(1) **Money Flow**. Models such as those used to illustrate this article must be extended to include money flow and costs in addition to information and material flow. It will then be possible to see how the financial position of the company and its profitability depend on operating decisions.

(2) **Consumer Market and Advertising**. The way now seems clear to establish an orderly behavioral structure to explain our experiences and our empirical knowledge of marketing, advertising, and consumer behavior. We should work toward a single

unified market model to cover the entire life cycle of a product from the period of launching through sales and profit growth, profitability decline and sales volume saturation, and finally sales decline as the product is overtaken by new technological advances. The growth and decay curves of different products would, of course, be different; but large classes of products can probably be represented by the same model of marketing and technological relationships. In such a model account will be taken not only of the action of the company and its competitors but also of the flow of consumers from possible users to prospective buyers, owners, repeat buyers (based on life and usage characteristics of the product), and secondhand buyers (if any).

(3) **Capital Investment**. A long-range representation of company growth must include the motivations and decisions which lead to construction, use, and discard of capital equipment. It must include sources of investment funds whether equity, debt, or retained earnings. Potential productivity increases from modern capital equipment must be related to older equipment already in use. The tendencies during the early growth phase of a product to overexpand capital investment must be related to the product life cycle.

(4) **Research and Development**. While not so well understood as other aspects of company operations, research and development are crucial to company growth and to a succession of profitable products. A clear understanding of the relationship of research to profit and capital investment is necessary for proper representation of long-term forces on company growth. The very long lead times, longer than generally supposed, between research policy decisions and the resulting influence on company profit must be properly related to current financial resources, production capacity, and market development.

(5) **Economic Conditions**. Although, as the exhibits show, major company crises are often of internal origin, others are initiated by outside conditions and are amplified internally. Ultimately, a truly satisfactory explanation and prediction of company evolution must incorporate the influences of national and international economics. We now know much about the separate parts of the economy, but we should not trust our intuitive estimate of what a development here or a change there implies for the behavior of the economic system as a whole.

(6) **Case Studies**. As the methods earlier described are developed, they must be tested in real situations. This will help us develop the means to cope with the complexities of actual management problems; it will also help us to guide research to areas of greatest payoff.

The six preceding areas are being incorporated in the Industrial Dynamics Program at the Industrial Management School of Massachusetts Institute of Technology. The research is being sponsored by foundation funds and also through cooperative research arrangements with interested industrial companies. To provide men competent to carry forward this work, both in research and in companies, industrial dynamics graduate courses and thesis research have already been inaugurated. Also, direct participation of men from cooperating companies will provide nuclei of talent in industry.

The industrial dynamics program at M.I.T. is aimed at four goals:

(1) To develop in the manager a better intuitive feel for the time-varying behavior of industrial and economic systems. The study of particular situations, as in the exhibits in this article, improves one's judgment about the factors influencing company success. The results are beneficial even before one feels that he has reached an accurate quantitative formulation of company behavior that will have reliable predictive value.

(2) To provide a background showing how the major aspects of a company are related to one another, so that the developing manager can derive the greatest benefit from his work experience. His operating experience should be more meaningful if he has a better understanding of how his immediate environment is related to the other company functions.

(3) To help predict the future course of an existing organization. Very often present decisions determine future company welfare five years or more ahead. We have seen in the exhibits how underlying, unchanging characteristics of a company can cause erratic manifestations in operations. We can hope, eventually, for better ability to see where present company practices will lead.

(4) To improve the future prospects of a company. Beyond prediction lies the ability to redesign an organization and its policies so that it stands a better chance of success. We have seen in the exhibits a preview of how one might explore and evaluate proposed changes in the company.

Effect on Manager

What do these new developments mean to the manager? How will they affect his position in the company?

Just as automation requires new skills at the worker level, so will improved methods require new abilities at the management level. The executive of the future will be concerned not so much with actual operating decisions as with the **basis** for wise operating decisions. He will be concerned not so much with day-to-day crises as with the establishment of policies and plans that minimize emergencies.

He will be able to do these jobs with a considerable amount of sophistication and skill. He will have a basis for understanding the implications of market trends and the probable behavior of consumer demand. He will recognize the ebb and flow of forces which interact to generate fluctuating economic conditions. He will be able to relate the changing factors in research, investment, and marketing to the life cycle of a product. He will be able to relate different financing methods with their relative advantages and risks to the uncertainties of market and economic conditions. He will know how to devise organizational forms that are efficient and encourage the creativity of people.

We can expect changes in various management responsibilities. There will be a merging of many line-and-staff functions. Today, the kind of systems studies discussed in this article would be the province of staff specialists. Yet the results will determine the future success of the company; the conclusions will be only as valid as the assumptions on which the studies are based, the answers no more pertinent than the questions asked. Therefore, systems planning must, in time, be the tool of the responsible manager. It is by nature a thoughtful process of weighing the past and present, not to reach the immediate decision of the moment but to derive guiding principles for the future.

All of this will lead to a more decisive separation of policy making from operations, with the dividing line much lower in the organization than at present. New ways to predict the interaction between company functions will speed up the movement toward decentralized management. In the past power has been concentrated at the top in order to improve over-all integration and control. With increased understanding of company operation, we can expect that improved definition of objectives and more pertinent standards for the measurement of managerial success will permit managers at the lower levels to take on more operating responsibility. Senior executives will then be free to give more attention to product innovation, economic conditions, and the organizational changes that will enhance man's creativity.

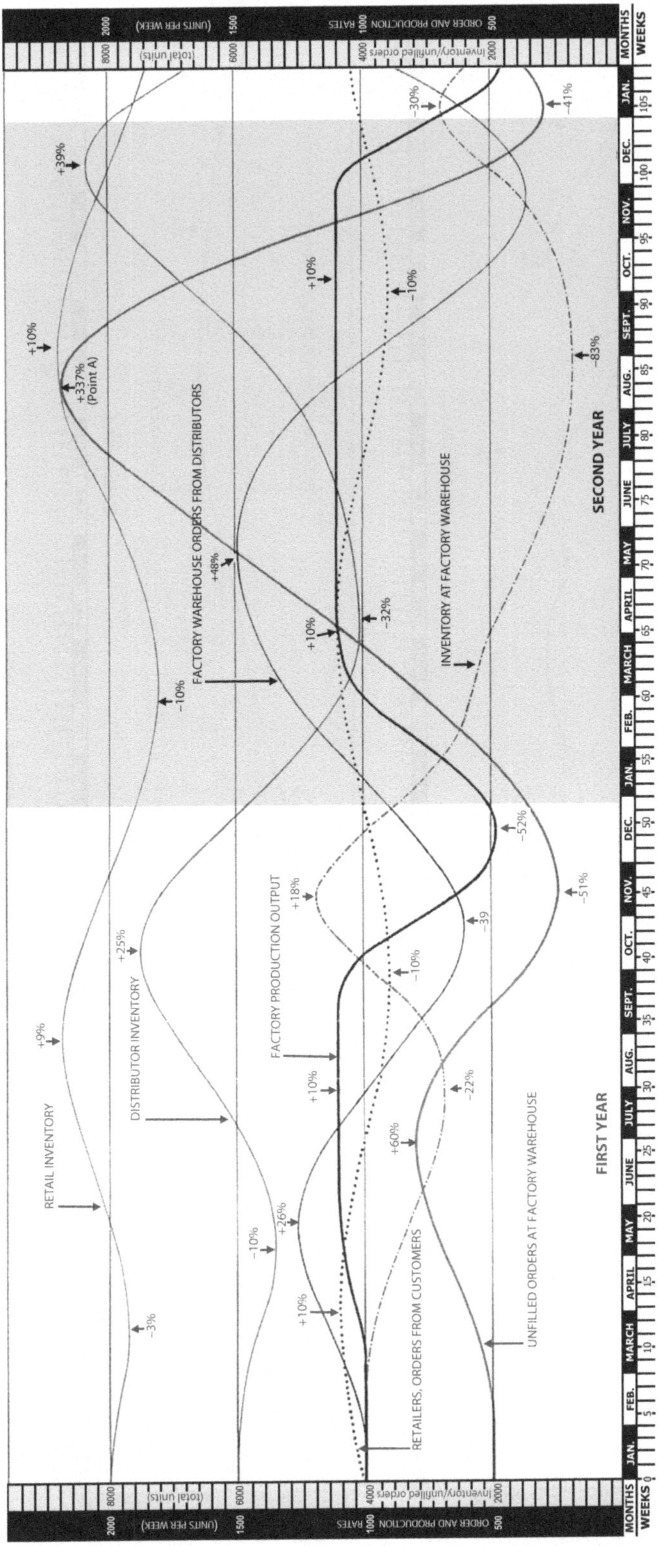

Figure A: Effect of fluctuating retail sales on factory with manufacturing capacity limited to 10% above average sales level

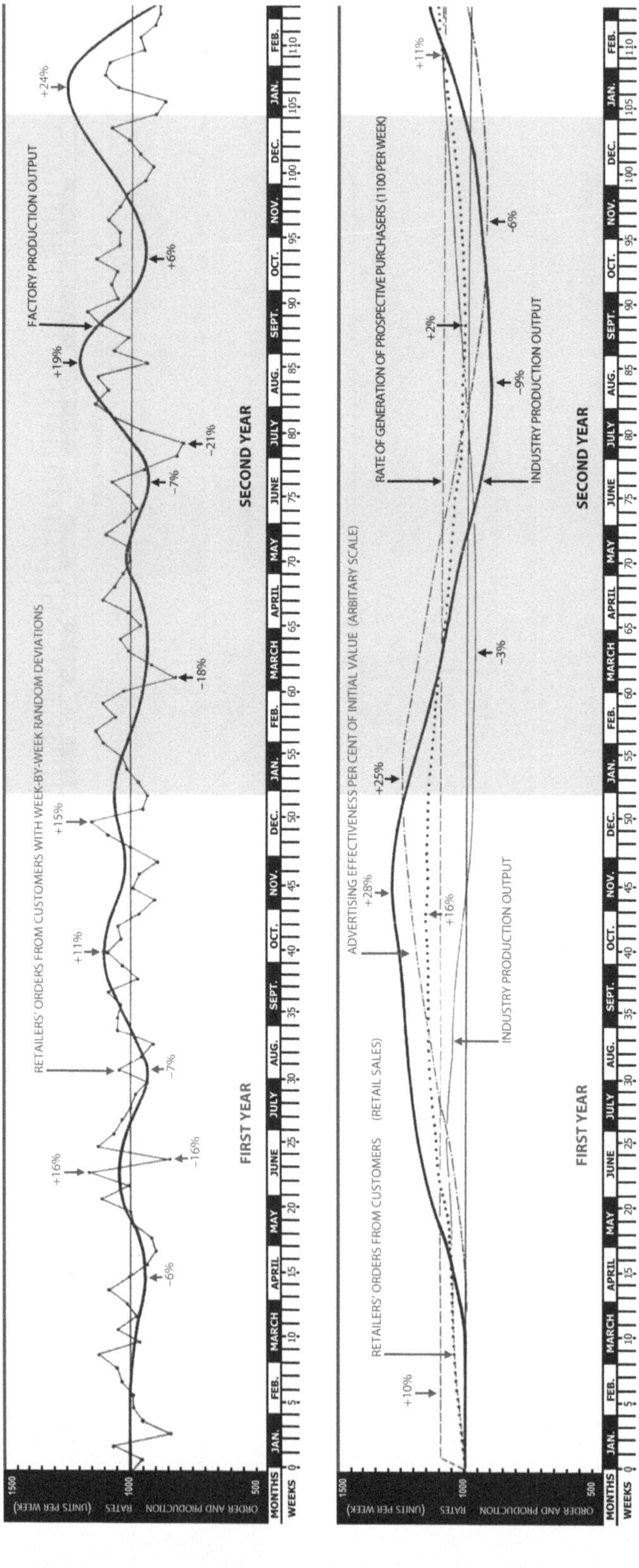

Figure B: Effect of random deviations in retail sales on factory production

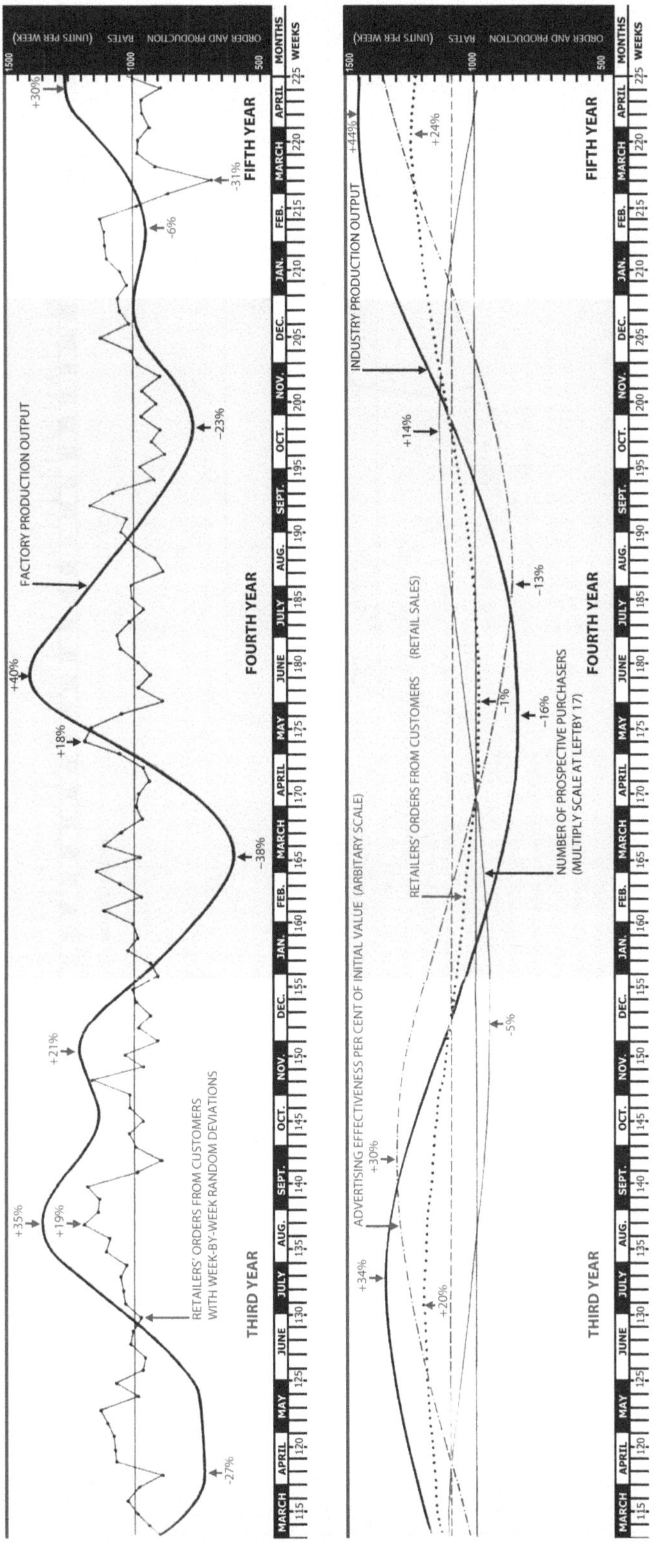

FIGURE C: EFFECT OF ADVERTISING WHEN GEARED TO SALES

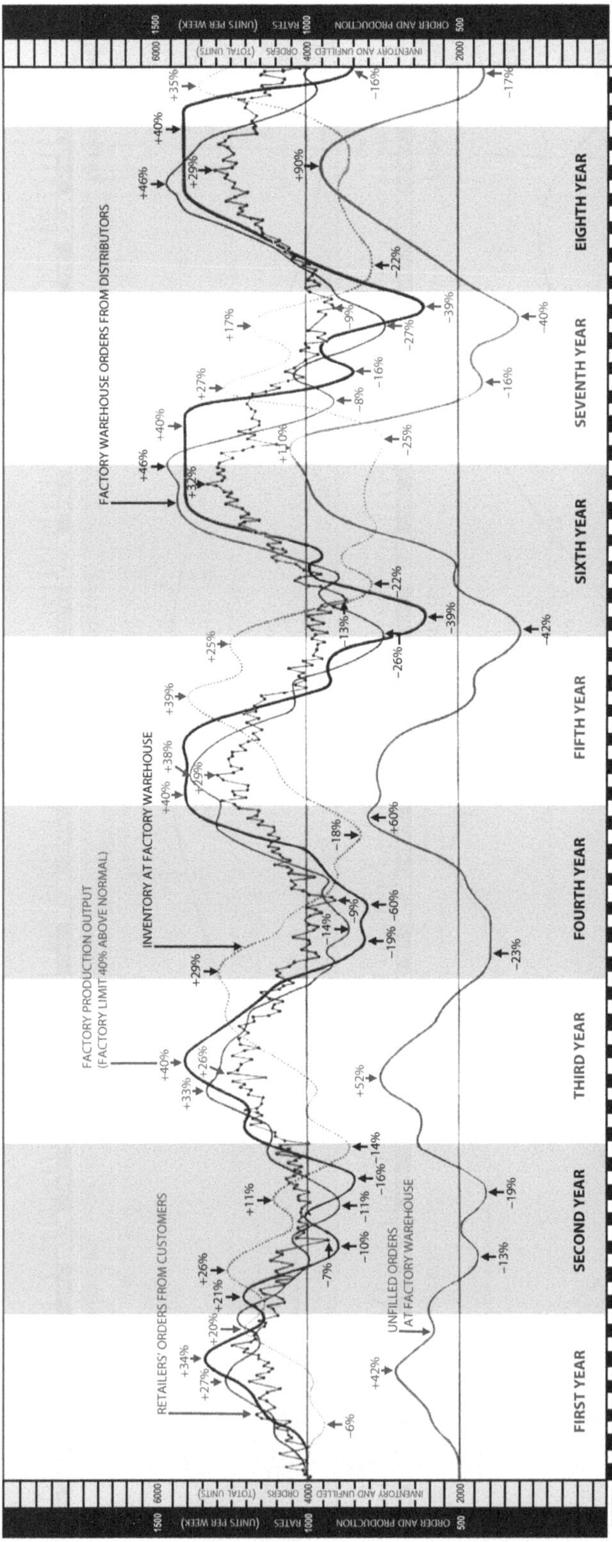

FIGURE D: COMPOSITE MODEL WITH ADVERTISING, A FACTORY LIMIT 40% ABOVE NORMAL, AND RANDOMIZED RETAIL SALES

Taiichi Ohno (1978/1988)

TOYOTA PRODUCTION SYSTEM

It was one man and his contribution to the practice of modern industrial production who revolutionized logistics management and probably contributed more to today's importance of logistics than any other individual person: Taiichi Ohno. He was a factory manager and vice-president for all of his professional life at Toyota Motor Co.

His contributions to the field were several: He and his company became the pioneers in the successful, large scale introduction of the "just-in-time" concept, meaning "in a flow process, the right parts needed in assembly reach the assembly line at the time they are needed and only in the amount needed. A company establishing this flow throughout can approach zero inventory".

In his 1978 book on the *Toyota Production System*, which immediately became a bestseller in Japan and had to be reprinted 20 times within the first two years, he described several other important ideas and concepts. Among them was the kanban system, as a critical, simple, partially self-controlling technique for implementing a smooth flow of production processes. The ingeniously simple idea behind it is a mobilization of orders and replenishment based on the "pull principle". The inspiration to this idea came from Ohno's visit of early American supermarkets in the 1950s. It implies that a workplace releases an order once having identified a demand thus "pulling" material instead of an external instance accomplishing an overall plan thus "pushing" material into the system. Other important experiences and innovations for efficient "flow management" which he described relate to the intelligent combination of automation technology and a respectful attitude towards the factory worker's ability to contribute to the continuous improvement and flexibility of their activities, and the importance of "leveling" flows. Selected quotes relating to these topics are reprinted in the following excerpt.

Ohno's and Toyota's example apparently inspired many other Japanese, now famous experts and developments such as Shingo (1986) and Imai (1986). It took years until experts and academics in the West such as Schonberger (1982), Wildemann (1984), Deming (1986) and Womack/Jones/Roos (1990) discovered and then gradually transferred and popularized the ideas of the Toyota Production System in the US, Europe, and eventually all over the world. Ohno's 1978 had not been translated into English before 1988.

With Ohno and his colleagues' and followers' work the focus in the field of Business Logistics widened from attention to a demand-side and marketing-related issues to those of production, supply, and the role of engineering, organizational and human resource issues.

P. Klaus and S. Müller (eds.), *The Roots of Logistics*, DOI 10.1007/978-3-642-27922-5_14,

Reference:

Taiichi Ohno: *Toyota Production System: Beyond Large-Scale Production*. Cambridge/MA: Productivity Press, 1988.

Further reading:

Deming, W. Edwards: *Out of the Crisis*. Cambridge/MA: MIT Press, 1986.

Imai, Masaaki: *Kaizen: The Key to Japan's Competitive Success*. McGraw-Hill/Irwin, 1986.

Schonberger, Richard: *Japanese Manufacturing Techniques: Nine Hidden Lessons in Simplicity*. New York: Free Press, 1982.

Shingo, Shigeo: *Zero Quality Control: Source Inspection and the Poka-Yoke System*. Cambridge/MA: Productivity Press, 1986.

Wildemann, Horst: Kanban-Systeme. In: *Die Betriebswirtschaft*, Vol. 44, No. 2, 1984, pp. 313–314.

Womack, James P., Daniel T. Jones, and Daniel Roos: *The Machine That Changed the World*. New York: Scribner, 1990.

Taiichi Ohno: Toyota Production System

JUST-IN-TIME[1]

The basis of the Toyota Production System is the absolute elimination of waste. The two pillars needed to support the system are:

- **just-in-time**
- **autonomation**, or automation with a human touch.

Just-in-time means that, in a flow process, the right parts needed in assembly reach the assembly line at the time they are needed and only in the amount needed. A Company establishing this flow throughout can approach zero inventory.

From the standpoint of production management, this is an ideal state. However, with a product made of thousands of parts, like the automobile, the number of processes involved is enormous. Obviously, it is extremely difficult to apply just-in-time to the production plan of every process in an orderly way.

An upset in prediction, a mistake in the paperwork, defective products and rework, trouble with the equipment, absenteeism — the problems are countless. A problem early in the process always results in a defective product later in the process. This will stop the production line or change a plan whether you like it or not.

By disregarding such situations and only considering the production plan for each process, we would produce parts without regard to later processes. Waste would result — defective parts on one hand, huge inventories of parts not needed immediately on the other. This reduces both productivity and profitability.

Even worse, there would be no distinction between normal and abnormal states on each assembly line. When there is a delay in rectifying an abnormal state, too many workers would make too many parts, a situation not quickly corrected.

Therefore, to produce using just-in-time so that each process receives the exact item needed, when it is needed, and in the quantity needed, conventional management methods do not work well.

---

1 Excerpt from Ohno (1988), p. 4ff.

USING A COMMON-SENSE IDEA

I am fond of thinking about a problem over and over. I kept thinking about how to supply the number of parts needed just-in-time. The flow of production is the transfer of materials. The conventional way was to supply materials from an earlier process to a later process. So, I tried thinking about the transfer of materials in the reverse direction.

In automobile production, material is machined into a part, the part is then assembled with others into a unit part, and this flows toward the final assembly line. The material progresses from the earlier processes toward the later ones, forming the body of the car.

Let's look at this production flow in reverse: a later process goes to an earlier process to pick up only the right part in the quantity needed at the exact time needed. In this case, wouldn't it be logical for the earlier process to make only the number of parts withdrawn? As far as communication between the many processes is concerned, wouldn't it be sufficient to clearly indicate what and how many are needed?

We will call this means of indication **kanban** (sign board) and circulate it between each of the processes to control the amount of production — that is, the amount needed. This was the beginning of the idea.

We experimented with this and finally decided on a system. The final assembly line is taken as the starting point. On this basis, the production plan, indicating the desired types of cars with their quantity and due date, goes to the final assembly line. Then the method of transferring the materials is reversed. To supply parts used in assembly, a later process goes to an earlier process to withdraw only the number of parts needed when they are needed. In this reverse way, the manufacturing process goes from finished product back to the earliest materials-forming department. Every link in the just-in-time chain is connected and synchronized. By this, the management work force is also reduced drastically. And kanban is the means used for conveying information about picking up or receiving the production order.

Kanban will be described later in detail. Here, I want the reader to understand the basic posture of the Toyota Production system. The System is supported by the just-in-time system, already discussed, and autonomation, described in the next section. The kanban method is the means by which the Toyota Production System moves smoothly.

GIVE THE MACHINE INTELLIGENCE

The other pillar of the Toyota Production System is called autonomation – not to be confused with simple automation. It is also known as automation with a human touch.

Many machines operate by themselves once the switch is turned on. Today's machines have such high performance capabilities, however, that a small abnormality, such as a piece of scrap falling into the machine, can damage it in some way. The dies or taps break, for instance. When this happens, tens and soon hundreds of defective parts are produced and quickly pile up. With an automated machine of this type, mass production of defective products cannot be prevented. There is no built-in automatic checking system against such mishaps.

This is why Toyota emphasizes autonomation – machines that can prevent such problems "autonomously" – over simple automation. The idea originated with the invention of an auto-activated weaving machine by Toyoda Sakichi (1867–1930), founder of the Toyota Motor Company.

The loom stopped instantly if any one of the warp or weft threads broke. Because a device that could distinguish between normal and abnormal conditions was built into the machine, defective products were not produced.

At Toyota, a machine automated with a human touch is one that is attached to an automatic stopping device. In all Toyota plants, most machines, new or old, are equipped with such devices as well as various safety devices, fixed-position stopping, the full-work system, and **baka-yoke**[2] foolproofing systems to prevent defective products [...]. In this way, human intelligence, or a human touch, is given to the machines.

Autonomation changes the meaning of management as well. An operator is not needed while the machine is working normally. Only when the machine stops because of an abnormal situation does it get human attention. As a result, one worker can attend several machines, making it possible to reduce the number of operators and increase production efficiency.

Looking at this another way, abnormalities will never disappear if a worker always attends to a machine and stands in for it when an abnormality does occur. An old Japanese saying mentions hiding an offensively smelly object by covering it up. If materials or machines are repaired without the managing supervisor's being made aware of it, improvement will never be achieved and costs will never be reduced.

Stopping the machine when there is trouble forces awareness on everyone. When the problem is clearly understood, improvement is possible. Expanding this thought, we established a rule that even in a manually operated production line, the workers themselves should push the stop button to halt production if any abnormality appears.

In a product like the automobile, safety must always be of primary importance. Therefore, on any machine on any production line in any plant, distinctions between normal and abnormal operations must be clear and countermeasures always taken to prevent recurrence. This is why I made autonomation the other pillar of the Toyota Production System.

---

[2] A more common term for "foolproofing" systems is "poka-yoke" (comment by the editors).

ESTABLISHING A PRODUCTION FLOW[3]

It is never easy to break the machine-shop tradition in which operators are fixed to jobs, for example, lathe operators to lathe work and welders to welding work. It worked in Japan only because we were willing to do it. The Toyota Production System began when I challenged the old system.

With the outbreak of the Korean War in June 1950, Japanese industry recovered its vigor. Riding this wave of growth, the automobile industry also expanded. At Toyota, it was a busy and hectic year, beginning in April with a three-month labor dispute over manpower reduction, followed by President Toyoda Kiichiro's assuming responsibility for the strike and resigning. After this, the Korean War broke out.

Although there were special wartime demands, we were far from mass production. We were still producing small quantities of many models.

At this time, I was manager of the machine shop at the Koromo plant. As an experiment, I arranged the various machines in the sequence of machining processes. This was a radical change from the conventional system in which a large quantity of the same part was machined in one process and then forwarded to the next process.

In 1947, we arranged machines in parallel lines or in an L-shape and tried having one worker operate three or four machines along the processing route. We encountered strong resistance among the production workers, however, even though there was no increase in work or hours. Our craftsmen did not like the new arrangement requiring them to function as multi-skilled operators. They did not like changing from "one operator, one machine" to a system of "one operator, many machines in different processes."

Their resistance was understandable. Furthermore, our efforts revealed various problems. For example, a machine must be set up to stop when machining is finished; sometimes there were so many adjustments that an unskilled operator found the job difficult to handle.

As these problems became clearer, they showed me the direction to continue moving in. Although young and eager to push, I decided not to press for quick, drastic changes, but to be patient.

PRODUCTION LEVELING

In business, nothing is more pleasing than customer orders. With the labor dispute over and the special demands of the Korean War beginning, a lively tension fueled

---

3 Excerpt from Ohno (1988), p. 10ff.

the production plant. How would we handle the demand for trucks? People in the production plant were frantic.

There was a shortage of everything from crude materials to parts. We could not get things in the quantity or at the time needed. Our parts suppliers were also short on equipment and manpower.

Because Toyota made chassis, when many parts did not arrive on time or in the right amounts, assembly work was delayed. For this reason, we could not do assembly during the first half of the month. We were forced to gather the parts that were arriving intermittently and irregularly and do the assembly work at the end of the month. Like the old song "**dekansho**" that tells of sleeping half the year, this was **dekansho** production and the approach nearly did us in.

If a part is needed at the rate of 1,000 per month, we should make 40 parts a day for 25 days. Furthermore, we should spread production evenly throughout the workday. If the workday is 480 minutes, we should average one piece every 12 minutes. This idea later developed into production leveling.

Establishing (1) a production flow and (2) a way to maintain a constant supply of raw materials from outside for parts to be machined was the way the Toyota, or Japanese, production system should be operated. Our minds were filled with ideas.

Because there were shortages of everything, we must have thought it all right to increase manpower and machines to produce and store items. At the time, we were making no more than 1,000 to 2,000 cars a month, and keeping a one-month inventory in each process. Except for needing a large warehouse, this did not seem too big a burden. We did foresee a big problem, however, if and when production increased.

To avoid this potential problem, we looked for ways to level all production. We wanted to get away from having to produce everything around the end of the month, so we started by looking inside Toyota itself. Then, when outside suppliers were needed, we first listened to their needs and then asked them to cooperate in helping us achieve leveled production. Depending on the situation, we discussed the supplier's cooperation in terms of manpower, materials, and money.

AN IDEA FROM THE U.S. SUPERMARKET[4]

[...] The two pillars of the Toyota Production System are just-in-time and automation with a human touch, or autonomation. The tool used to operate the system is kanban, an idea I got from American supermarkets.

---

[4] Excerpt from Ohno (1988), p. 25ff.

Following World War II, American products flowed into Japan – chewing gum and Coca-Cola, even the jeep. The first U.S.-style supermarkets spread in the mid-1950s.

And, as more and more Japanese people visited the United States, they saw the intimate relationship between the supermarket and the style of daily life in America. Consequently, this type of store became the rage in Japan due to Japanese curiosity and fondness for imitation.

In 1956, I toured U.S. production plants at General Motors, Ford, and other machinery companies. But my strongest impression was the extent of the supermarket's prevalence in America. The reason for this was that by the late 1940s, at Toyota's machine shop that I managed, we were already studying the U.S. supermarket and applying its methods to our work.

Combining automobiles and supermarkets may seem odd. But for a long time, since learning about the setup of supermarkets in America, we made a connection between supermarkets and the just-in-time system.

A supermarket is where a customer can get (1) what is needed, (2) at the time needed, (3) in the amount needed. Sometimes, of course, a customer may buy more that he or she needs. In principle, however, the supermarket is a place where we buy according to need. Supermarket operators, therefore, must make certain that customers can buy what they need at any time.

Compared to Japan's traditional, turn-of-the-century merchandising methods such as peddling medicines door to door, going around to customers to take orders, and hawking wares, America's supermarket system is more rational. From the seller's viewpoint, labor is not wasted carrying items that may not sell, while the buyer does not have to worry about whether to buy extra items.

From the supermarket we got the idea of viewing the earlier process in a production line as a kind of store. The later process (customer) goes to the earlier process (supermarket) to acquire the required parts (commodities) at the time and in the quantity needed. The earlier process immediately produces the quantity just taken (restocking the shelves). We hoped that this would help us approach our just-in-time goal and, in 1953, we actually applied the system in our machine shop at the main plant.

In the 1950s, American-style supermarkets appeared in Japan, bringing the object of our research even closer. And when in America in 1956, I finally fulfilled my desire to visit a supermarket firsthand.

Our biggest problem with this system was how to avoid throwing the earlier process into confusion when a later process picked up large quantities a time. Eventually, after trial and error, we came up with production leveling [...].

WHAT IS KANBAN?

The operating method of the Toyota Production System is **kanban**. Its most frequently used form is a piece of paper contained in a rectangular vinyl envelope.

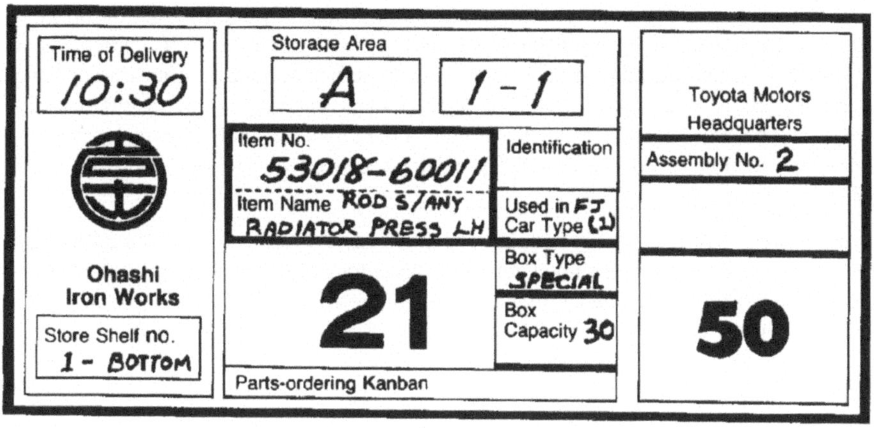

EXHIBIT: A SAMPLE OF KANBAN

This piece of paper carries information that can be divided into three categories: (1) pickup information, (2) transfer information, and (3) production information. The kanban carries the information vertically and laterally within Toyota itself and between Toyota and the cooperating firms.

As I said earlier, the idea came from the supermarket. Suppose we take kanban into the supermarket. How would it work?

Commodities purchased by customers are checked out through the cash register. Cards that carry information about the types and quantities of commodities bought are then forwarded to the purchasing department. Using this information, commodities taken are swiftly replaced by purchasing. These cards correspond to the withdrawal kanban in the Toyota Production System. In the supermarket, the commodities displayed in the store correspond to the inventory at the production plant.

If a supermarket had its own production plant nearby, there would be production kanban in addition to the withdrawal kanban between the store and the production department. From the directions on this kanban, the production department would produce the number of commodities picked up.

Of course, supermarkets have not gone that far. In our production plant, however, we have been doing this from the beginning.

The supermarket system was adopted in the machine shop around 1953. To make it work, we used pieces of paper listing the part number of a piece and other information related to machining work. We called this "kanban".

Subsequently, this was called the "kanban system". We felt that if this system were used skillfully, all movements in the plant could be unified or systematized. After all, one piece of paper provided at a glance the following information: production quantity, time, method, sequence or transfer quantity, transfer time, destination, storage point, transfer equipment, container, and so on. At the time, I did not doubt that this means of conveying information would certainly work.

Generally in a business, **what**, **when**, and **how many** are generated by the work planning section in the form of a work start plan, transfer plan, production order, or delivery order passed through the plant. When this system is used, "when" is set arbitrarily and people think it will be all right whether parts arrive on time or early. Managing parts made too early, however, means carrying a lot of intermediate workers. The word "just" in "just-in-time" means exactly that. If parts arrive anytime prior to their need – not at the precise time needed – waste cannot be eliminated.

In the Toyota Production System, overproduction is completely prevented by kanban. As a result, there is no need for extra inventory and, consequently, there is no need for the warehouse and its manager. A generation of countless paper slips also becomes unnecessary.

R. Keith Oliver/Michael D. Webber (1982)

SUPPLY-CHAIN MANAGEMENT: LOGISTICS CATCHES UP WITH STRATEGY

Management innovations and new concepts may develop in various ways: One possibility is that theoreticians develop and suggest something new which – if it is found to be practicable – is adopted by practitioners and applied in their daily business activities. But more often innovation in business and management emerges "backwards": A new method or concept arises in everyday practice, and some time later it will be noticed by the academics, to be reinterpreted conceptually and enhanced. This happened in case of Kanban and the Toyota production system, as Ohno's (1978) report included in this volume illustrated. It also happened in the case of the – now so important and much discussed – concept of Supply Chain Management:

American business consultants formed this phrase in the 1980s and implemented the ideas behind it in several projects. A first extensive report was published in the Booz, Allen and Hamilton company magazine Outlook in 1982. The authors were two Booz consultants – R. Keith Oliver and Michael D. Webber. In their paper, entitled *Supply-Chain Management: Logistics Catches up with Strategy* they were the first to extensively use and define Supply Chain Management:

> "First, it views the supply chain as a single entity rather than relegating fragmented responsibility for various segments in the supply chain to functional areas such as purchasing, manufacturing, distribution, and sales. [...] 'Supply' is a shared objective of practically every function on the chain and is of particular strategic significance because of its impact on overall costs and market share. Third, supply-chain management provides a different perspective on inventories, which are used as a balancing mechanism of last, not first, resort..."

The early notion of Supply Chain Management, which Oliver/Webber suggest, focuses on the task of balancing inventories and functional integration in the spirit of logistics as a "science of coordination" (cf. to the Sheffi/Klaus commentary in this volume). But it also anticipates ideas of the "flow system" paradigm of logistics, for which Quesnay, Nordsieck, Forrester and Ohno had paved the way (cf. the respective selections in this volume and, in place for innumerable other publications, Bechtel/Jayaram (1997) and the references given there).

P. Klaus and S. Müller (eds.), *The Roots of Logistics*, DOI 10.1007/978-3-642-27922-5_15,

Some time after the publication of Oliver and Webber's paper, practitioners have eagerly taken up the concept: Many companies have established supply chain management departments or are employing supply chain managers. An increasing number of academic institutes and organizations are doing research with regard to supply chain management, even replacing the "Logistics" term by "Supply Chain Management" (e.g. Cooper/Lambert/Pagh (1997)) and the American Council of Logistics Management's 2005 decision to rename the organization the "Council of Supply Chain Management Professionals").

Reference:

Keith Oliver and Michael D. Webber: Supply-Chain Management: Logistics Catches up with Strategy. In: *Outlook*; Booz, Allen and Hamilton, 1982.

Further reading:

Bechtel, Christian and Jayanth Jayaram: Supply Chain Management: A Strategic Perspective. In: *International Journal of Logistics Management*, Vol. 8, No. 1, 1997, pp.15–34.

Cooper, Martha C., Douglas M. Lambert, and Janus D. Pagh: Supply Chain Management: More than a New Name for Logistics. *International Journal of Logistics Management*, Vol. 8, Nr. 1, 1997, pp. 1–14.

R. Keith Oliver/Michael D. Webber: Supply-Chain Management: Logistics Catches up with Strategy

In the world of the logistics manager ten years ago – another era altogether in terms of business economics – the mission, while perhaps not always readily achieved, was at least clear: balancing-inventories between both production capacity and the demands of customer service. While the manager might have understood intellectually that assets should be employed to make the most of both factors, it was also accepted that hidden costs were bound to creep into even the best-managed system – and that these could be borne.

Now all that has changed. The job of the logistics manager has begun not only to overwhelm him or her, but also to have an increasing significance for overall corporate health. What were once simple trade-offs seem to result, with increasing frequency, in what is not merely a suboptimal utilization of assets, but, in the worst cases, a no-win materials management policy. Companies are faced with the prospect of incurring double or even triple costs in order to build up inventories to support marginal increases in customer service. But at the same time, more intensive competition in slow-growth markets, combined with the rising costs of other production and supply factors, means that many firms can't afford not to supply a level of service that results in a competitive edge.

Over the last few years, our analyses of the particular problems of multinational companies in the context of today's more challenging economic environment resulted in a fundamental shift in our own perceptions about materials management. What were hitherto considered 'mere' logistics problems have now emerged as much more significant issues of **strategic management**. Through our study of firms in a variety of industries in the USA, Japan, and western Europe, we found that the traditional approach of seeking trade-offs among the various conflicting objectives of key functions – purchasing, production, distribution, and sales – along the supply chain no longer worked very well. We needed a new perspective and, following from it, a new approach: **supply-chain management**.

1. CHALLENGES OF THE MARKETPLACE

To appreciate the urgency in the move toward adoption of new approaches to logistics, executives need look no further than the current economic and competitive environment. Within the macro-economic sphere, the cost of capital alone stands out as a major concern. Beween 1977 and 1981, the prime rate tripled in the United States and the relevance of that to European financial health is seen not just in parallel high European rates but in the acrimonious international debates surrounding the issue. This factor, coupled with generally slower growth and uncertainty in demand, makes investment decisions all along the supply chain – whether in capacity, systems, or inventories – riskier than ever before.

Economic uncertainty and, in particular, the swings in and out of the recession are, by themselves, problematic from the viewpoint of materials planning. The natural inertia of the supply chain itself and of its decision-making mechanisms magnifies the impact of such changes, producing major inventory surpluses or shortages in sharp succession. So wrenching are these swings that even politicians have started to monitor national inventory trends. Add to this problem the tendency of management to overreact to these ups and downs in the hope of providing a safety net for the next swing, and the incurring of still further costs and risks is almost guaranteed.

Evidence of such behaviour can be found in various functions all along the supply chain. Marketing, for example, may boost its forecasts in order to secure large allocations from manufacturing so as not to be caught short in a potential upswing. In response, the manufacturing and distribution functions may develop their own independent forecasts or try to second-guess actual sales and inventories. Functions all along the supply chain tend to exhibit a certain possessiveness, and as a result inventories accumulate like snowdrifts at organizational fences. Finally, both short-term and long-term decisions on capacity, lagging as they do behind the market and inevitably reflecting the tendency to build in a safety net, as described above, very often make the firm even more vulnerable to subsequent changes in demand – if not actually influencing the market.

To a large extent, the driving force behind these conflicts and responses is another reality of the current business environment: increased competition in slower growth markets. Higher investment and operating costs cry out for increased market share as a means of assuring high (or at least adequate) return on investment. But as ever-stronger competitors vie for a larger part of the same (and in some cases shrinking) pie, the overall cost implications of increased performance in the distribution and marketing functions may be overshadowed by the desire to enhance the firm's competitive position through greatly increased customer service.

The trouble is, certain competitors out there are doing a better job of not only providing a high level of customer service, but also simultaneously keeping inventory costs down. Evidence of such performance showed up in a Booz Allen survey, completed in mid-1980, in which inventory turnover and facility utilization in some 1,100 companies representing 18 industries were examined. The study showed that inventory turnover among Japanese firms had increased 31% over a nine-year period, compared with an increase of only 20% in the USA and a drop of 2% for European firms (see EXHIBIT 1). And because the Japanese were starting from a higher base, they came out the undisputed winner. This study showed, quite tangibly, why the "Japanese challenge" should not be discounted merely because it has, by now, become a cliché.

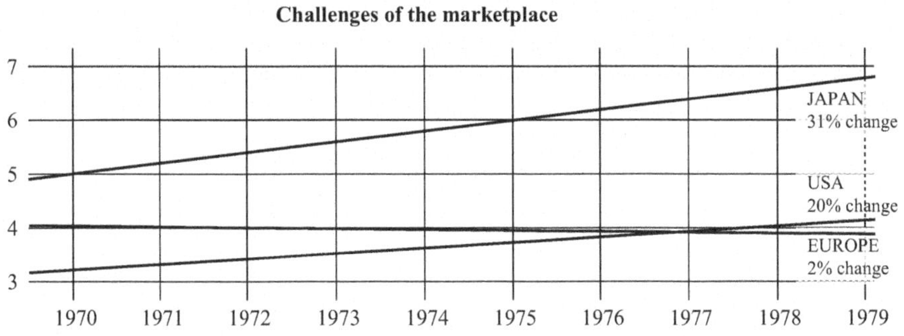

EXHIBIT 1: COMPARATIVE RATIOS OF TOTAL COSTS TO INVENTORY (TOTAL COSTS OF GOODS DIVIDED BY INVENTORY = INVENTORY TURNOVER).

It is not feasible, of course, to suggest that Western firms slavishly emulate the Japanese, either in terms of the particular features of their production operation (deintegration, smaller plants, narrowly focused facilities, etc.) or in terms of their particular managerial style (consensus und "bottom-up" decision making). Nor is it logical to suggest that the historical and environmental factors that have shaped Japanese industry – almost complete dependence on imported raw materials, a limited domestic market, and as a consequence of the two, the need to build a strong presence internationally – have much relevance for the majority of Western firms today. However, the study does indicate certain characteristics that American and European firms successful in supply chain management have in common with Japanese firms; these can, and probably should, be emulated. The strategic balance of supply and demand, based on firmwide objectives, and more particularly, its support by a systems approach that places a premium on the fast transfer and accessibility of information across functional barriers are all highly relevant. While "natural" conflicts between functions may be inevitable, common and shared data encourage the development of a broader perspective on supply-chain management and foster decision making that is more likely to be keyed to the overall objectives of a business rather than to the local or parochial functional objectives that have tended to dominate in the past.

2. FUNDAMENTALS OF SUPPLY CHAIN MANAGEMENT

Supply-chain management differs significantly from classic materials and manufacturing control in four respects. First, it views the supply chain as a single entity rather than relegating fragmented responsibility for various segments in the supply chain to functional areas such as purchasing, manufacturing, distribution, and sales (see Exhibit 2, on the next page). The second distinctive feature of supply-chain management flows directly from the first: It calls for – and in the end, depends upon – strategic decision making. "Supply" is a shared objective of practically every function on the chain and is of particular strategic significance because of its impact on overall costs and market share. Third, supply-chain management provides a different perspective on inventories, which are used as a balancing mechanism of last, not first, resort. Finally, supply-chain management required a new approach to systems: Integration, not simply interface, is the key.

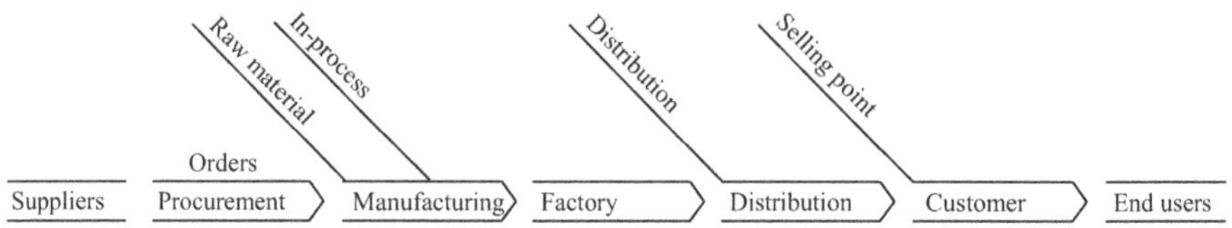

EXHIBIT 2: THE SCOPE OF SUPPLY-CHAIN MANAGEMENT. SUPPLY-CHAIN MANAGEMENT COVERS THE FLOW OF GOODS FROM SUPPLIER THROUGH MANUFACTURING AND DISTRIBUTION CHAINS TO THE END USER.

All of these features, and all of the challenges of the business environment that lie behind the move towards supply-chain management, point in one direction: to the top. Only top management can assure that conflicting functional objectives along the supply chain are reconciled and balanced, that inventories assume their proper role as a mechanism for dealing with inevitable residual imbalances, and, finally, that an integrated systems strategy that reduces the level of business vulnerability is developed and implemented. Logistics and materials managers will continue to play important roles, but only top management can be expected to have the perspective to recognize the significance of supply-chain management, and only top management can provide the impetus for adopting this new approach.

3. BALANCING FUNCTIONAL OBJECTIVES

The conflicting objectives of marketing, sales, manufacturing, and distribution are a fact of business life. The imbalances resulting from these conflicts have become almost structural in nature and traditionally have been bridged by inventory. It is not necessary to challenge the direction of the individual strategies of each of those functions. What is needed, rather, is a critical evaluation of the opportunities for trade-offs between the key elements of these strategies with significant cross-functional implications, such as:

1) The demand characteristics, lead time, reliability, and responsiveness implications of the marketing strategy;
2) The lead time, flexibility, minimum run size, and changeover implications of the manufacturing strategy;
3) The variety, range, and likely demand implications of the product strategy.

Some companies tend to resist the notion that a better balance can indeed be achieved, claiming to be victims of their own structures, culture, and environment.

Many companies also find it difficult to articulate their objectives at this level, particularly in manufacturing, where traditionally reactive positioning precludes "objectives" and where management tends to talk in terms of "constraints", in sharp contrast to the Japanese view of manufacturing as a "competitive weapon". Customer service objectives are frequently taken as givens, and the utilization of certain supply points and lead times are considered rigid. So, the traditional approach to managing these imbalances is via inventory and sophisticated control systems.

Supply-chain management suggests a quite different approach, however: addressing the imbalances directly and evaluating opportunities for minimizing them. In our diagnostic studies for international companies, we found – sometimes to management's genuine surprise – that key functional strategies affecting logistics are indeed negotiable. It may be possible, for example, that delivery reliability to customers can, in some instances, be traded against lead times: A 99% reliability on a four-week lead time is frequently more acceptable to customers than a 85–90% reliability on a two-week lead time. Moreover, such supply policies are usually negotiable by individual item based on relative volume or price.

In addition, the tendency of manufacturing executives to use the longest procurement time for materials as the basis for quoting standard lead times can be challenged – as can many attempts to protect manufacturing costs via batch sizes, particularly in times of low demand. We can see that decisions at these levels do have a fundamental and significant impact on inventory requirements. And as a consequence, the failure to achieve a balance of objectives at the strategic level will tend to put the burden on inventories alone, inevitably resulting in greater working capital requirements and increased costs.

4. OPTIMIZATION OF INVENTORY

This kind of approach – analysing in detail the individual functional strategies and attempting to reduce structural imbalance to a minimum – leads to a whole new role for inventory and supply policies. They become the mechanisms by which the inevitable residual imbalances are bridged, rather than the primary tool for managing the supply chain. The deployment of inventory as an asset, however, remains a strategic rather than an operational issue. At this level it should not focus on planning and batching rules or on systems tools, but on sourcing and supply policies, customer service and delivery performance objectives, and demand and supply decisions. This is the only way in which structural reductions in inventory requirements are likely to be achieved. The superior performance of Japanese industry – five and even ten times that of Western firms – highlights the opportunities for reducing the dependence of high inventory levels in the logistics systems of most European and US companies. This performance does not emanate from knowing how much is where but from why is it there? How many so-called inventory control systems in Europe and the USA today advance beyond mere "status" into the areas of simulation and goal setting?

5. CONTROL SYSTEMS STRATEGY

Once this strategic framework has been established, then management and control mechanisms can be developed. It is only at this point that the issues of organization, tools, techniques and support systems should be addressed. The basic architecture of most systems today follows the traditional functional and organizational divisions of the supply chain – purchasing systems, production control systems, distributions systems, and sales order processing systems. This structure may do a fine job of supporting the hierarchical upward flow of data within any single function, but it is not very effective in the context of cross-functional or total supply-chain communication. The same holds true for "interfaces" (frequently a synonym for time-consuming meetings), which by now have become a commonly accepted means of providing the necessary cross-functional communications.

The systems interface approach has several shortcomings, not the least of which is its hidden cost in terms of indirect manpower, and the related delay and distortion in the transfer of information. In addition, this approach frequently results in strategic and directional decision-making at too low a level. Further, systems groups either directly or indirectly "sell" their services separately to functional vice-presidents or directors. Typically, there is no one "client" for cross-functional communications, although the information-processing technology needed to support supply-chain management has been available for some time. The end result is that systems modules have tended to reflect organization segmentation with little or no emphasis on integration across the total chain.

Probably the best example of this problem is the master scheduling or master production planning "interface" – actually one of the most critical tactical activities in any manufacturing business. Until very recently, the level of systems support available to this function was minimal. To be sure, materials requirement planning (MRP), which assumes the availability of a master schedule, has been around for some 15 years. Real-time order processing and sophisticated sales-forecasting systems for the front end of the chain are equally mature. However, the conversion process involving orders and forecasts and MRP occurs at a functional interface, and those conversions have been almost devoid of systems support. What is lacking here is a unified, overall perspective on the entire supply chain: who takes responsibility for systems interface in a predominantly functional organization structure?

Here, again, management must lead the way, laying the groundwork for integration of the various systems throughout the organization's supply chain rather than simply accepting the traditional forms of interaction among them.

There can be no doubt that the supply-chain management approach places additional burdens on top management: it requires the incorporation of a logistics focus into the strategic decision-making mechanisms of the business. It demands the rejection of inventory as the easy "buy-out" option to many of the troublesome balancing and trade-off decisions. It implies an approach to control systems which supersedes the traditional functional divisions and which is likely to have significant organizational implications in the longer term. None of these are easy challenges, but the competitive environment of the 1980s for Europe and for the USA puts a high premium on meeting them.

6. MAKING SUPPLY-CHAIN MANAGEMENT WORK – TWO EXAMPLES

Supply-chain management opens up new vistas when it comes to both diagnosing and correcting inefficiencies. Most important, top management is heavily involved in supply-chain management; what previously were merely logistical issues are now strategic ones. Then, too, the broader view provided by supply-chain management makes it possible to find solutions that directly reflect the concerns of the customer and the technological boundaries within which a firm operates.

This is evident from the example of two companies, one in the United States and one in the United Kingdom, that Booz Allen has counselled in the implementation of supply-chain management. In both cases, the application of traditional techniques might well have resulted in an all too familiar no-win trade-off: a cost premium for higher capacity or inventory investment versus shortfalls in customer service, with real costs showing up eventually as lost business. With the broader perspective of supply-chain management, however, it became evident that these were two quite different problems requiring two quite different solutions.

6.1 Alternatives to improved customer service

A US-based multinational pharmaceutical company faced market-growth problems in several divisions; these were exacerbated by customers demanding better service than they were receiving. Each division defined its problems somewhat differently, but it was clear that no one functional manager could solve the problem independently. Eventually the supply-chain management concept was used in four divisions, but its application in two of them highlights the effectiveness of the approach.

One division produced a highly specialized pharmaceutical product whose lead time had grown to three years and whose delivery performance was so poor that senior management was handling customer complaints. Only 30–35% of the orders were being delivered on time. As a result, customers were forced to stockpile inventory at a time when interest rates were rising and square foot costs were increasing. Manufacturing responded by blaming unreliable forecasts and long purchasing lead times. While these were partially to blame, this explanation ignored the internal scheduling problems that compounded that of poor delivery performance.

A traditional approach to solving these problems might have started with an attempt to reduce total lead time by eliminating capacity bottlenecks. However, the supply-chain management method called for a different first step: analysing the benefits that would result from shorter lead times or improved delivery reliability. Once it was demonstrated to management that the market really wanted greater reliability – not necessarily shorter lead times – and that increased reliability could result in increased volume, then the decision to invest in alternative solutions was considerably easier to make.

The primary recommendation arising from our approach involved the development of an improved manufacturing system – one that was tied to forecasting and that incorporated a mechanism which offered a certain level of protection against forecast deviations. On-time reliability increased to more than 95% within two to three months. Reducing work-in-process inventory via smaller in-process buffer stocks resulted in a 15% decrease in overall inventory investment in the first year and a 30% decrease in the following year, as market volume increased.

The second division was coping with another set of problems. Externally, it faced a retail and wholesale trade group that was becoming increasingly sophisticated in its inventory management policies. As a result, the group was operating with less inventory, ordering more frequently, and demanding that orders be filled completely from regional distribution centres. Internally, the division was faced with conflicting views on its performance. According to marketing and sales, management had three problems: a stagnant market, stiff competition, and uncompetitive service levels. Manufacturing, on the other hand, believed that poor forecasting and antiquated distribution methods were to blame and that no one had demonstrated that improved service would sell more products. In the meantime, management was being evaluated on return on assets and was therefore reluctant to do much more than squeeze inventories.

A traditional logistics approach would have tried to reduce raw material and work-in-progress inventory to allow some incremental investment in finished goods. However, with the supply-chain management approach, management agreed to involve all the functional groups in solving the problem and not to assume that improved service was the answer until it could evaluate its cost implications.

A customer service survey of the trade indicated that service was becoming an important criterion in the purchasing decision and that there was a correlation between increased market share and high levels of service. As a result, programmes aimed at improved sales forecasting, new distribution policies, and plant scheduling methods were developed and cost implications were weighed against the service targets.

Implementation proceeded on all the programmes and within six months, service levels rose from 60% order fill to 75%, management's target. At the same time, inventory turnover did not change and in fact is expected to improve by 10% in the first full year of implementation. Finally, management was left with a system that could respond quickly to market demands for service as well as internal pressures to minimize inventory investment at certain points during the year.

6.2 Complementing manufacturing and marketing strategies

At the confectionery division of a UK-based food and beverage producer we found a vastly different operating environment and different market requirements. Not surprisingly, the application of supply-chain management here resulted in a radically different solution.

When we began our diagnostic study, the firm was almost three years into a major overhaul designed to strengthen its competitive position. The effects of the ambitious programme – which included modernizing and rationalizing production facilities, restructuring the product range, and refocusing marketing programmes – were already visible at that time. Market share was rising, manufacturing productivity was improving, and inventory levels had been reduced. In addition, rationalization of the new product line, coupled with the introduction of new size ranges, had provided a strong foundation for an upgraded advertising programme that capitalized on the brand franchise. All this showed up in the bottom line: Sales, profits, and return on assets were all improving.

Despite that obvious success, management sensed that fundamental structural issues had not been faced squarely. Their particular concern was one which supply-chain management addresses: the balance between manufacturing strategy and inventory investment as a means of meeting the challenges of the marketplace.

Booz Allen's diagnostic study let to a more precise identification of supply-chain management issues that validated management's concern:

1) High customer service and product availability ex-stock were givens, not open to significant negotiation;

2) Forecast accuracy at the item level on a month-to-month basis was inaccurate (not unusual for consumer products businesses).

The new manufacturing strategy, focusing on four principal manufacturing lines, had succeeded in reducing costs, but was considerably less flexible than the previous system which had been built around 18 different lines.

The results of the three-year programme thus carried a certain irony. The company had succeeded both in rationalizing product lines and manufacturing and in streamlining inventories, yet these successes worked against each other. All "stack" or flexibility had been removed from the system.

One obvious area for potential improvement was forecasting accuracy; a more traditional approach might have focused on this, probably without success, given the challenges of consumer product market places. Perhaps an even more obvious response might have been to reinstall the inventory buffers that had so recently been eliminated. But in this case, quite aside from the issue of carrying costs, this option was particularly unappealing for several reasons. First, brand franchise depended largely upon quality, and chocolate and candy plainly and simply taste better if they have not spent excessive time in a warehouse. Second, EEC regulations requiring open-date coding were imminent, and could be expected to draw consumer attention to the product age – yet another reason not to adopt a strategy that would have added a month of finished inventory to the supply-chain time frame.

Strategically, management faced the challenge of absorbing the high level of forecast error and seasonality while providing the desired level of customer service through an appropriate, cost-effective mix of capacity, flexibility, and inventory. That challenge was made even more formidable by an ever-changing environment – in particular, by continuing shifts in overall volume levels and in product mix coupled with changes in interest rates and hence investment or carrying costs. A one-time exercise would not be sufficient; the firm needed a framework within which periodic reappraisal of capacity and inventory costs could be measured and adjustments made. Working with management, we developed a model in which all relevant elements – including forecast error and inventory costs – could be converted to standard hour equivalents. Using this model, management could determine the appropriate method for meeting the demands of the market place, e.g. high inventory/less capacity, greater capacity/less inventory.

The initial appraisal exercise showed that the environment at that time militated against inventory build-up: the inventory investment would have to double in order to reach the desired level of customer service. The solution lay in stretching production capacity – specifically, by investing more in the packaging process. This basic strategic decision was complemented by the introduction of a new, capacity-oriented approach to production scheduling. Installation of tighter controls over the total number of changeovers and associated down-time were built into the scheduling system, allowing a limited number of product-mix decisions to be made over a short horizon. As a result, the impact of near-term forecast error was considerable softened in terms of manufacturing, economics while allowing customer service and supply objectives to be maintained.

These two examples provide only a small window on the benefits that flow from effective supply-chain management. Still, it is evident that such management is a far cry from simple manipulation of inventories and more than improved materials management, both of which may be inadequate – not inappropriate – responses to an unstated economic environment. Unlike these approaches, supply-chain management requires an underlying strategic focus, along with the involvement and commitment of the top management. And because supply-chain management can provide the framework needed to meet the demands of today's uncertain economy, top management is likely to find that commitment worthwhile.

George Stalk jr. (1988)

TIME – THE NEXT SOURCE OF COMPETITIVE ADVANTAGE

In search of the conceptual roots of modern logistics – the logistics of flows and flow systems – one further important contribution is the explicit introduction of the dimension of time. It was a consultant, George Stalk, long time senior associate and fellow with Boston Consulting Group, who in 1988 published his seminal article *Time – The Next Source of Competitive Advantage* in Harvard Business Review.

With his contribution he directed management attention to the phenomena of time and to the importance which the intelligent use of the resource of time has for gaining and maintaining competitive advantage. From this insight and mastering the rather elementary logistics function of compensating time discrepancies between production and consumption through inventories, it is just a small step to the recognition of time-related flow phenomena such as flow velocity, flow regularity, and flow adaptivity to changing demands, which are at the centre of successful logistics and supply chain management today.

Reference:

George Stalk: Time – The Next Source of Competitive Advantage. In: *Harvard Business Review*, Vol. 65, July-August 1988, pp. 41–51.

P. Klaus and S. Müller (eds.), *The Roots of Logistics*, DOI 10.1007/978-3-642-27922-5_16,

George Stalk jr.: Time – The Next Source of Competitive Advantage[1]

Like competition itself, competitive advantage is a constantly moving target. For any company in any industry, the key is not to get stuck with a single simple notion of its source of advantage. The best competitors, the most successful ones, know how to keep moving and always stay on the cutting edge.

Today, **time** is on the cutting edge. The ways leading companies manage time – in production, in new product development and introduction, in sales and distribution-represent the most powerful new sources of competitive advantage. Though certain Western companies are pursuing these advantages, Japanese experience and practice provide the most instructive examples – not because they are necessarily unique but because they best illustrate the evolutionary stages through which leading companies have advanced.

In the period immediately following World War II, Japanese companies used their low labor costs to gain entry to various industries. As wage rates rose and technology became more significant, the Japanese shifted first to scale-based strategies and then to focused factories to achieve advantage. The advent of just-in-time production brought with it a move to flexible factories, as leading Japanese companies sought both low cost and great variety in the market. Cutting-edge Japanese companies today are capitalizing on time as a critical source of competitive advantage: shortening the planning loop in the product development cycle and trimming process time in the factory-managing time the way most companies manage costs, quality, or inventory.

In fact, as a strategic weapon, time is the equivalent of money, productivity, quality, even innovation. Managing time has enabled top Japanese companies not only to reduce their costs but also to offer broad product lines, cover more market segments, and upgrade the technological sophistication of their products. These companies are time-based competitors.

From Low Wages to Variety Wars

Since 1945, Japanese competitors have shifted their strategic focus at least four times. These early adaptations were straightforward; the shift to time-based competitive advantage is not nearly so obvious. It does, however, represent a logical evolution from the earlier stages.

---

[1] George Stalk is vice president of the Boston Consulting Group, where he specializes in time-based competition. With James C. Abbeglen, he wrote Kaisha, the Japanese Corporation: The New Competitors in World Business (Basic Books, 1985).

In the immediate aftermath of World War II, with their economy devastated and the world around them in a shambles, the Japanese concentrated on achieving competitive advantage through low labor costs. Since Japan's workers were still productive and the yen was devalued by 98.8% against the dollar, its labor costs were extraordinarily competitive with those of the West's developed economies.

Hungry for foreign exchange, the Japanese government encouraged companies to make the most of their one edge by targeting industries with high labor content: textiles, shipbuilding, and steel-businesses where the low labor rates more than offset low productivity rates. As a result, Japanese companies took market share from their Western competition.

But this situation did not last long. Rising wages, caused by high inflation, combined with fixed exchange rates to erode the advantage. In many industries, manufacturers could not improve their productivity fast enough to offset escalating labor costs. By the early 1960s, for instance, the textile companies – comprising Japan's largest industry – were hard-pressed. Having lost their competitive edge in world markets, they spiraled downward, first losing share, then volume, then profits, and finally position and prestige. While the problem was most severe for the textile business, the rest of Japanese industry suffered as well.

The only course was adaptation: in the early 1960s, the Japanese shifted their strategy, using capital investment to boost work-force productivity. They inaugurated the era of scale-based strategies, achieving high productivity and low costs by building the largest and most capital-intensive facilities that were technologically feasible. Japanese shipbuilders, for example, revolutionized the industry in their effort to raise labor productivity. Adapting fabrication techniques from mass production processes and using automatic and semiautomatic equipment, they constructed vessels in modules. The approach produced two advantages for the Japanese. It drove up their own productivity and simultaneously erected a high capital-investment barrier to others looking to compete in the business.

The search for ways to achieve even higher productivity and lower costs continued, however. And in the mid-1960s, it led top Japanese companies to a new source of competitive advantage – the focused factory. Focused competitors manufactured products either made nowhere else in the world or located in the high-volume segment of a market, often in the heart of their Western competitors' product lines. Focusing of production allowed the Japanese to remain smaller than established broad-line producers, while still achieving higher productivity and lower costs-giving them great competitive power.

Factory costs are very sensitive to the variety of goods a plant produces. Reduction of the product-line variety by half, for example, raises productivity by 30%, cuts costs 17%, and substantially lowers the break-even point. Cutting the product line in half again boosts productivity by 75%, slashes costs 30%, and diminishes the break-even point to below 50%. (See EXHIBIT 1, "The Benefits of Focus", next page.)

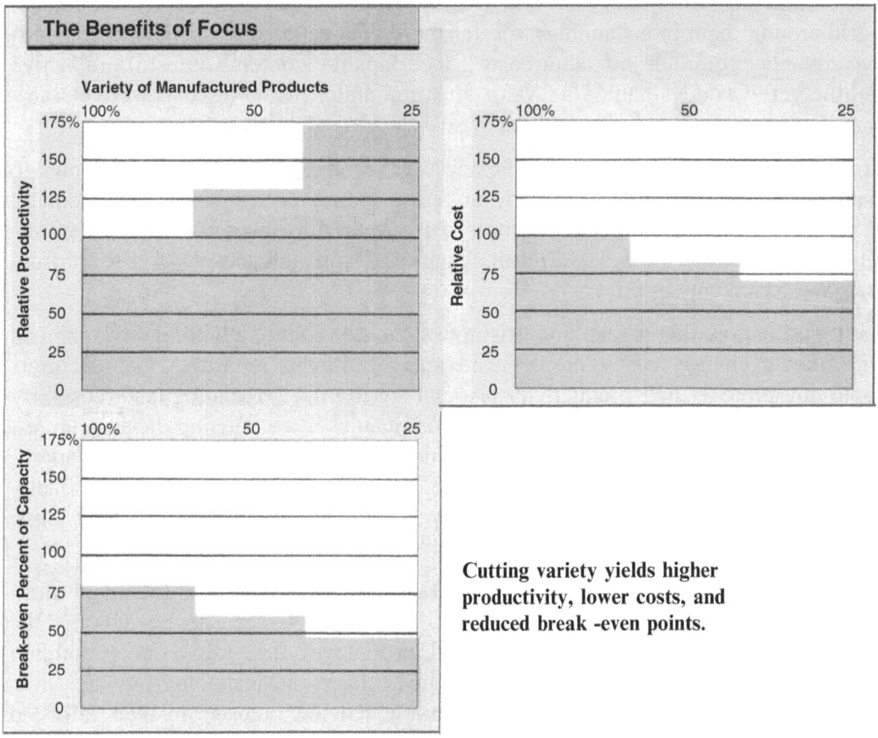

EXHIBIT 1: THE BENEFITS OF FOCUS

In industries like bearings, where competition was fierce in the late 1960s, the Japanese fielded product lines with one-half to one-quarter the variety of their Western competitors. Targeting the high-volume segments of the bearing business – bearings for automobile applications was one – the Japanese used the low costs of their highly productive focused factories to undercut the prices of Western competitors.

SKF was one victim. With factories scattered throughout Europe, each geared to a broad product line for the local market, the Swedish company was a big target for the Japanese. SKF reacted by trying to avoid direct competition with the Japanese: it added higher margin products to serve specialized applications. But SKF did not simultaneously drop any low-margin products, thereby complicating its plant operations and adding to production costs. In effect, SKF provided a cost umbrella for the Japanese. As long as they operated beneath it, the Japanese could expand their product line and move into more varied applications.

Avoiding price competition by moving into higher margin products is called margin retreat – a common response to stepped-up competition that eventually leads to corporate suicide. As a company retreats, its costs rise as do its prices, thus "subsidizing" an aggressive competitor's expansion into the vacated position. The retreating company's revenue base stops growing and may eventually shrink to the point where it can no longer support the fixed cost of the operation. Retrenchment, restructuring, and further shrinkage follow in a cycle that leads to inevitable extinction.

SKF avoided this fate by adopting the Japanese strategy. After a review of its factories, the company focused each on those products it was best suited to manufacture. If a product did not fit a particular factory, it was either placed in another, more suitable plant or dropped altogether. This strategy not only halted SKF's retreat but also beat back the Japanese advance.

At the same time, however, leading Japanese manufacturers began to move toward a new source of competitive advantage – the flexible factory. Two developments drove this move. First, as they expanded and penetrated more markets, their narrow product lines began to pinch, limiting their ability to grow. Second, with growth limited, the economics of the focus strategy presented them with an unattractive choice: either reduce variety further or accept the higher costs of broader product lines.

In manufacturing, costs fall into two categories: those that respond to volume or scale and those that are driven by variety. Scale-related costs decline as volume increases, usually falling 15% to 25% per unit each time volume doubles. Variety-related costs, on the other hand, reflect the costs of complexity in manufacturing: setup, materials handling, inventory, and many of the overhead costs of a factory. In most cases, as variety increases, costs increase, usually at a rate of 20% to 35% per unit each time variety doubles.

The sum of the scale- and variety-related costs represents the total cost of manufacturing. With effort, managers can determine the optimum cost point for their factories – the point where the combination of volume and variety yields the lowest total manufacturing cost for a particular plant. When markets are good, companies tend to edge toward increased variety in search of higher volumes, even though this will mean increased costs. When times are tough, companies pare their product lines, cutting variety to reduce costs.

In a flexible factory system, variety-driven costs start lower and increase more slowly as variety grows. Scale costs remain unchanged. Thus the optimum cost point for a flexible factory occurs at a higher volume and with greater variety than for a traditional factory. A gap emerges between the costs of the flexible and the traditional factory – a cost/variety gap that represents the competitive advantage of flexible production. Very simply, a flexible factory enjoys more variety with lower total costs than traditional factories, which are still forced to make the trade-off between scale and variety. (See EXHIBIT 2, "The Advantage of Flexible Manufacturing", next page.)

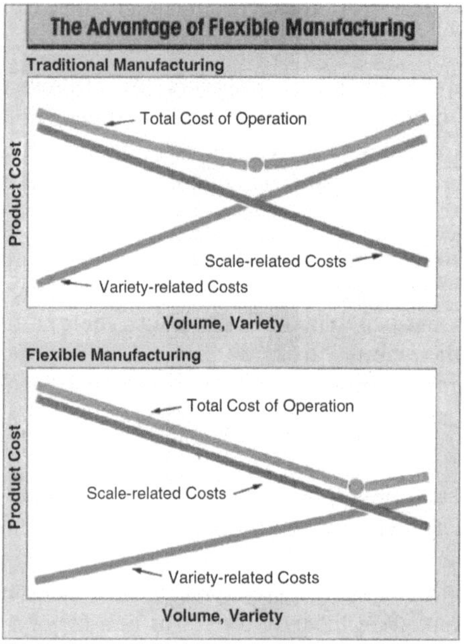

For flexible factories, the optimum cost points occur at a higher volume and with higher variety than for traditional factories

EXHIBIT 2: THE ADVANTAGE OF FLEXIBLE MANUFACTURING

Yanmar Diesel illustrates how this process works. In 1973, with the Japanese economy in recession, Yanmar Diesel was mired in red ink. Worse, there was no promise that once the recession had passed, the existing strategy and program would guarantee real improvement in the company's condition.

As a Toyota supplier, Yanmar was familiar with the automaker's flexible manufacturing system. Moreover, Yanmar was impressed with the automaker's ability to weather the recession without losing money. Yanmar decided to install the Toyota procedure in its own two factories. The changeover took less than five years and produced dramatic results: manufacturing costs declined 40% to 60%, depending on the product; factory break-even points dropped 80% to 50%; total manufacturing labor productivity improved by more than 100%.

But it was Yanmar's newfound capability in product variety that signaled the arrival of a unique strategic edge: during the restructuring Yanmar more than quadrupled its product line. With focused factories, Yanmar could have doubled productivity in such a short time only by reducing the breadth of the product line by 75%. The Toyota system made Yanmar's factories more flexible, reducing costs and producing a greater variety of products.

As its inventor, Taiichi Ohno, said, the Toyota production system was "born of the need to make many types of automobiles, in small quantities with the same manufacturing process." With its emphasis on just-in-time production, total quality control, employee decision making on the factory floor, and close supplier relations, the Toyota system gave the many Japanese manufacturers who adopted it in the mid-1970s a distinct competitive advantage.

A comparison of a U.S. company with a Japanese competitor in the manufacture of a particular automotive suspension component illustrates the nature and extent of the Japanese advantage. The U.S. company bases its strategy on scale and focus: it produces 10 million units per year – making it the world's largest producer – and offers only 11 types of finished parts. The Japanese company's strategy, on the other hand, is to exploit flexibility. It is both smaller and less focused: it manufactures only 3.5 million units per year but has 38 types of finished parts.

With one-third the scale and more than three times the product variety, the Japanese company also boasts total labor productivity that is half again that of its American competitor. Moreover, the unit cost of the Japanese manufacturer is less than half that of the U.S. company. But interestingly, the productivity of the Japanese direct laborers is not as high as that of the U.S. workers, a reflection of the difference in scale. The Japanese advantage comes from the productivity of the overhead employees: with one-third the volume and three times the variety, the Japanese company has only one-eighteenth the overhead employees. (See EXHIBIT 3, "Flexible Manufacturing's Productivity Edge.")

Flexible Manufacturing's Productivity Edge
(Automobile Suspension Component)

	U.S. Competitor	*Japanese Competitor*
Annual Volume	10M	3.5M
Employees		
Direct	107	50
Indirect	135	7
Total	242	57
Annual Units/Employee	43,100	61,400
Types of Finished Parts	11	38
Unit Cost for Comparable Part (index)	$100	$49

(1987 figures)

EXHIBIT 3: FLEXIBLE MANUFACTURING'S PRODUCTIVITY EDGE

In the late 1970s, Japanese companies exploited flexible manufacturing to the point that a new competitive thrust emerged – the variety war. A classic example of a variety war was the battle that erupted between Honda and Yamaha for supremacy in the motorcycle market, a struggle popularly known in Japanese business circles as the H-Y War. Yamaha ignited the H-Y War in 1981 when it announced the opening of a new factory which would make it the world's largest motorcycle manufacturer, a prestigious position held by Honda. But Honda had been concentrating its corporate resources on the automobile business and away from its motorcycle operation. Now, faced with Yamaha's overt and public challenge, Honda chose to counterattack.

Honda launched its response with the war cry, "Yamaha wo tsubusu!" ("We will crush, squash, slaughter Yamaha!") In the no-holds-barred battle that ensued, Honda cut prices, flooded distribution channels, and boosted advertising expenditures. Most important – and most impressive to consumers – Honda also rapidly increased the rate of change in its product line, using variety to bury Yamaha. At the start of the war, Honda had 60 models of motorcycles. Over the next 18 months, Honda introduced or replaced 113 models, effectively turning over its entire product line twice. Yamaha also began the war with 60 models; it was able to manage only 37 changes in its product line during those 18 months.

Honda's new product introductions devastated Yamaha. First, Honda succeeded in making motorcycle design a matter of fashion, where newness and freshness were important attributes for consumers. Second, Honda raised the technological sophistication of its products, introducing four-valve engines, composites, direct drive, and other new features. Next to a Honda, Yamaha products looked old, unattractive, and out of date. Demand for Yamaha products dried up; in a desperate effort to move them, dealers were forced to price them below cost. But even that didn't work. At the most intense point in the H-Y War, Yamaha had more than 12 months of inventory in its dealers' showrooms. Finally Yamaha surrendered. In a public statement, Yamaha President Eguchi announced, "We want to end the H-Y War. It is our fault. Of course there will be competition in the future but it will be based on a mutual recognition of our respective positions."

Honda didn't go unscathed either. The company's sales and service network was severely disrupted, requiring additional investment before it returned to a stable footing. However, so decisive was its victory that Honda effectively had as much time as it wanted to recover. It had emphatically defended its title as the world's largest motorcycle producer and done so in a way that warned Suzuki and Kawasaki not to challenge that leadership. Variety had won the war.

Time-Based Competitive Advantage

The strength of variety as a competitive weapon raises an interesting question. How could Japanese companies accommodate such rapid rates of change? In Honda's case, there could be only three possible answers. The company did one of the following:

1. Began the development of more than 100 new models 10 to 15 years before the attack.
2. Authorized a sudden, massive spending surge to develop and manufacture products on a crash basis.
3. Used structurally different methods to develop, manufacture, and introduce new products.

In fact, what Honda and other variety-driven competitors pioneered was time-based competitiveness. They managed structural changes that enabled their operations to execute their processes much faster. As a consequence, time became their new source of competitive advantage.

While time is a basic business performance variable, management seldom monitors its consumption explicitly – almost never with the same precision accorded sales and costs. Yet time is a more critical competitive yardstick than traditional financial measurements.

Today's new-generation companies compete with flexible manufacturing and rapid-response systems, expanding variety and increasing innovation. A company that builds its strategy on this cycle is a more powerful competitor than one with a traditional strategy based on low wages, scale, or focus. These older, cost-based strategies require managers to do whatever is necessary to drive down costs: move production to or source from a low-wage country; build new facilities or consolidate old plants to gain economies of scale; or focus operations down to the most economic subset of activities. These tactics reduce costs but at the expense of responsiveness.

In contrast, strategies based on the cycle of flexible manufacturing, rapid response, expanding variety, and increasing innovation are time based. Factories are close to the customers they serve. Organization structures enable fast responses rather than low costs and control. Companies concentrate on reducing if not eliminating delays and using their response advantages to attract the most profitable customers.

Many – but certainly not all – of today's time-based competitors are Japanese. Some of them are Sony, Matsushita, Sharp, Toyota, Hitachi, NEC, Toshiba, Honda, and Hino; time-based Western companies include Benetton, The Limited, Federal Express, Domino's Pizza, Wilson Art, and McDonald's. For these leading competitors, time has become the overarching measurement of performance. By reducing the consumption of time in every aspect of the business, these companies also reduce costs, improve quality, and stay close to their customers.

Breaking the Planning Loop

Companies are systems; time connects all the parts. The most powerful competitors understand this axiom and are breaking the debilitating loop that strangles much of traditional manufacturing planning.

Traditional manufacturing requires long lead times to resolve conflicts between various jobs or activities that require the same resources. The long lead times, in turn, require sales forecasts to guide planning. But sales forecasts are inevitably wrong; by definition they are guesses, however informed. Naturally, as lead times lengthen, the accuracy of sales forecasts declines. With more forecasting errors, inventories balloon and the need for safety stocks at all levels increases. Errors in forecasting also mean more unscheduled jobs that have to be expedited, thereby crowding out scheduled jobs. The need for longer lead times grows even greater and the planning loop expands even more, driving up costs, increasing delays, and creating system inefficiencies.

Managers who find themselves trapped in the planning loop often respond by asking for better forecasts and longer lead times. In other words, they treat the symptoms and worsen the problem. The only way to break the planning loop is to reduce the consumption of time throughout the system; that will, in turn, cut the need for lead time, for estimates, for safety stocks, and all the rest. After all, if a company could ever drive its lead time all the way to zero, it would have to forecast only the next day's sales. While that idea of course is unrealistic, successful time-based competitors in Japan and in the West have kept their lead times from growing and some have even reduced them, thereby diminishing the planning loop's damaging effects.

Thirty years ago, Jay W. Forrester of MIT published a pioneering article in HBR, "Industrial Dynamics: A Major Breakthrough for Decision Makers" (July-August 1958), which established a model of time's impact on an organization's performance.[2] Using "industrial dynamics" – a concept originally developed for shipboard fire control systems – Forrester tracked the effects of time delays and decision rates within a simple business system consisting of a factory, a factory warehouse, a distributors' inventory, and retailers' inventories. The numbers in the illustration "Time in the Planning Loop" (see EXHIBIT 4) are the delays in the flow of information or product, measured in weeks. In this example, the orders accumulate at the retailer for three weeks, are in the mail for half a week, are delayed at the distributor for two weeks, go back into the mail for another half a week, and need eight weeks for processing at the factory and its warehouse. Then the finished product begins its journey back to the retailer. The cycle takes 19 weeks.

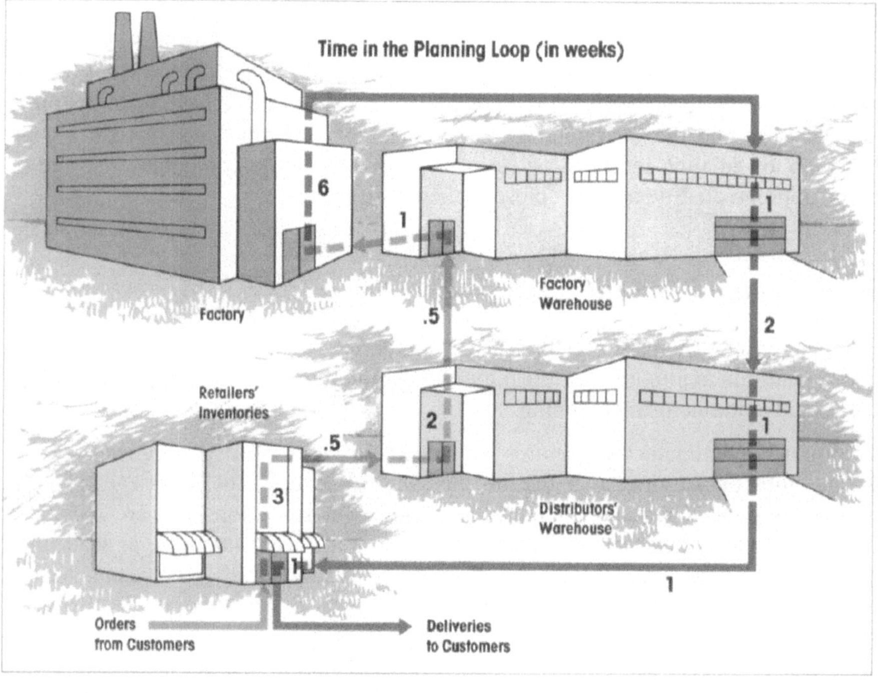

EXHIBIT 4: TIME IN THE PLANNING LOOP

The system in this example is very stable – as long as retail demand is stable or as long as forecasts are accurate 19 weeks into the future. But if unexpected changes

[2] A reprint of Forrester's article is to be found in this volume (comment by the editors).

occur, the system must respond. The chart, also taken from the Forrester article, shows what happens to this system when a simple change takes place: demand goes up 10%, then flattens. Acting on new forecasts and seeking to cut delivery delays, the factory first responds by ramping up production 40%. When management realizes – too late – that it has overshot the mark, it cuts production 30%. Too late again it learns that it has overcorrected. This ramping up and cutting back continue until finally the system stabilizes, more than a year after the initial 10% increase.

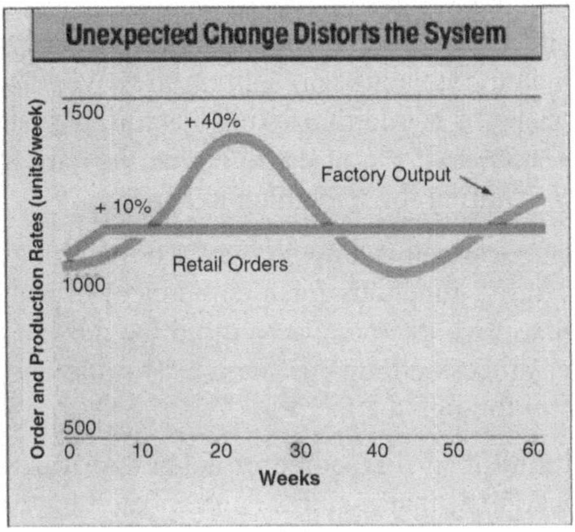

EXHIBIT 5: UNEXPECTED CHANGE DISTORTS THE SYSTEM

What distorts the system so badly is time: the lengthy delay between the event that creates the new demand and the time when the factory finally receives the information. The longer that delay, the more distorted is the view of the market. Those distortions reverberate throughout the system, producing disruption, waste, and inefficiency.

These distortions plague business today. To escape them, companies have a choice: they can produce to forecast or they can reduce the time delays in the flow of information and product through the system. The traditional solution is to produce to forecast. The new approach is to reduce time consumption.

Because time flows throughout the system, focusing on time-based competitive performance results in improvements across the board. Companies generally become time-based competitors by first correcting their manufacturing techniques, then fixing sales and distribution, and finally adjusting their approach to innovation. Ultimately, it becomes the basis for a company's overall strategy.

Time-Based Manufacturing

In general, time-based manufacturing policies and practices differ from those of traditional manufacturers along three key dimensions: length of production runs, organization of process components, and complexity of scheduling procedures.

When it comes to lot size, for instance, traditional factories attempt to maximize production runs while time-based manufacturers try to shorten their production runs as much as possible. In fact, many Japanese companies aim for run lengths of a single unit. The thinking behind this is as simple as it is fundamental to competitive success: reduced run lengths mean more frequent production of the complete mix of products and faster response to customers' demands.

Factory layout also contributes to time-based competitive advantage. Traditional factories are usually organized by process technology centers. For example, metal goods manufacturers organize their factories into shearing, punching, and braking departments; electronic assemblers have stuffing, wave soldering, assembly, testing, and packing departments. Parts move from one process technology center to the next. Each step consumes valuable time: parts sit, waiting to move; then move; then wait to be used in the next step. In a traditional manufacturing system, products usually receive value for only .05% to 2.5% of the time that they are in the factory. The rest of the time products sit waiting for something to happen.

Time-based factories, however, are organized by product. To minimize handling and moving of parts, the manufacturing functions for a component or a product are as close together as possible. Parts move from one activity to the next with little or no delay. Because the production process eliminates the need to pile and repile parts, they flow quickly and efficiently through the factory.

In traditional factories, scheduling is also a source of delay and waste. Most traditional factories use central scheduling that requires sophisticated materials resource planning and shop-floor control systems. Even though these systems are advanced, they still waste time: work orders usually flow to the factory floor on a monthly or weekly basis. In the meantime, parts can sit idle.

In time-based factories, local scheduling enables employees to make more production control decisions on the factory floor, without the time-consuming loop back to management for approval. Moreover, the combination of the product-oriented layout of the factory and local scheduling makes the total production process run more smoothly. Once a part starts through the production run, many of the requirements between manufacturing steps are purely automatic and require no intermediate scheduling.

These differences between traditional and time-based factories add up. Flexible factories enjoy big advantages in both productivity and time: labor productivity in time-based factories can be as much as 200% higher than in conventional plants; time-based factories can respond eight to ten times faster than traditional factories. Flexible production means significant improvements in labor and net-asset productivity. These, in turn, yield reductions of up to 20% in overall costs and increases in growth for much less investment.

Toyota offers a dramatic example of the kinds of improvements that leading time-based competitors are making. Dissatisfied with the response time of a supplier, Toyota went to work. It took the supplier 15 days to turn out a component after arrival of the raw materials at its factory. The first step was to cut lot sizes, reducing response time to 6 days. Next Toyota streamlined the factory layout, reducing the number of inventory holding points. The response time fell to 3 days. Finally Toyota eliminated all work-in-progress inventories at the supplier's plant. New response time: 1 day.

Toyota, of course, is not alone in improving manufacturing response times. Matsushita cut the time needed to make washing machines from 360 hours to just 2; Honda slashed its motorcycle fabricating time by 80%; in North America, companies making motor controllers and electrical components for unit air conditioners have improved their manufacturing response times by 90%.

Time-Based Sales and Distribution

A manufacturer's next challenge is to avoid dissipation of factory performance improvements in other parts of the organization. In Jay Forrester's example of the planning loop, the factory and its warehouse accounted for roughly one-half of the system's time. In actuality today, the factory accounts for one-third to one-half of the total time – often the most "visible" portion of time. But other parts of the system are just as important, if less apparent. For example, in the Forrester system, sales and distribution consume as much or more time than manufacturing.

What Forrester modeled, the Japanese experienced. By the late 1970s, leading Japanese companies were finding that inefficient sales and distribution operations undercut the benefits of their flexible manufacturing systems. Toyota, which at that time was divided into two separate companies, Toyota Motor Manufacturing and Toyota Motor Sales, again makes this point. Toyota Motor Manufacturing could manufacture a car in less than 2 days. But Toyota Motor Sales needed from 15 to 26 days to close the sale, transmit the order to the factory, get the order scheduled, and deliver the car to the customer. By the late 1970s, the cost-conscious, competition-minded engineers at Toyota Manufacturing were angry at their counterparts at Toyota Motor Sales, who were frittering away the advantage gained in the production process. The sales and distribution function was generating 20% to 30% of a car's cost to the customer – more than it cost Toyota to manufacture the car!

Finally, in 1982 Toyota moved decisively to remedy the problem. The company merged Toyota Motor Manufacturing and Toyota Motor Sales. The company announced that it wanted to become "more marketing driven". While Toyota assured the public that the reorganization only returned it to its configuration in the 1950s, within 18 months all the Toyota Motor Sales directors retired. Their jobs were left vacant or filled by executives from Toyota Motor Manufacturing.

The company wasted no time in implementing a plan to cut delays in sales and distribution, reduce costs, and improve customer service. The old system, Toyota found, had handled customer orders in batches. Orders and other crucial information would accumulate at one step of the sales and distribution process before dispatch to the next level, which wasted time and generated extra costs.

To speed the flow of information, Toyota had to reduce the size of the information batches. The solution came from a company-developed computer system that tied its salespeople directly to the factory scheduling operation. This link bypassed several levels of the sales and distribution function and enabled the modified system to operate with very small batches of orders.

Toyota expected this new approach to cut the sales and distribution cycle time in half – from four to six weeks to just two to three weeks across Japan. (For the Tokyo and Osaka regions, which account for roughly two-thirds of Japan's population, the goal was to reduce cycle time to just two days.) But by 1987 Toyota had reduced system responsiveness to eight days, including the time required to make the car. In the Forrester example, this achievement is equivalent to cutting the 19-week cycle to 6 weeks. The results were predictable: shorter sales forecasts, lower costs, happier customers.

Time-Based Innovation

A company that can bring out new products three times faster than its competitors enjoys a huge advantage. Today, in one industry after another, Japanese manufacturers are doing just that to their Western competition:

- In projection television, Japanese producers can develop a new television in one-third the time required by U.S. manufacturers.
- In custom plastic injection molds, Japanese companies can develop the molds in one-third the time of U.S. competitors and at one-third the cost.
- In autos, Japanese companies can develop new products in half the time – and with half as many people – as the U.S. and German competition.

To accomplish their fast-paced innovations, leading Japanese manufacturers have introduced a series of organizational techniques that precisely parallel their approach to flexible manufacturing:

- In manufacturing, the Japanese stress short production runs and small lot sizes. In innovation, they favor smaller increments of improvement in new products, but introduce them more often – versus the Western approach of more significant improvements made less often.
- In the organization of product development work, the Japanese use factory cells that are cross-functional teams. Most Western new product development activity is carried out by functional centers.
- In the scheduling of work, Japanese factories stress local responsibility, just as product development scheduling is decentralized. The Western approach to both requires plodding centralized scheduling, plotting, and tracking.

The effects of this time-based advantage are devastating; quite simply, American companies are losing leadership of technology and innovation – supposedly this country's source of long-term advantage.

Unless U.S. companies reduce their new product development and introduction cycles from 36–48 months to 12–18 months, Japanese manufacturers will easily out-innovate and outperform them. Taking the initiative in innovation will require even faster cycle times.

Residential air conditioners illustrate the Japanese ability to introduce more technological innovation in smaller increments – and how in just a few years these improvements add up to remarkably superior products. The Japanese introduce innovations in air conditioners four times faster than their American competitors; in technological sophistication the Japanese products are seven to ten years ahead of U.S. products.

Look at the changes in Mitsubishi Electric's three-horsepower heat pump between 1975 and 1985. From 1975 to 1979, the company did nothing to the product except change the sheet metal work, partly to improve efficiency but mostly to reduce materials costs. In 1979, the technological sophistication of the product was roughly equal to that of the U.S. competition. From this point on, the Japanese first established, and then widened the lead.

In 1980, Mitsubishi introduced its first major improvement: a new product that used integrated circuits to control the air-conditioning cycle. One year later, the company replaced the integrated circuits with microprocessors and added two important innovations to increase consumer demand. The first was "quick connect" freon lines. On the old product (and on the U.S. product), freon lines were made from copper tubing and cut to length, bent, soldered together, purged, and filled with freon – an operation requiring great skill to produce a reliable air conditioner. The Japanese substituted quick-connect freon lines – precharged hoses that simply clicked together. The second innovation was simplified wiring. On the old product (and still today on the U.S. product) the unit had six color-coded wires to connect. The advent of microprocessors made possible a two-wire connection with neutral polarity.

These two changes did not improve the energy-efficiency ratio of the product; nor were they intended to. Rather, the point was to fabricate a unit that would be simpler to install and more reliable, thereby broadening distribution and increasing demand. Because of these innovations, white-goods outlets could sell the new product, and local contractors could easily install it.

In 1982, Mitsubishi introduced a new version of the air conditioner featuring technological advances related to performance. A high-efficiency rotary compressor replaced the outdated reciprocating compressor. The condensing unit had louvered fins and inner fin tubes for better heat transfer. Because the balance of the system changed, all the electronics had to change. As a result, the energy-efficiency ratio improved markedly.

In 1983, Mitsubishi added sensors to the unit and more computing power, expanding the electronic control of the cycle and again improving the energy-efficiency ratio.

In 1984, Mitsubishi came out with another version of the product, this time with an inverter that made possible an even higher energy-efficiency ratio. The inverter, which requires additional electronics for the unit, allows unparalleled control over the speed of the electric motor, dramatically boosting the appliance's efficiency.

Using time-based innovation, Mitsubishi transformed its air conditioner. The changes came incrementally and steadily. Overall they gave Mitsubishi – and other Japanese companies on the same track – the position of technological leadership in the global residential air-conditioning industry.

In 1985, a U.S. air-conditioner manufacturer was just debating whether to use integrated circuits in its residential heat pump. In view of its four- to five-year product development cycle, it could not have introduced the innovation until 1989 or 1990 – putting the American company ten years behind the Japanese. Faced with this situation, the U.S. air-conditioner company followed the example of many U.S. manufacturers that have lost the lead in technology and innovation: it decided to source its air conditioners and components from its Japanese competition.

Time-Based Strategy

The possibility of establishing a response time advantage opens new avenues for constructing winning competitive strategies. At most companies, strategic choices are limited to three options:

1. Seeking coexistence with competitors. This choice is seldom stable, since competitors refuse to cooperate and stay put.
2. Retreating in the face of competitors. Many companies choose this course, the business press fills its pages with accounts of companies retreating by consolidating plants, focusing their operations, outsourcing, divesting businesses, pulling out of markets, or moving upscale.
3. Attacking, either directly or indirectly. The direct attack involves the classic confrontation – cut price and add capacity, creating head-on competition. Indirect attack requires surprise. Competitors either do not understand the strategies being used against them or they do understand but cannot respond – sometimes because of the speed of the attack, sometimes because of their inability to mount a response.

Of the three options, only an attack creates the opportunity for real growth. Direct attack demands superior resources; it is always expensive and potentially disastrous. Indirect attack promises the most gain for the least cost. Time-based strategy offers a powerful new approach for successful indirect attacks against larger, established competitors.

Consider the remarkable example of Atlas Door, a ten-year-old U.S. company. It has grown at an average annual rate of 15% in an industry with an overall annual growth rate of less than 5%. In recent years, its pretax earnings were 20% of sales, about five times the industry average. Atlas is debt free. In its tenth year the company achieved the number one competitive position in its industry.

The company's product: industrial doors. It is a product with almost infinite variety, involving limitless choices of width and height and material. Because of the importance of variety, inventory is almost useless in meeting customer orders; most doors can be manufactured only after the order has been placed.

Historically, the industry had needed almost four month to respond to an order for a door that was out of stock or customized. Atlas's strategic advantage was time: it could respond in weeks to any order. It had structured its order-entry, engineering, manufacturing, and logistics systems to move information and products quickly and reliably.

First, Atlas built just-in-time factories. These are fairly simple in concept. They require extra tooling and machinery to reduce changeover times and a fabrication process organized by product and scheduled to start and complete all of the parts at the same time. But even the performance of the factory – critical to the company's overall responsiveness – still only accounted for 2 ½ weeks of the completed product delivery cycle.

Second, Atlas compressed time at the front end of the system, where the order first entered and was processed. Traditionally, when customers, distributors, or salespeople called a door manufacturer with a request for price and delivery, they would have to wait more than one week for a response. If the desired door was not in stock, not in the schedule, or not engineered, the supplier's organization would waste even more time, pushing the search for an answer around the system.

Recognizing the opportunity to cut deeply into the time expenditure in this part of the system, Atlas first streamlined, then automated its entire order-entry, engineering, pricing, and scheduling processes. Today Atlas can price and schedule 95% of its incoming orders while the callers are still on the telephone. It can quickly engineer new special orders because it has preserved on computer the design and production data of all previous special orders – which drastically reduces the amount of re-engineering necessary.

Third, Atlas tightly controlled logistics so that it always shipped only fully complete orders to construction sites. Orders require many components. Gathering all of them at the factory and making sure that they are with the correct order can be a time-consuming task. It is even more time-consuming, however, to get the correct parts to the job site **after** they have missed the initial shipment. Atlas developed a system to track the parts in production and the purchased parts for each order, ensuring arrival of all necessary parts at the shipping dock in time – a just-in-time logistics operation.

When Atlas started operations, distributors were uninterested in its products. The established distributors already carried the door line of a larger competitor, they saw no reason to switch suppliers except, perhaps, for a major price concession. But as a start-up, Atlas was too small to compete on price alone. Instead, it positioned itself as the door supplier of last resort, the company people came to if the established supplier could not deliver or missed a key date.

Of course, with industry lead times of almost four months, some calls inevitably came to Atlas. And when it did get a call, Atlas commanded a higher price because of its faster delivery. Atlas not only got a higher price but its time-based processes also yielded lower costs: it thus enjoyed the best of both worlds.

In ten short years, the company replaced the leading door suppliers in 80% of the distributors in the country. With its strategic advantage the company could be selective, becoming the house supplier for only the strongest distributors.

In the wake of this indirect attack, the established competitors have not responded effectively. The conventional view is that Atlas is a "garage shop operator" that cannot sustain its growth: competitors expect the company's performance to degrade to the industry average as it grows larger. But this response – or nonresponse – only reflects a fundamental lack of understanding of time as the source of competitive advantage. The extra delay in responding only adds to the insurmountable lead the indirect time-based attack has created. While the traditional companies track cost and size, the new competitor drives advantage from time, staying on the cutting edge, leaving its rivals behind.

PART C: THE INSTRUMENTATION OF LOGISTICS: TERMINOLOGY, CONCEPTS AND METHODOLOGICAL APPROACHES

Wilhelm Launhardt (1882)

THE DETERMINATION OF THE MOST EXPEDIENT LOCATION OF AN INDUSTRIAL PLANT

The choice of an appropriate location for a logistics facility (i.e. a warehouse, a transshipment point or a production site) is one of the first and most important planning tasks when designing and configuring a logistics network. The choice of location requires the facility's geographic positioning relative to its suppliers, its customers, and its service providers. Decisions about locations affect transport, warehousing, and transshipment costs and conditions.

The support of location planning decisions, based on systematic, often quantitative analysis, has a long tradition: Johann Heinrich von Thünen, a German economist, published a first monograph on economic location theory in 1826, using primarily qualitative and graphical arguments. The German economist, engineer and mathematician William Launhardt (1832–1918) in his 1882 article on *The Determination of the Most Expedient Location of an Industrial Plant* (Die Bestimmung des zweckmäßigsten Standortes einer gewerblichen Anlage) discussed a series of considerations for the determination of optimal locations for industrial plants, focusing primarily on the optimization of transportation cost. He uses both mathematical calculations and graphical representations of the problem. While using assumptions which appear fairly simple from today's perspective, it is an impressive example of early model-based quantitative analysis of an important type of logistics decision.

Von Thünen's and Launhardt's works on location decisions were followed by a long and rich stream of work, including Alfred Weber's (1909) famous, still quoted *Theory of Industrial Location* up to today's Operations Research based network planning, design and location ranking models (e.g. Magnanti/Wong (1984)).

Reference:

Wilhelm Launhardt: Die Bestimmung des zweckmäßigsten Standortes einer gewerblichen Anlage. In: *Zeitschrift des Vereines Deutscher Ingenieure*, Vol. XXVI, No. 3, March 1882, pp. 105–116.

Further reading:

Magnanti, Thomas L. and R.T. Wong: Network Design and Transportation Planning: Models and Algorithms. In: *Transportation Science*, Vol. 18, No. 1, 1984, pp. 1–55.

Thünen, Heinrich von: *Der isolirte Staat.* Hamburg: Perthes, 1826.

Weber, Alfred: *Über den Standort der Industrien. Reine Theorie des Standorts.* Tübingen: Mohr, 1909.

P. Klaus and S. Müller (eds.), *The Roots of Logistics*, DOI 10.1007/978-3-642-27922-5_17,

Wilhelm Launhardt: The Determination of the Most Expedient Location of an Industrial Plant[1]

As a general rule, the determinants of the location of an industrial plant are dependent on the location of the sources of raw and auxiliary materials and of the points of use of the plant's products. Next to the consideration of the availability and capability of transport facilities, in some instances still further circumstances need to be taken into consideration such as, for example, differences in the prices for the acquisition of land, the availability of hydroelectric power, differences in the cost of living and wages for workers, or the availability of workers trained for the jobs required, etc. However, the latter circumstances can only be taken into account after the most expedient location of the plant has been determined with respect to the availabilities and conditions for transportation.

If, for example, A is the point of origin of one raw material (see FIGURE 1), e.g. of clay for a brick-making plant or of iron ore, if point B is the sourcing location of coal, and C, finally, is the location of the finished product, i.e. of the fired bricks or the pig iron; furthermore, if A also stands for the tonnage volume of raw material, B for the volume of the auxiliary material, and C for the weight of the finished product then the transport costs for an assumed location of the plant (be it the brick-making plant or the blast furnace) are easily calculated. If a, b, and c are the distances between the three locations in kilometres, and if Ψ is the freight rate per ton-kilometre, then the transport costs amount to:

FIGURE 1

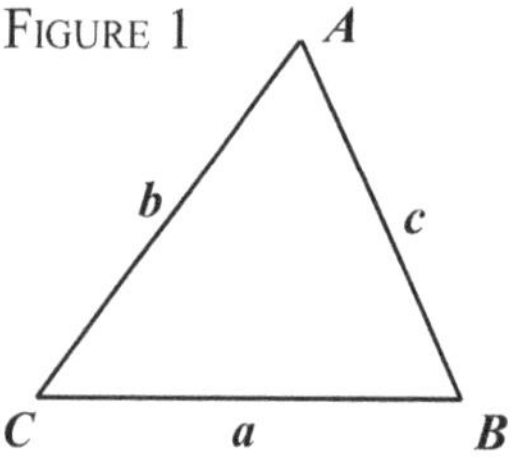

- if A is chosen as the location: $K_i = (Bc + Cb)\,\Psi$,
- if the plant is at B: $K_{ii} = (Ac + Ca)\,\Psi$,
- if the plant is utilized in C: $K_{iii} = (Ab + Ba)\,\Psi$.

On the basis of this simple calculations it is quickly determined which of these three locations would be the lowest cost one from the perspective of transportation cost. This consideration shall be exemplified through the following calculation:

A be a blast furnace plant where the ore is extracted, which is $c = 10$ km away from the coal sourcing location B, and $b = 12$ km away from the destination of the manufactured pig iron (for instance the closest railway station). B, the sourcing location of the coals is $a = 8$ km distant to C. If 32,000 tons of pig iron are obtained annually through smelting 100,000 tons of iron ore under the consumption of 80,000 tons of coal, then by installing a blast furnace in A, an annual volume of 1,184,000 ton-kilometres (80,000 tons × 10 kilometres + 32,000 tons × 12 kilometres) is generated. Installing the plant in B will annually generate 1,256,000 ton-kilometres (100,000 tons × 12 kilometres + 32,000 tons × 8 kilometres). A decision for plant C will lead

[1] Translation by the editors.

to 1,840,000 ton-kilometre units of transportation performance (100,000 tons × 12 kilometres + 80,000 tons × 8 kilometres).

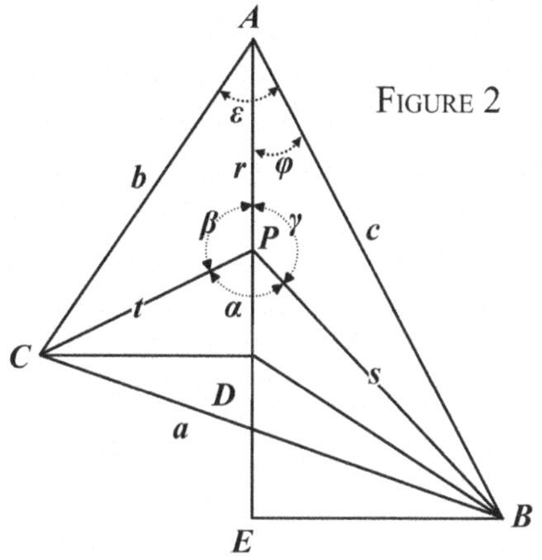

FIGURE 2

This consideration, however, is in no way exhaustive since the location decision is not restricted to one of the three existing locations. Instead, any intermediate point P could also be selected alternatively. According to FIGURE 2, the total transport costs that are to be spent are determined by the transport costs per kilometre and the transport quantity from A, B, and C (as introduced above), considering the three access routes r, s, and t.

Hence, the total transport costs can now be formulated as follows:

$S = Ar + Bs + Ct$ or

$$S = Ar + B\,[r^2 + c^2 - 2\,r\,c\cos\varphi]^{1/2} + C\,[r^2 + b^2 - 2\,r\,b\cos(\varepsilon - \varphi)]^{1/2} \qquad (1)$$

The transport costs in this model depend on the values of the variables r and φ. A differentiation according to these variables results in the conditions for the minimum transport costs:

$$A + B\,\frac{r - \cos\varphi}{[r^2 + c^2 - 2\,r\,c\cos\varphi]^{1/2}} + C\,\frac{r - b\cos(\varepsilon - \varphi)}{[r^2 + b^2 - 2\,r\,c\cos(\varepsilon - \varphi)]^{1/2}} = 0 \qquad (2)$$

and

$$\frac{B\,r\,c\sin\varphi}{[r^2 + c^2 - 2\,r\,c\cos\varphi]^{1/2}} - \frac{C\,r\,b\sin(\varepsilon - \varphi)}{[r^2 + b^2 - 2\,r\,b\cos(\varepsilon - \varphi)]^{1/2}} = 0 \qquad (3)$$

If the ray $r = AP$ is extended beyond P and if two right-angled lines CD and BE are drawn from C and B against these extensions then; EB = $c \sin \varphi$, CD = $b \sin(\varepsilon - \varphi)$, PD = $b \cos(\varepsilon - \varphi) - r$ and PE = $c \cos\varphi - r$. Inserting these relations into the equations (2) and (3) and reintroducing the side lengths of the triangle ABC into the denominator results in:

$$A - B\,\frac{\overline{PE}}{c} - C\,\frac{\overline{PD}}{b} = 0 \qquad (4)$$

and

$$B\,\frac{\overline{EB}}{c} - C\,\frac{\overline{CD}}{b} = 0 \qquad (5)$$

or, if the three angles at point P are correspondingly denominated as α, β and γ:

$$A + B\cos\gamma + C\cos\beta = 0 \qquad (6)$$

and

$$B\sin\gamma - C\sin\beta = 0 \qquad (7)$$

These equations imply that the angles α, β and γ at point P must be equal to the exterior angles of a triangle whose side lengths are equal to the kilometrical transport costs A, B, and C.

The solution for the most favourable location of the production site can also be expressed as follows: **The kilometrical transport costs A, B, and C, regarded as forces, need to be in balance at the resulting location of the production site.**

Another way to find the most favourable location of the production site could be through a simple experiment: Three threads are tied together with a knot. The ends of these threads are connected to the points A, B, and C, which are marked on a horizontal board. Each of the threads is strained with a weight that represents the kilometrical transport costs.[2]

In still another very simple way the solution could also be found by a geometrical construction, as shown in FIGURE 3. Arcs are plotted around point A with the length $B/C \times c$ and around B with the length $A/C \times c$. Their intersection in point O results in the triangle ABO. Its sides are in the ratio of the kilometrical transport costs A, B, and C in such a way that the angles of this triangle α, β and γ are the complementary angles of the angles α, β and γ at the location P. Accordingly, the location must lie on the circle orbiting the triangle ABO. In fact, it must lie on the point, which is cut through by the straight line from C to O. The circle is described as the "**location circle**" and point O as "**pole**". A blast furnace for smelting the iron ore of A while using the coals of B should be on the location circle. Appointing the site on the circle depends on the location of the utilization site C. A randomly located utilization site must always lie on a straight line between the utilization site and the poles.

FIGURE 3

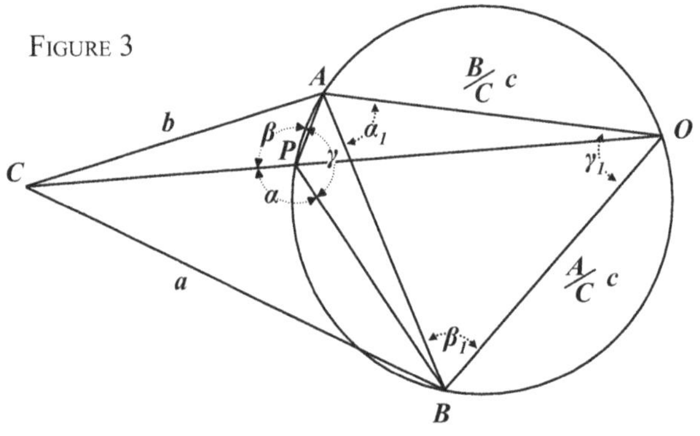

2 The procedure Launhardt describes here was introduced already in the 17th century by the French mathematician Pierre de Varignon. It is known as "Varignon's apparatus" (comment by the editors).

For pole A, a further implication can be derived. According to Ptolemy's theorem for a quadrangle that is surrounded by a circle applies that the product of the diagonals equals the sum of the products of the opposite sides. This can be transferred to the above considerations by the following equations:

$$\overline{AP} \times \overline{BO} + \overline{BP} \times \overline{AO} = \overline{AB} \times \overline{PO}$$

that is:

$$r\,\frac{A}{C}\,c + s\,\frac{B}{C}\,c = c \times \overline{PO}$$

or

$$rA + sB = c \times \overline{PO} \tag{8}$$

The total transport costs amount to:

$$S = Ar + Bs + Ct$$

that is, considering equation (8):

$$S = C \times \overline{PO} + C \times t$$

or

$$S = C \times \overline{CO} \tag{9}$$

This means: **The total transport costs comprise an amount as if there was merely a transport of the finished product from pole *O* to destination *C*.** Hence, the pole totally replaces the locations *A* and *B*, regarding both – the direction of the access route as well as the amount of the transport costs. In case of the utilization site *C* lying between the extended pole rays *AD* and *BE*, the access route must then always take the direction of *C* to *O*. In this case point *C* is also to be chosen as the location of the intersection point of the ray *CO* with the location circle where the total transport costs amount to $C \times CO$. However, if the plant was located on the opposite side of the line *AB*, then the pole would move to the other side from *AB* to O_1. This second pole then would serve as the target point for all utilization sites between the pole rays *AF* and *BG*. Conversely, if the destination point *C* lay between the pole rays *AD* and *AF*, then *A* would have to be chosen as the production site. In this case, the total transport costs would have a level as if the finished product was to be transported from *O* via *A* to *C*. If the utilization site lay within the pole rays *BE* and *BG* then *B* would need to be selected as the relevant location. Finally, if the utilization site was located within both of the location circles *APB* and AP_1B (which is the case

FIGURE 4

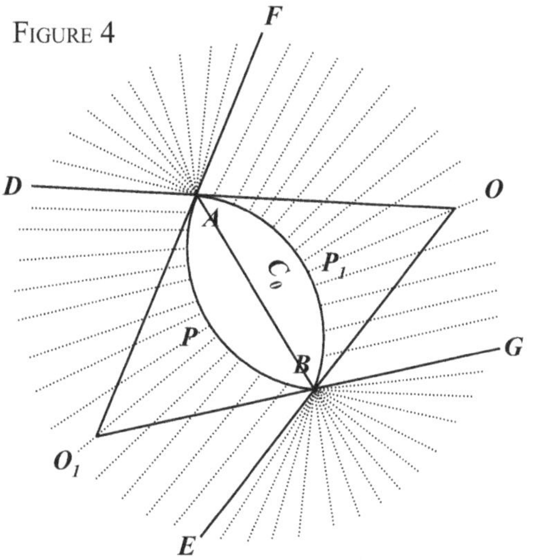

e.g. at C_0 in FIGURE 4), then the utilization site would have to be determined as the location of the production. Accordingly, FIGURE 4 shows the various directions of the access routes from the utilization sites to those locations that lie on location circles *APB* or AP_1B.

In a similar way to how the pole *O* for the locations *A* and *B* was found, also pole *Q*, concerning the locations *A* and *C*, can be determined (compare FIGURE 5). This can be done by plotting one arc around location *A* with the length $C/B \times b$ and one around location *C* with the length $A/B \times b$. Finally, pole *R* for the locations *B* and *C* can be constructed. Location *P* is then determined through the intersection point of the three vector rays *AR*, *BQ*, and *CO*. For the construction of these, only two rays are required; also the location circles are dispensable.

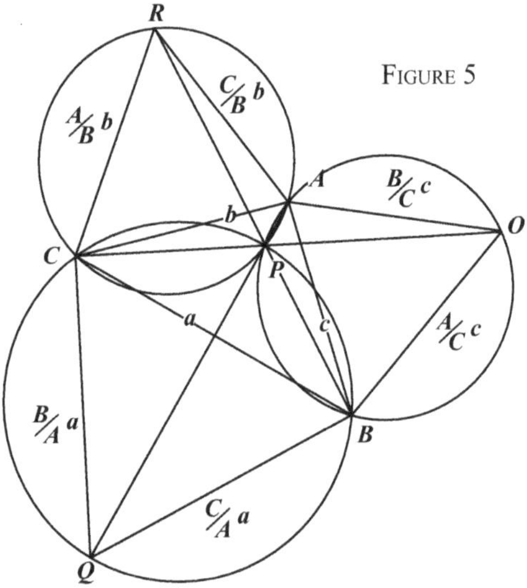

FIGURE 5

If the determination of the most expedient location is conducted for the above example, the lengths of the three access routes are found to be $r = 3.81$, $s = 6.28$, and $t = 8.96$. Accordingly, the units of transportation performance in ton-kilometres adds up to

$$100{,}000 \times 3.81 + 80{,}000 \times 6.28 + 32{,}000 \times 8.96 = 1{,}170{,}000.$$

The distance of point *C* to the junction *O* amounts to 36.59 kilometres. This leads to a transport performance of $36.59 \times 32{,}000 = 1{,}170{,}880$ ton-kilometre units.

Compared with the situation of establishing the blast furnace at the location of the iron extraction, *A*, this most expedient location therefore results in savings of 13,000 ton-kilometres.

Of course, in the assumed example the attained reduction of transport costs is so low that other changes could be considered instead. E.g., even regarding the savings in administration and supervision costs it could be expedient to operate the plant directly from the ore extraction site. Before a decision about the matter can be made, the total possible minimum of transport costs is to be figured out by applying the method described above.

Until now, the same topographical conditions of the terrain and transport infrastructure was assumed for any place. However, if transport costs are to be figured out realistically, as it is always required for practical applications, then the site's optimal location will be affect more strongly.

If, e.g., the blast furnace with the ore mines in point *A* would be linked with the sourcing location of coal in point *B* by constructing of a railway connection whose investment costs amount to 80,000 Marks ($\mathcal{M}$), then the kilometrical transport costs for the ore transport can be calculated as follows:

Interest of investment capital in railway $0.05 \times 80{,}000$	$= 4{,}000\ \mathcal{M}$
Railway maintenance and supervision	$= 2{,}000\ \mathcal{M}$
Transport costs 0.02 per ton-kilometre $900{,}000 \times 0.02$	$= 2{,}000\ \mathcal{M}$
total	$A = 8{,}000\ \mathcal{M}$

In the same way the kilometrical transport costs for the procurement of the coal can be calculated:

Interest of investment capital in railway	$= 4{,}000\ \mathcal{M}$
Railway maintenance	$= 2{,}000\ \mathcal{M}$
Transport costs $0.02 \times 80{,}000$	$= 1{,}600\ \mathcal{M}$
	$B = 7{,}600\ \mathcal{M}$

If a road was constructed from the blast furnace plant P to the utilization site, then the kilometrical transport costs for the dispatch of the pig iron would be imposed in the following way:

Interest of investment capital in road $8{,}000 \times 0.05$	$= 400\ \mathcal{M}$
Road maintenance $200 + 0.03 \times 32{,}000$	$= 1{,}160\ \mathcal{M}$
Transport costs on the road $0.17 \times 32{,}000$	$= 5{,}440\ \mathcal{M}$
	$C = 7{,}000\ \mathcal{M}$

If these restated values of the kilometrical transport costs are used the location for the blast furnace results at a distance of $r = 7.7$ kilometres from the ore mines in A. The blast furnace location is $s = 3.3$ km distant to the coal sourcing site in B and $t = 6.0$ km to the utilization site C. The total transport costs then amount to:

$$S = 7.7 \times 8{,}000 + 3.3 \times 7{,}600 + 6.0 \times 7{,}000 = 128{,}680\ \mathcal{M}.$$

If, instead, the blast furnace had been established at the ore extraction site A, which is the most expedient location among the sites A, B, and C, then the transport costs would have been $10 \times 7{,}600 + 12 \times 7{,}000 = 160{,}000\ \mathcal{M}$. Thus the selection of the most advantageous location, compared with selecting site A as the location would earn savings of totally 31,320 $\mathcal{M}$ transport costs per annum.

With these considerations the importance of selecting the most expedient location for the industrial plant has been highlighted sufficiently.

Now follows the consideration of an instance, where more than three sources of materials exist, from where transport-moves leave. For example, iron ore smelting requires a supplementary material of 300,000 tons that is to be procured from point D. This point is 8 kilometres distant from A and 4 kilometres distant from B, as shown in FIGURE 6. The procedure for identifying the most expedient location then is the following.

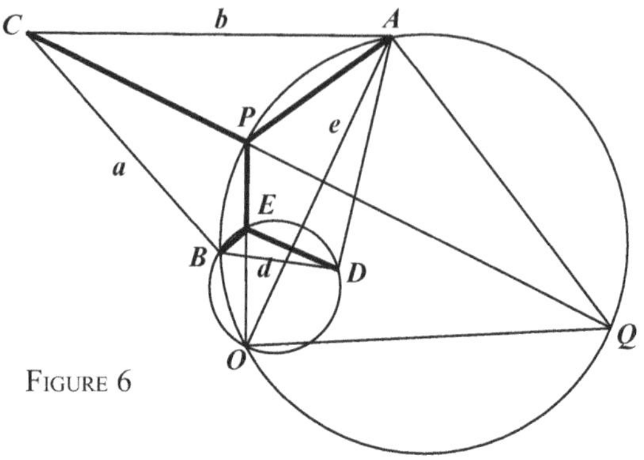

FIGURE 6

At first, the pole for two of the sourcing areas is determined, e.g. pole *O* for points *B* and *D*, and the surrounding location circle of *B*, *D*, and *O* is created. Pole *O* now acts in place of the related points *B* and *D* and is combined with points *A* and *C* for identifying a further expedient location. For this purpose the procedure is repeated: The new pole, *Q*, is searched for points *O* and *A*, and *O*, *A*, and *Q* are connected by a location circle. Hereafter, a straight line is drawn from *C* to *Q* through the location circle in point *P*, where the site is to be installed. Then two connections from *P* to *A* and from *P* to *O* are drawn. The last-mentioned line is only outlined to *E*, which lies on the intersection of the location circle of *O*, *B*, and *D*. *E* then branches into *B* and *D*. The access routes *B* and *D* join at *E* from where the transport to the location *P* is jointly carried out on the line *EP*, on which accordingly a transport of 80,000 + 30,000 = 110,000 tons takes place. Assuming a railway connection to point *D* and the previously introduced cost parameters, then the kilometrical transport costs on the three rays *BE*, *DE*, and *EP* to $B = 6{,}000 + 0.02 \times 80{,}000 = 7{,}600\ \mathcal{M}$; $D = 6{,}000 + 0.02 \times 30{,}000 = 6{,}600\ \mathcal{M}$ and $E = 6{,}000 + 0.02 \times 110{,}000 = 8{,}200\ \mathcal{M}$. At the same time, the kilometrical transport costs on the rays *CP* and *AP* remain the same, respectively 7,000 and 8,000 $\mathcal{M}$. Hence, pole *O* is to be found starting from line *BD* (named *d*), if two arcs are drawn from point *B* with a length of $\frac{D}{E} d = \frac{6{,}600}{8{,}200} \times 4 = 3.22$ and from point *D* with a length of $\frac{B}{E} d = \frac{7{,}600}{8{,}200} \times 4 = 3.71$. Pole *O* is to be found at the intersection of the arcs and is at a distance of $e = 11.6$ kilometres from *A*. In order to now find pole *Q*, circle arcs are drawn from *O* with a length of $\frac{A}{C} e = \frac{8{,}000}{7{,}000} \times 11.6 = 13.24$ and from *A* with a length of $\frac{E}{C} e = \frac{8{,}200}{7{,}000} \times 11.6 = 13.59$. In the solution, the individual straight lines amount to the lengths of $CP = 7.4$, $AP = 7.0$, $EP = 1.5$, $BE = 2.0$, and $DE = 3.1$ kilometres.

It is to be pointed out, however, that a second deviating solution also exists, if, according to FIGURE 7, the search of pole *O* starts from

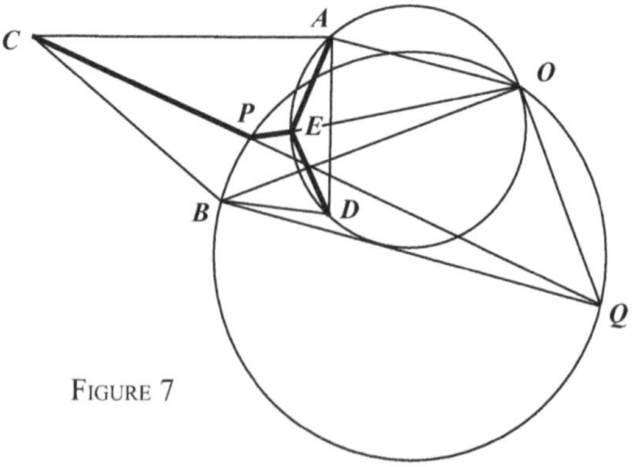

FIGURE 7

points *D* and *A* and location *P* is determined thereupon for the remaining points *C*, *B*, and *O*.

In most cases, it can be concluded immediately which of the possible solutions is the best; in ambiguous cases, both solutions are to be constructed and the costs are to be compared.

Probably it is not necessary to mention that for any other number of points it is possible to proceed in the described way – by gradually substituting each two points by their respective pole.

Finally, the question needs to be considered which deviations for the most expedient location result from the utilization of already existing traffic routes (roads, railroads, or waterways). For this purpose, a general consideration is to be preceded.

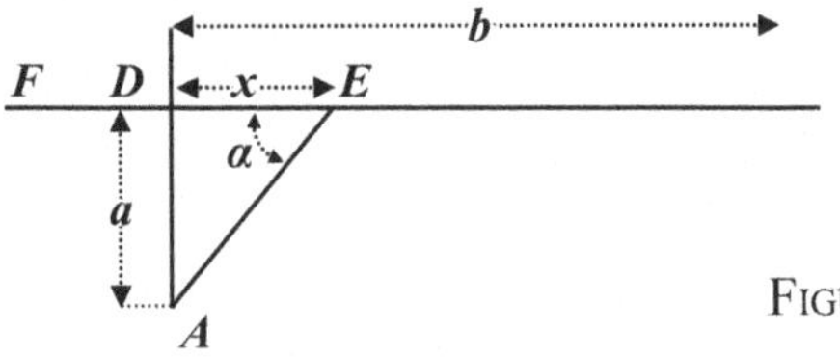

FIGURE 8

It is assumed that a transport from point *A* to point *G* (see FIGURE 8) is to take place. In the direction of *F* a traffic route already exists. In this case it will probably not be favourable to construct a new traffic route from *A* to *G* or to do the transport from *A* to *G* on a route without a developed route infrastructure. Instead, one will try to reach the existing route *FG* via any point *E* and continue the transport on the existing route. The location of point *E* is determined through the following calculation.

If the right-angled distance between point *A* and straight line *FG* = *AD* = *a* and the distance *DG* = *b*, if moreover the kilometrical transport costs on the access route is *AE* = *A* and the costs on the existing route *EG* = *E*, then the costs for a transport from *A* via *E* to *G* are calculated by

$$S = A\,(a^2 + x^2)^{\frac{1}{2}} + E\,(b - x).$$

The condition for the cost minimum is then:

$$\frac{dS}{dx} = \frac{Ax}{(a^2 + x^2)^{\frac{1}{2}}} - E = 0$$

or, since $\frac{x}{(a^2 + x^2)^{\frac{1}{2}}} = \cos\alpha$, the minimum condition is expressed by:

$$\cos\alpha = \frac{E}{A} \qquad (10)$$

This simple equation for the direction through which the access route has to lead is succeeded by a further relation. If a straight line *GHJ* is drawn from point *G* including an angle of 90°– *α* (see FIGURE 9), the extension of the access route *AE* will hit this straight line *GHJ* rectangularly in point *H*. In this consideration the rectangular distance of *AH* to straight line *GHJ* – which may be called "**access front**"

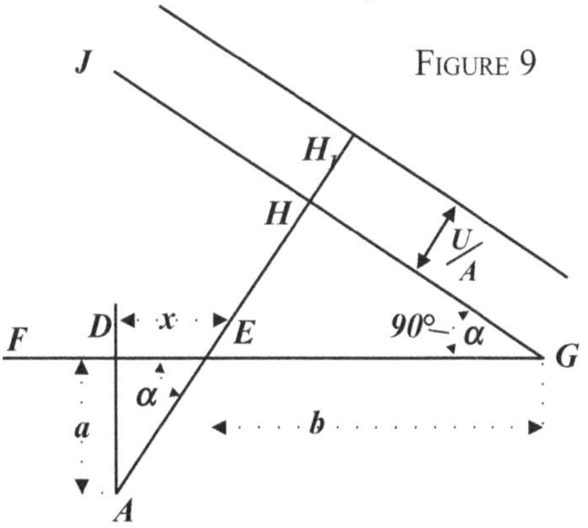

FIGURE 9

provides a measure for the transport costs between *A* and *G*.

$$S = A\ (a^2 + x^2)^{½} + E\ (b - x)$$

or

$$S = A\ [(a^2 + x^2)^{½} + \frac{E}{A}(b - x)]$$

and since $\frac{E}{A} = \cos\alpha$ and furthermore $\cos\alpha\,(b - x) = \overline{EH}$:

$$S = A\ [(a^2 + x^2)^{½} + \overline{EH}]\,,$$

that is:

$$S = A \times \overline{AH}$$

Hence, the access to a cheaper traffic route must be positioned rectangularly against the access front, whereas the total transport costs would be as high as if the access route to the access front had to be covered.

If the transition to the more advantageous traffic route entails transshipment of goods in *E*, the costs of which amount to the cost parameter *U* per ton, then a parallel translation of the access front by the measure U/A has to take place and the transport costs are quantified by the distance *AH* (see FIGURE 9).

FIGURE 10

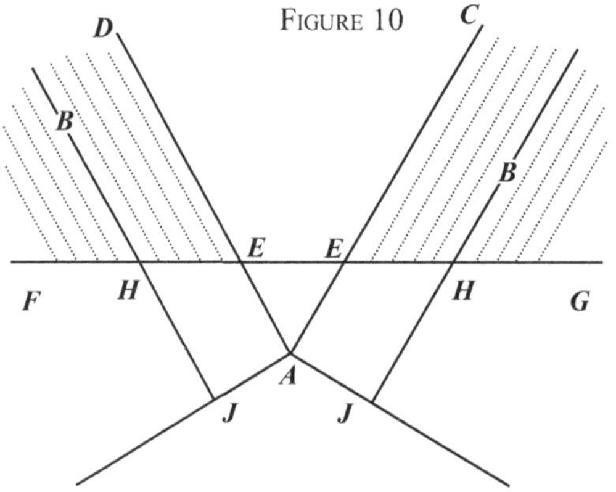

If there is no transshipment required upon transition to the cheaper traffic route *FE*, then the traffic from *A* to all locations *B*, which lie between the extended access routes *EC* and *EG* or *DE* and *FE* (see FIGURE 10) will not be conducted via the direct path in the direction of *AB*, but via the combined line course *AEHB*. In this case the costs would be as high as if path *BHJ* with the kilometrical transport costs of the access routes had to be covered. The straight lines *JA* are access fronts which are directed rectangularly against the access routes *AE*.

On the other hand, if there is transshipment necessary upon transition to the cheaper traffic route in *E* and upon leaving this route in *H* with transshipment costs of *U* per ton, then the area that is expedient for the transition to the cheaper traffic route is restricted by a parabola E_1B (see FIGURE 11). This parabola

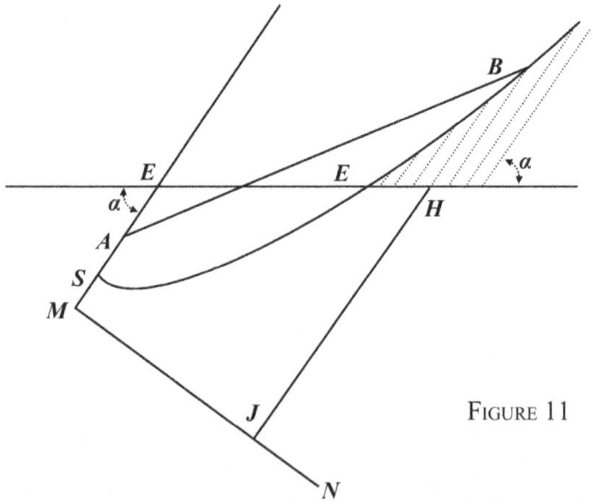

FIGURE 11

is determined in the following way. The straight line *AE* is extended beyond *A* by the measure AM = 2U/A and the access front MN is drawn rectanularly from *M* to *EAM*. If furthermore *AM* is bisected in *S* and a parabola is constructed based on the parameter *AM* with *S* being the vertex and *ME* being the axis, then for any point *B* of the parabola it applies that the ray *BA* equals the straight line *BJ* that is rectangularly drawn toward the access front. For all points that lie between the line E_1G and the parabola E_1B traffic to A will not be conducted via the straight connection *BA*, but via the combined line course *BHEA*.

For the current purpose there are no further considerations regarding this type of investigations that need to be made. Rather we can return to determining the most expedient location of the production site through using the existing traffic routes.

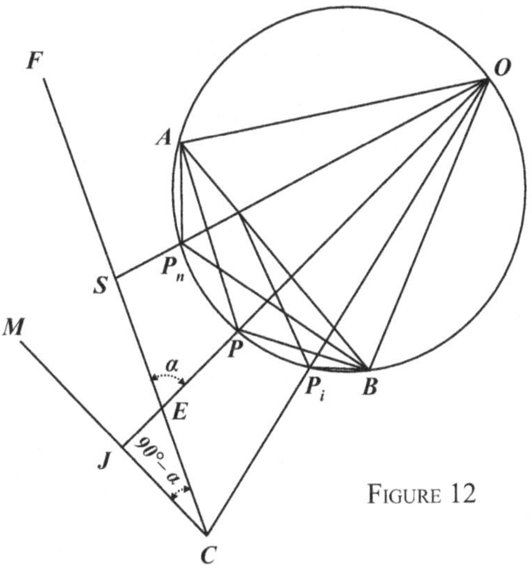

FIGURE 12

Be *FC* (in FIGURE 12) a road leading to utilization site *C*. On this road one ton-kilometre can be transported at a rate of 0.2 $\mathcal{M}$. Now a highway is to be constructed from production site *P* to *C* or for the connection to the road in *E*. In this situation, the costs of transporting a goods volume of 32,000 tons on the finished route from *E* to *C* to *E* = 0.2 × 32,000 = 7,000 $\mathcal{M}$. On the connection route *EP* (that is to be constructed) transport costs amount to 0.2 × 32,000 + 600 $\mathcal{M}$ = 6,400 $\mathcal{M}$, if the interest of the investment capital amounts to 600 $\mathcal{M}$. The connection angle α must therefore be: $\alpha = \arc\cos\frac{6{,}400}{7{,}000}$. If, according to the previously stated method, the pole *O* was determined for the locations for *A* and *B* and, moreover, the access from *MC* is drawn with the angle $90°-\alpha$, then a straight line would have to be drawn from pole *O* rectangularly towards the access front in order to obtain point *P* as the location and point *E* as the connection to the existing road. If the existing traffic road *FC* was a waterway which only requires 0.02 $\mathcal{M}$ transport costs per ton-kilometre, then the connection angle would have to be less acute and would have to be $\alpha = \arc\cos\frac{640}{7{,}000}$. However, in this case the transshipment costs would have to be added and in a comparative calculation it would have to be examined if generating a direct access from *O* to *C* by relocating the plant to P_1 would not be cheaper than utilizing a waterway.

Finally, if the existing traffic route was a railroad then the most expedient access to this route would have to be found in the same way. However, then the access route would have to be accordingly translated to the next existing railway station *S*. Also, in this instance a comparison with the direct connection of the access route on *C* thus avoiding transshipment costs would need to be undertaken.

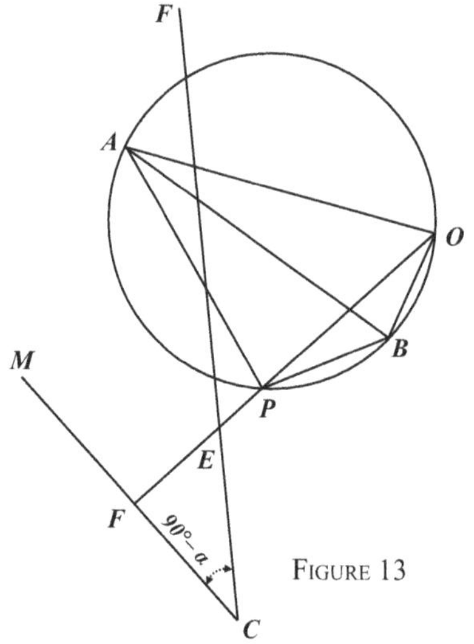

FIGURE 13

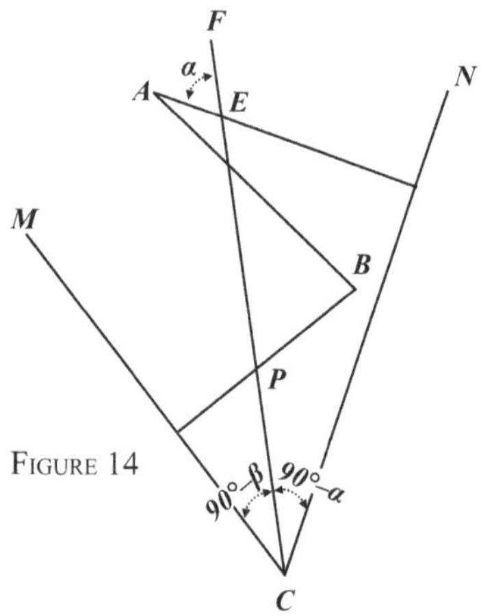

FIGURE 14

If the two locations *A* and *B* lay on opposite sides of the existing traffic route *FC*, as shown in FIGURE 13, then it could occur the case that the transport route *AP* cuts the route *FC* with such an acute angle that a transition of the traffic from *A* to route *FC* is advantageous. In this case, the entire solution alters. The traffic that previously transited from *A* to line *FC* will not leave this route again, as easily conceivable. Moreover, also the traffic from *B* will have to be connected to this route and the location must also lie on this latter access point. Therefore, the access front *CN* for the traffic from *A* would have to be drawn starting from point *C*, as stated in FIGURE 14. The same is then to be done with the access front *CM* for the traffic from *B*; it also had to start from *C*. Thereupon the access paths *AE* and *BP* would be directed towards the access fronts in a right-angled way thus resulting in *P* as the location of the production site.

At this stage it appears to be superfluous to go into the detail of still further alternatives.

If the extraction sites of the raw materials and auxiliary materials are in long distance to each other, transport planning is dependent on the existing railroads and waterways. In this case, by considering the network density of the railway system, it is adequate – at least for the initial attempt – to determine the most expedient location of the production site on the assumption of a consistent passability of terrain in all directions. This may be still illustrated by a further example.

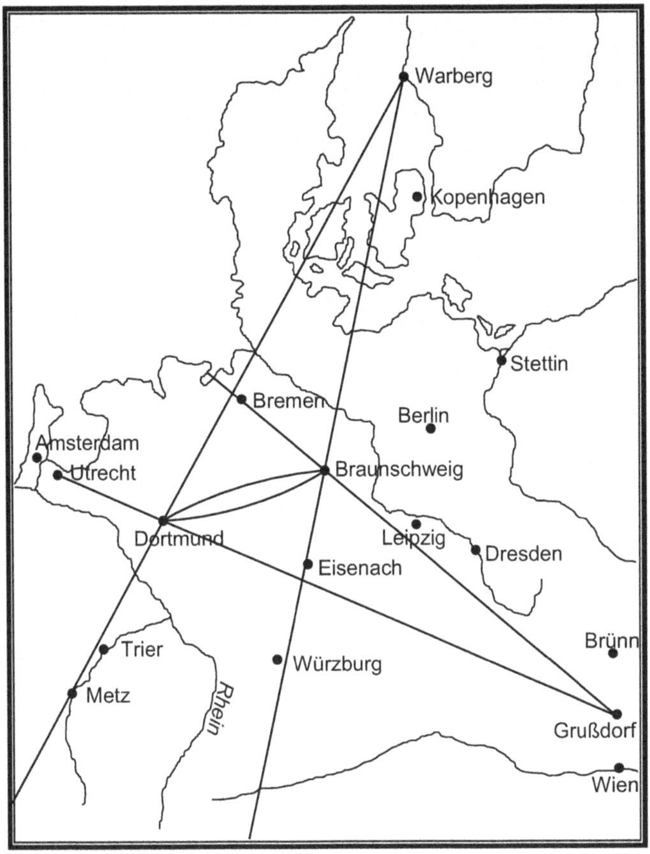

FIGURE 15

Brunswick (Braunschweig) be the location where iron ore is extracted; its smelting is carried out with coals procured from Dortmund. If for ore, coal and pig iron a quantity proportion of 100:80:32 is assumed like previously discussed, then the kilometrical transport costs comply with these figures, since all three goods belong to the same railway tariff class. For the determination of the pole, arcs are to be plotted from Brunswick to Dortmund (both places being in 225 kilometres air-line distance from each other). The arc from Brunswick has a length of

$\frac{80}{32} \times 225 = 562\ ½\ \text{km}$; the one from Dortmund amounts to $\frac{100}{32} \times 225 = 703\ ⅛\ \text{km}$.

A southern pole is to be found at the station of Grußdorf, situated at the railroad line Vienna (Wien) – Brno (Brünn), 70 kilometres north of Vienna. This pole functions as the target point for all sourcing sites of pig iron that lie between the pole rays Brunswick – Bremen and Dortmund – Utrecht.

The northern pole for the extraction sites Brunswick and Dortmund is to be found at Warberg, situated at the Norwegian coast at the Kattegat, approximately 170 kilometres north of Copenhagen. This pole is the target point for those utilization sites of pig iron that lie between the pole rays Brunswick – Eisenach – Würzburg and Dortmund – Trier – Metz. For all shipping locations east of the line Bremen – Brunswick and Brunswick – Eisenach – Würzburg, Brunswick should be chosen as the location of the blast furnace. In contrast, for all utilization sites that lie between the rays Utrecht – Dortmund and Dortmund – Trier – Metz, the smelting of the ore should take place in Dortmund.

The northern location circle touches upon the places Dortmund – Hamm – Gütersloh, Hameln, Nordstemmen, Hildesheim, Brunswick. The southern location circle leads from Dortmund via Soest, Driburg, Alfeld and Derneburg to Brunswick. If the utilization site of the pig iron would lie within both of these circle arcs, then the iron smelting would have to take place there. If the pig iron was destined for an ironworks near the utilization site in Osnabrück, then the blast furnace would have to be situated on that point, where the line Osnabrück – Grußdorf intersects the northern location circle. This point would then be close to Bielefeld. With regard to the existing railway connections, the nearby railway junction Löhne would be suitable.

This example emphasizes how much the cost of the production is affected by the distance to the extraction sites of the raw materials and auxiliary materials. The airline distance from Brunswick to Dortmund amounts to 225 km. If iron ore and coal were available next to each other at the pole Grußdorf, which is 700 km away from Dortmund, then it would be possible to obtain the pig iron cheaper than from the site in Brunswick – Dortmund. Even from north of the line Brunswick – Dortmund, the iron from Grußdorf would only cost the same price.

The method for the determination of the most expedient location of an industrial plant represents, as it can finally be pointed out, the application of the theory for a commercial rail track routing, which was originally published in 1872 in the journal of Hanoverian Architects and Engineers' Society in which I have previously elaborated on this subject.

G. Clarke/J.W. Wright (1964)

SCHEDULING OF VEHICLES FROM A CENTRAL DEPOT TO A NUMBER OF DELIVERY POINTS

Decisions about the location of facilities ("nodes") are one fundamental task in building logistical networks, which economists and mathematicians started to tackle in the 19th century already (see the article by Launhardt (1882) in this volume). Decisions relating to the assignment of vehicles, orders, possible crews and container equipment to routes between the nodes of a network, are a fundamental task of network operations.

The development and application of mathematical tools for finding optimal solutions to the rather complex, dynamic issues of network operations started in the 1960s only. One of the pioneering contributions was G. Clarke and J.W. Wright's development of the "savings algorithm" – introduced in their paper *Scheduling of Vehicles from a Central Depot to a Number of Delivery Point* of 1964, which is reprinted here.

The savings algorithm is based on a comparison of serving paired customers compared with serving individual customers from a central depot. This is analysed for varying pairs of customers and leads to savings being the higher the longer the customers' distance from the depot and the closer both customers are to each other.

With this algorithm Clarke and Wright introduced the first systematic solution method for the vehicle routing problem. Many other contributions followed to solve more complex scheduling and routing problems and provide alternative solution approaches, such as the sweep algorithm (proposed by Gillett/Miller in 1974, also Fisher/Jaikumar (1981)).

Reference:

G. Clarke; J.W. Wright: Scheduling of Vehicles from a Central Depot to a Number of Delivery Points. In: *Operations Research*, Vol. 12, 1964, pp. 568–581.

Further reading:

Fisher, Marshall L. and Ramchandran Jaikumar: A Generalized Assignment Heuristic for the Vehicle Routing Problem. In: *Networks*, Vol. 11, No. 2, 1981, pp. 109–124.

Gillett, Billy E. and Leland R. Miller: Heuristic Algorithm for the Vehicle-Dispatch Problem. In: *Operations Research*, Vol. 22, No. 2, March-April 1974, pp. 340–349.

P. Klaus and S. Müller (eds.), *The Roots of Logistics*, DOI 10.1007/978-3-642-27922-5_18,

G. Clarke/J.W. Wright: Scheduling of Vehicles from a Central Depot to a Number of Delivery Points

The optimum routing of a fleet of trucks of varying capacities from a central depot to a number of delivery points may require a selection from a very large number of possible routes, if the number of delivery points is also large. This paper, after considering certain theoretical aspects of the problem, develops an iterative procedure that enables the rapid selection of an optimum or near-optimum route. It has been programmed for a digital computer but is also suitable for hand computation.

The paper is concerned with the optimum routing of a fleet of trucks of varying capacities used for delivery from a central depot to a large number of delivery points. The merchandise is homogeneous with respect to the unit of capacity. The shortest route between every two points in the system is given. It is desired to allocate loads to trucks in such a manner that all the merchandise is assigned and the total mileage covered is a minimum. The procedure given is simple but effective in producing a near-optimal solution and has been programmed for several digital computers. This truck dispatching (as a mathematical) problem was first formulated by Dantzig and Ramser,[1] who obtained a method of solution. Multiple demand and multiple truck capacity were considered as a further formulation. The formulation in this paper is essential similar, but a restriction was first imposed to meet a particular practical application for which the method was being developed. Their method has been developed and a new solution found.

FORMULATION

A number of trucks x_i of capacity C_i $(i = 1...n)$ are available and loads q_j are required to be delivered to points P_j $(j = 1...M)$ from a depot P_0. Given the distance $d_{y,z}$ between all such points it is required to minimize the total distance covered by the trucks.

For convenience of computation C_i are ordered such that $C_{i-1} < C_i$ $(i = 2...n)$ and it is assumed that

---

1 Cf. George B. Dantzig and J.H. Ramser: The Truck Dispatching Problem, in: *Management Science*, Vol. 6, 1959, pp. 80–91; George B. Dantzig and J.H. Ramser: Optimum Routing of Gasoline Delivery Trucks, in: *Proc. Fifth World Petroleum Cong.*, Section VIII, 1959, p. 19.

$$C_1 \ll \sum\nolimits_{j=1}^{j=m} q_j.$$

Since in the solution some trucks may only be partially loaded, x_1 needs to be sufficiently large to ensure that all loads are allocated. For purpose of computation it is therefore made infinite.

It might be noted that if $C_n \geq \sum\nolimits_{j=1}^{j=m} q_j$, then the problem becomes the traveling salesman problem.

COMMENTS ON THE DANTZIG AND RAMSER METHOD

This method only considers the state $C_2 = C_3 = \ldots C_n = 0$.

Due to the restriction that, in the first of N stages, only customers whose combined load does not exceed $C_1/2^{N-1}$ are permitted to be linked, points may be linked that are far apart, and may be virtually on opposite ends of a straight line through the depot. Although obviously long links may be excluded in the initial stages by "rapid corrections", when two points become linked in an "aggregation" they remain aggregated.

As a result, this method tends to lay more emphasis on filling trucks to near capacity than on minimizing distance. The distance table in the Nth stage could require each cell to contain the shortest distance from the depot through 2^N points, i.e., the "traveling salesman problem" must be solved; this can be extremely time consuming for only a few points, e.g. supposing after two stages 100 customers are aggregated into 25 groups of four customer each, then the mileage table for the next stage requires the solution of 300 traveling salesman problems each of nine points. These can of course be approximated too graphically.

A MODIFIED PROCEDURE BASED ON THE DANTZIG AND RAMSER METHOD

If the restriction is removed that in the rth stage aggregations, only customers whose combined load does not exceed $C_1/2^{N-r}$ may be joined, the "traveling salesman problem" will still be encountered, but it is now permissible in each stage to join any two points whose combined load does not exceed C_1. This method has been found to give better results than the Dantzig and Ramser method in a number of cases tested. The relative merit of the two methods depends on the variability of the customer loads q_j. In the numerical example cited in reference 1 the original method gives 294 units, while this method gives 312 units.

This led the authors to seek another method of solution.

THEORETICAL ASPECTS OF THE PROBLEM

Consider a feasible allocation of trucks to loads. In all cases each customer point will be linked to two other points, one or both of which may be the depot P_0. Consider the two points P_y and P_z. Let the two points linked to P_y be $P_{y\pm 1}$ and similarly for P_z. The effect of linking P_y and P_z will be calculated. It is assumed that P_y and P_z are on separate "runs" from P_0. If they are on the same "run" the same considerations apply except that one of the four cases is not permissible.

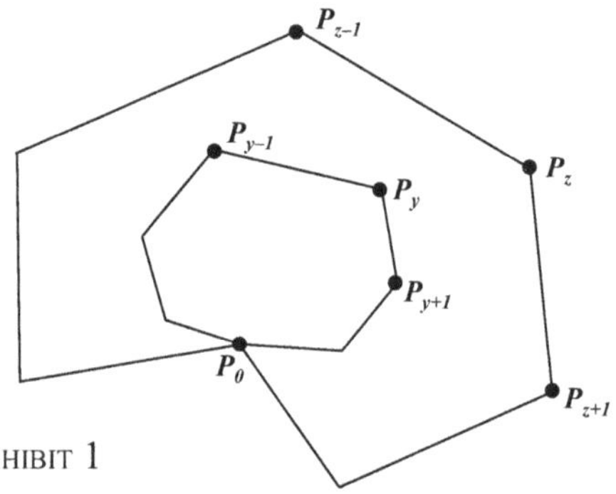

EXHIBIT 1

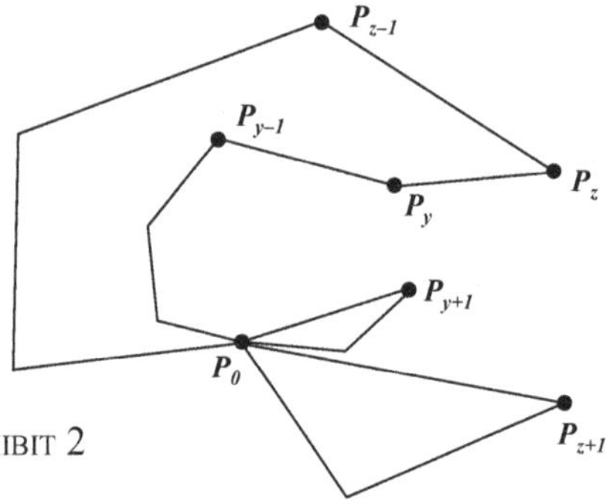

EXHIBIT 2

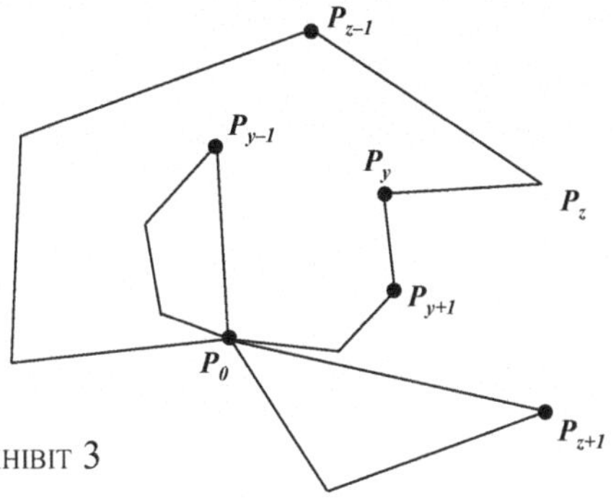

EXHIBIT 3

EXHIBIT 1 shows the positions of P_y and P_z in the feasible allocation. EXHIBITS 2, 3, 4, 5 show the four possible decompositions of these runs caused by joining P_y and P_z. These consist of the severing of links $P_{y-1}\,P_y$ or $P_y\,P_{y+1}$ with the severing of links $P_{z-1}\,P_z$ or $P_z\,P_{z+1}$. The distances saved by each of these decompositions are as follows:

(2) $d_{y,y+1} - d_{0,y+1} + d_{z,z+1} - d_{0,z+1} - d_{y,z}$,

(3) $d_{y-1,y} - d_{0,y-1} + d_{z,z+1} - d_{0,z+1} - d_{y,z}$,

(4) $d_{y,y+1} - d_{0,y+1} + d_{z,z-1} - d_{0,z-1} - d_{y,z}$,

(5) $d_{y-1,y} - d_{0,y-1} + d_{z,z-1} - d_{0,z-1} - d_{y,z}$.

These four "savings" are calculated for each pair of customers.

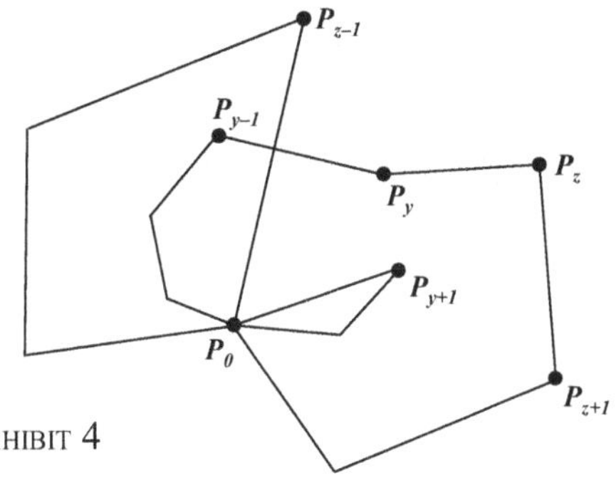

EXHIBIT 4

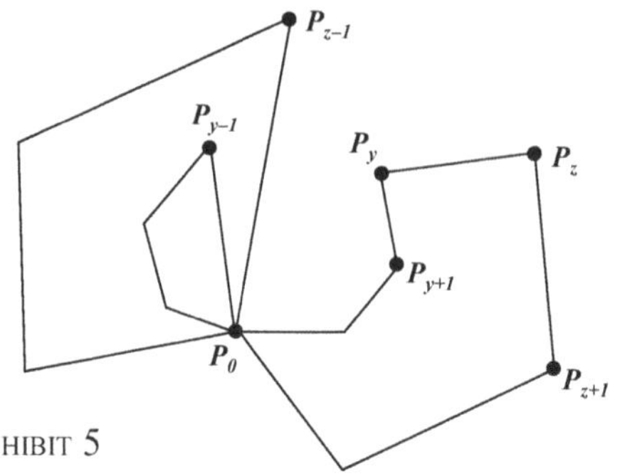

EXHIBIT 5

The maximum of these is selected that would, if linked, produce feasible routes consistent with truck availabilities and capacities. These two customers are now linked and the "savings" recalculated. In linear programming terminology this method is equivalent to allotting two "shadow costs" to a customer, the shadow cost for customer P_y being $d_{y-1,y} - d_{0,y-1}$ and $d_{y,y+1} - d_{0,y+1}$. The four "evaluations" for cell $(y{:}z)$ are then enumerated above. When a link is severed, the appropriate "shadow cost" is reduced by the value of the cell causing the severance. Since this cell value is a maximum value, whenever a point is linked to two others (not P_0) all its cell values become negative, and this point will not be considered again for linking.

As a result of this, the only links that will be severed will be those of points linked to P_0, if these were involved in the linking of cell $(y{:}z)$, the saving would be $d_{0,y} + d_{0z} - d_{yz}$.

TABLE 1

Q	P_0												
1200	2 9	P_1											
1700	2 14	18 5	P_2										
1500	2 21	18 12	28 7	P_3									
1400	2 23	10 22	20 17	34 10	P_4								
1700	2 22	10 21	20 16	22 21	26 19	P_5							
1400	2 25	10 24	16 23	16 30	20 28	38 9	P_6						
1200	2 32	10 31	20 26	26 27	30 25	44 10	50 7	P_7					
1900	2 36	10 35	20 30	20 37	24 35	42 16	50 11	58 10	P_8				
1800	2 38	10 37	16 36	16 43	20 41	38 22	50 13	54 16	68 6	P_9			
1600	2 42	10 41	20 36	32 31	36 29	44 20	50 17	64 10	72 6	68 12	P_{10}		
1700	2 50	10 49	20 44	34 37	42 31	44 28	50 27	64 18	72 14	76 12	84 8	P_{11}	
1100	2 52	10 51	20 46	34 39	46 29	44 30	50 27	64 20	72 16	70 20	84 10	92 10	P_{12}

COMPUTATIONAL PROCEDURE

It is assumed that the values of q_j are such that an initial allocation of one vehicle to each customer is possible. If this is not true it is assumed that an allocation can be made by splitting a load into two (or more) full truckloads of the highest capacities available and only considering the remainder of that load, an amount less than a truckload of the highest capacity. E.g., q_j = 1700 gallons and the available trucks are one of 700 gallons, two of 600 gallons, three of 500 gallons then the 700 gallon and the 600 gallon trucks would be allocated to the customer and 400 gallons used as the value of q_j with an availability now of three 500 gallon trucks. Hence $q_j \leq C_n$ ($j = 1...M$).

For hand computation it is recommended that half matrix be used. The numerical example used here is the same as that used in Danzig/Ramser (The Truck Dispatching Problem (1959), see reference 1). In TABLE 1 the entries in the lower right-hand corner of each cell (y:z) are the appropriate distances $d(P_y:P_z)$ by the shortest practicable route. The entries in the lower left-hand corner of each cell are the "savings". For cell (y:z) with $y,z \geq 1$ and the $y \neq z$ this value is $d_{0,y} + d_{0,z} - d_{y,z}$. A column vector $Q = (Q_1...Q_M)$ is added on the left-hand side of the matrix. Initially this consists of the loads q_j required by customer P_j ($j = 1...M$). The entry in the middle of cell (y:z) is $t_{y,z} = 1$ if the two customers P_y and P_z are linked on a truck's route; otherwise $t_{y,z} = 0$ for $y,z > 0$. If a customer is served exclusively by a truck $t_{y,0} = 2$. The following relation always exists:

$$\sum_{z=0}^{y-1} t_{y,z} \sum_{z=y+1}^{z=M} t_{y,z} = 2...(A).$$

The initial basic solution is now entered as $t_{y,0} = 2$ ($y=1...M$).

Trucks	Up to 4000 gal	Over 4000 gal	Over 5000 gal	Over 6000 gal
Available	∞	7	4	0
Allocated	12	0	0	0

TABLE 2

TABLE 2 is a table showing the number of available trucks above each capacity level and the number of trucks already allocated.

In the numerical example shown, it is assumed that there is an unlimited supply of trucks of capacity 4,000 gallons, 3 trucks of capacity 5,000 gallons, and 4 trucks of capacity 6,000 gallons.

TABLES 1 and 2 show the initial feasible solution.

The rows and columns of the half matrix are searched for the maximum "saving", subject to the conditions that if this occurs in cell (y:z):

(I) $T_{y,0}$ and $t_{z,0}$ must be greater than zero.

(II) P_y and P_z are not already allocated on the same truck tun.

(III) Amending TABLE 2 by removing the trucks allocated to leads Q_y and Q_z and adding a truck to cover the load $Q_y + Q_z$ does not cause the trucks allocated to exceed the trucks available in any column in TABLE 2.

TABLE 3

Q	P_0												
1200	2	P_1											
1700	2	18	P_2										
2900	1	18	28*	P_3									
2900	1	10	20	1 34	P_4								
1700	2	10	20	22	26	P_5							
5100	1	10	16	16	20	38	P_6						
5600	1	10	20	26	30	44	50	P_7					
—		10	20	20	24	42	1 50	58	P_8				
5100	1	10	16	16	20	38	50	54	1 68	P_9			
5600	1	10	20	32	36	44	50	64	72	68	P_{10}		
—		10	20	34	42	44	50	1 64	72	76	84	P_{11}	
—		10	20	34	46	44	50	64	72	70	1 84	1 92	P_{12}

* Indicates maximum saving satisfying all conditions (I), (II), and (III) for use in next iteration

Trucks	Up to 4000 gal	Over 4000 gal	Over 5000 gal	Over 6000 gal
Available	∞	7	4	0
Allocated	4	2	2	0

TABLE 4

If these conditions hold, $t_{y,z}$ is made equal to 1 and other values of $t_{i,j}$ amended subject to relation (A). The vector Q is amended by firstly making all Q_j zero where $t_{j,0}$ is zero and making Q_j equal to the total load on the "run" for all other j. This completes the first iteration.

TABLE 5

Q	P_0												
5800	1	P_1											
—		1	P_2										
—			1	P_3									
5800	1			1	P_4								
1700	2					P_5							
5100	1						P_6						
5600	1							P_7					
—							1		P_8				
5100	1								1	P_9			
5600	1										P_{10}		
—								1				P_{11}	
—											1	1	P_{12}

Trucks	Up to 4000 gal	Over 4000 gal	Over 5000 gal	Over 6000 gal
Available	∞	7	4	0
Allocated	1	3	3	0

TABLE 6

If there are two or more equal maxima in the search it is suggested that one of these be selected randomly. The procedure is repeated until no more links are possible. TABLES 3 and 4 show an intermediate stage in the computation. TABLES 5 and 6 show the final solution. The routes are as follows: $P_0P_1P_2P_3P_4P_0$, $P_0P_6P_8P_9P_0$, $P_0P_{10}P_{12}P_{11}P_7P_0$, giving a total distance of 290 units, believed by Dantzig and Ramser to be optimum.

Although the improvement in this example is slight, in an example with 30 customers an improvement of 17 per cent on the earlier method was obtained. The results of this example with the initial mileage matrix are given in the appendices to this paper.

While the solution does give the order of visiting the customers it may be beneficial to solve the traveling salesman problem for each truck in the final allocation to obtain the true optimum order of visiting.

Practical limitations such as certain customers only accepting certain truck sizes or types and other priority treatments can be incorporated into the computation without much difficulty. Details of some of these restrictions together with computational methods for a digital computer will be found in a Case Study which will be published shortly.

Appendix I

MIDLANDS II

Proposed Runs for December 4, 1961, using Dantzig's Method

Code No.	Society	T	Cwt	Load		Mileage
				T	Cwt	
1	Stoke	1	4			
2	Burslem	1	14	2	18	82
3	Barrow on Soar		11			
4	Mount Sorrel		15			
5	Fleckney		11			
6	Stoughton		1			
7	Huncote		3			
8	Dudley	1	9			
9	Sapcote		6			
10	Tenacres	1	5	5	1	238
11	Whetstone		6			
12	Tamworth	1	5			
13	Enderby		2			
14	Nuneaton	1	8			
15	Broughton Astley		8			
16	Cosby		10	3	19	201
17	Silverdale		18			
18	Walsall	2	5			
19	Birmingham	1	13			
20	Stafford		17			
21	Market Drayton		9	6	2	183
22	Shepshed		16			
23	Coalville	1	15			
24	Loughborough		5			
25	Wolverhampton	3	0	5	16	201
26	Coventry	4				
27	Lockhurst Lane	1	19	5	19	197
28	Melton Mowbray	4	15	4	15	186
29	Leicester	4	10	4	10	178
30	Oakengates	6	3	6	3	136
19	Birmingham	7	0	7	0	164
		Total				1,766

MILEAGE TABLE

		0	1	2	3	4	5	6	7	8	9	10	11	12	13	14	15	16	17	18	19	20	21	22	23	24	25	26	27	28	29	30
0	Newton Heath																															
1	Stoke	41																														
2	Burslem	38	3																													
3	Barrow	80	54	56																												
4	Mt. Sorrel	80	54	56	3																											
5	Fleckney	97	64	67	19	16																										
6	Stoughton	92	59	62	13	10	7																									
7	Huncote	96	56	59	16	14	11	10																								
8	Dudley	78	39	41	54	54	53	57	48																							
9	Sapcote	98	59	62	20	17	12	13	4	39																						
10	Tenacres	87	52	50	47	46	46	42	35	12	33																					
11	Whetstone	95	58	61	15	12	8	8	3	45	6	39																				
12	Tamworth	77	38	41	30	29	34	32	25	24	21	18	28																			
13	Enderby	93	55	58	15	12	10	14	3	47	7	39	3	26																		
14	Nuneaton	91	52	53	25	22	24	19	12	30	9	24	15	14	14																	
15	Broughton	98	58	61	19	16	10	10	4	42	3	36	4	24	6	12																
16	Cosby	96	59	62	17	14	8	8	4	44	5	38	2	26	4	14	2															

		0	1	2	3	4	5	6	7	8	9	10	11	12	13	14	15	16	17	18	19	20	21	22	23	24	25	26	27	28	29	30
17	Silverdale	40	45	5	60	61	71	65	63	40	66	49	65	40	62	54	65	66														
18	Walsall	73	34	37	46	46	50	46	39	8	39	12	43	16	41	28	42	44	36													
19	Birmingham	82	48	46	44	44	45	42	33	9	31	4	36	18	36	21	34	35	45	9												
20	Stafford	55	16	19	54	54	58	55	48	22	45	30	53	24	51	38	48	50	18	22	31											
21	Market Drayton	52	16	17	68	68	77	72	69	36	65	46	71	44	68	57	67	69	14	36	45	21										
22	Shepshed	76	46	49	8	9	19	15	18	44	22	40	16	20	17	24	20	18	52	37	35	45	59									
23	Coalville	76	44	46	11	11	20	15	15	42	19	36	18	18	14	18	20	18	47	35	33	39	57	6								
24	Loughborough	76	50	53	4	4	20	14	18	50	21	43	17	25	16	28	20	19	57	41	40	50	64	5	10							
25	Wolverhampton	72	33	34	53	54	57	55	47	6	45	15	49	22	49	34	46	48	34	6	15	16	30	47	42	50						
26	Coventry	98	58	61	33	30	27	30	21	27	15	18	19	19	21	8	17	19	60	26	18	44	61	34	28	35	32					
27	Lockhurst Lane	98	58	61	32	29	26	29	20	28	15	20	18	19	20	7	16	18	60	26	19	44	61	34	28	34	34	3				
28	Melton Mowbray	93	66	68	14	12	23	18	22	65	25	54	20	41	20	32	24	22	72	57	54	61	79	20	26	15	64	39	39			
29	Leicester	89	55	58	10	9	8	5	7	48	10	39	5	29	5	18	9	7	62	44	39	51	69	15	12	11	52	24	23	15		
30	Oakengates	68	32	33	64	64	67	66	57	22	55	38	63	34	60	47	58	60	32	26	33	21	18	66	53	60	18	51	52	76	65	

Appendix II

MIDLANDS II

Proposed Runs for December 4, 1961, using proposed method with 7-ton Capacity

Code No.	Society	T	Cwt	Load		Mileage
				T	Cwt	
2	Burslem	1	14			
1	Stoke	1	4			
23	Coalville	1	15			
22	Shepshed	0	16	5	9	167
12	Tamworth	1	5			
14	Nuneaton	1	8			
10	Tenacres	1	5			
19	Birmingham	1	13			
8	Dudley	1	9	7	0	206
17	Silverdale	0	18			
27	Lockhurst Lane	1	19			
26	Coventry	4	0	6	17	201
20	Stafford	0	17			
25	Wolverhampton	3	0			
18	Walsall	2	5	6	2	150
24	Loughborough	0	5			
4	Mount Sorrel	0	15			
28	Melton Mowbray	4	15			
3	Barrow on Soar	0	11	6	6	186
6	Stoughton	0	1			
5	Fleckney	0	11			
11	Whetstone	0	6			
16	Cosby	0	10			
15	Broughton Astley	0	8			
9	Sapcote	0	6			
7	Huncote	0	3			
13	Enderby	0	2			
29	Leicester	4	10	6	17	215
30	Oakengates	6	3			
21	Market Drayton	0	9	6	12	138
19	Birmingham	7	0	7	0	164
		Total				1,427

ACKNOWLEDGMENT

The directors of the Cooperative Wholesale Society Limited are thanked for their permission to publish this work.

REFERENCES

[1] G.B. Dantzig and J.H. Rameser, The Truck Dispatching Problem, *Management Science* 6, 80–91 (1959)

[2] G.B. Dantzig and J.H. Ramser, Optimum Routing of Gasoline Delivery Trucks, *Proc. Fifth World Petroleum Cong.*, Section VIII, p. 19 (1959)

Arnoldo C. Hax/Harlan C. Meal (1973)

HIERARCHICAL INTEGRATION OF PRODUCTION PLANNING AND SCHEDULING

Next to location planning (cf. Launhardt (1882) in this volume) and routing (cf. Clarke/Wright (1964) in this volume) the fundamental issues in running logistical systems are: Setting the quantitative levels of production capacities, lot sizes, optimal inventories and their dynamic interactions. Order taking, production, and delivery operations need to be scheduled along time lines.

First efforts and classical contributions in dealing with these kinds of problems were, for example, Wagner/Whitin's (1958) method for determining optimal lot sizes. Finding appropriate solutions for more integrative Planning and Scheduling System problems in the context of modern, often multi-location industrial production networks, is becoming extremely complex.

Hax/Meal's 1973 paper on Hierarchical Integration of Production Planning and Scheduling (and follow-up work such as Bitran/Hax (1977), as well as the parallel developments of MRP systems (cf. Wallace (1985)) and constraint-based industrial planning systems such as Goldratt's (1984) and Wiendahl's (1995) are pioneering contributions and a basis for the development of advanced Planning and Scheduling systems, which have become core elements now of the ubiquitous computer-based Enterprise Resource Planning systems.

Reference:

Arnoldo C. Hax and Harlan C. Meal: *Hierarchical Integration of Production Planning and Scheduling*. Published as Sloan School of Management Working Paper No. 656-73, Massachusetts Institute of Technology (MIT), 1973, pp. 1–25.

Further reading:

Bitran, Gabriel R. and Arnoldo C. Hax: On the Design of Hierarchical Production Planning Systems. In: *Decision Science*, Vol. 8, No. 1, 1977, pp. 28–55.

Goldratt, Eliyahu M.: *The Goal. A Process of Ongoing Improvement.* Croton-on-Hudson: North River Press, 1984.

Wagner, Harvey M. and Thomas M. Whitin: Dynamic Version of the Economic Lot Size Model. In: *Management Science*, No. 5, 1958, pp. 89–96.

Wallace, Thomas F.: *MRP II: Making it Happen.* Essex/NH: Oliver Wight, 1985.

Wiendahl, Hans-Peter: *Load-oriented Manufacturing Control.* Berlin; Heidelberg: Springer, 1995.

P. Klaus and S. Müller (eds.), *The Roots of Logistics*, DOI 10.1007/978-3-642-27922-5_19,

Arnoldo C. Hax/Harlan C. Meal: Hierarchical Integration of Production Planning and Scheduling

ABSTRACT

This paper describes the development of a hierarchical planning and scheduling system for a multiple plant, multiple product, seasonal demand situation. In this hierarchical structure, optimal decisions at an aggregate level (planning) provide constraints for detailed decision making (scheduling).

Four levels of decision making are used: **first**, products are assigned to plants (using mixed-integer programming), making long-term capacity provision and utilization decisions; **second**, a seasonal stock accumulation plan is prepared (using linear programming), making allocations of capacity in each plant among product Types within which the products have similar inventory costs; **third**, detailed schedules are prepared for each product Family (using standard inventory control methods for items grouped for production since the Items in a Family share a major setup), allocating the type capacity among the product Families in the Type; **fourth**, individual run quantities are calculated for each Item in each Family, again using standard inventory methods.

The approximations used and the procedures developed are described in sufficient detail to guide a similar application. We also discuss the problems encountered in implementation and the approach used to resolve these problems. Finally, we estimate the costs and benefits of this system application.

I. INTRODUCTION

The objective of the present paper is to provide a framework for a hierarchical decision making approach in which the aggregate results of capacity planning provide constraints for detailed scheduling decisions. In order to illustrate, with specific examples, the design and implementation issues in such a hierarchical system, we describe the actual application of this approach to the development of a production planning and scheduling system for a firm producing many products in several plants.

We will first describe the characteristics of the production and distribution problems presented in the firm where the system design was carried out. Then, we will comment on the general nature of hierarchical planning and scheduling systems, and justify the specific approach followed. Subsequently, we will discuss the details of each of the components of the overall system. Finally, we will describe the implementation efforts, the difficulties encountered, how these were overcome and what costs and benefits can be derived from the system operation.

II. SITUATION

In order to protect the anonymity of this manufacturing firm we will describe it as a process manufacturing situation analogous in some ways to a chemical plant or a steel mill. There are some minor assembly operations; but for our purposes, these can be treated as though they were mixing operations in a batch process chemical plant.

1. **Multiple Plants.** There are four plants, geographically separated so as to service reasonably separated market areas in the north, south, east and west. The combination of manufacturing and transportation costs indicates that some items should be made in only one plant, some in two, and so on. The assignment of product to plants is an important problem faced by the firm.
2. **Seasonal Demand.** The customer demand for this set of products is seasonal, with three distinct seasonal demand patterns. Some products can be characterized as winter season and some as summer season, with significant differences in the size of these two markets. A third seasonal pattern is neither winter nor summer, but shows distinct variations through the year.
3. **Items Grouped for Production.** There are natural groupings of products in manufacturing. A major setup cost is incurred whenever a particular set of products is run, while minor setup costs are required in switching production from one of these products to another. Moreover, production of any of the group in a plant requires a fixed capital investment (tooling cost)

which then permits production of the remainder of the group without additional tooling cost.

4. **Level Production.** To a greater extent than is common in most manufacturing operations, there is a strong incentive to maintain a nearly level manufacturing rate. This arises from two primary considerations: a) The capital cost of equipment is very high compared with the cost of shift premiums for labor, and the system normally operates three shifts on a five-day week. Occasionally a sixth and a seventh day are worked. b) The labor union is very strong and demands that production levels be kept nearly constant throughout the year. Thus, even though there may be a tradeoff between investment in capital equipment and investment in seasonal inventories, the question is largely moot, since the labor union is strongly opposed to substantial swings in work force level.

The planning of inventories, particularly seasonal stocks, is difficult in this situation. There is a need to build seasonal stocks, but very often the time at which one should build them does not seem to be reasonable, given the immediate demand problems. Also, as in most seasonal stock situations, it is hard to decide in which items to build seasonal stocks.

The list of symptoms exhibited by the production and distribution activities in this company was almost a classic set:

1. Poor customer service
2. High inventory
3. High production cost.

The poor customer service was exhibited in two ways, which are related. Some promised shipping dates were missed on make-to-order items for which a specific promise was given to the customer. In addition to these, shortages of stock items had risen to a level which caused the Sales Department a good deal of concern. These two were related because attempts to correct the former led to expediting and short runs, which resulted in reduced production output and stock shortages.

While high inventory in the face of poor customer service seems to be an inconsistency, it is a common occurrence. Poor control leads to excessive seasonal stock accumulation in some items and shortages in others. The excess inventories cannot be used to improve the service in the short items.

The high production costs are primarily a consequence of runs which are uneconomically short with consequent high setup cost and low productivity associated with the high frequency of changeover. These, in turn, resulted from the high runout rate and the consequent need to produce a small amount of each of many items in order to clear up some of the backorders created. This situation often arises when a simple order point, order quantity system is used to control inventories in an environment characterized by limited capacity and strong seasonalities. At the beginning of the peak season, when the demand starts to increase, many items trigger the order point simultaneously, creating a surge in manufacturing load and thus forcing the normal order quantities to be reduced because of the limited production capacity available. At the end of the season, if no seasonal stock limits are built into the order quantity procedures, the last production run will exceed the demand requirement to the end of the season; this leads to an inventory which is inactive until the beginning of the new season.

The system to be described here was developed to help management solve these problems. The center of the difficulties seemed to lie in the planning of aggregate production levels, particularly in allocating available production capacity among several product types with differing seasonal demand patterns, and in the subsequent detailed scheduling of each item belonging to a product type. The result of these efforts is an integrated production planning and scheduling system.

III. STRUCTURE OF THE PLANNING AND SCHEDULING SYSTEM

We decided upon a hierarchical system, one which makes decisions in sequence, with each set of decisions at an aggregate level providing constraints within which more detailed decisions must be made. We did this because we found that we could not, with available analytic methods and data processing capability, develop an optimization of the entire system. However, even if the current state of the art allowed solution of a detailed integrated model of the production process we would have rejected that approach because it would have prevented management involvement at the various stages of the decision-making process. A model that facilitates overall planning can only become effective if it helps in establishing at the various organizational levels subgoals which are consistent with the management responsibilities at each level. The model should allow for corrections to be made to these subgoals by the managers at each level, and for coordination among the decisions made at each level. This is the essential characteristic of hierarchical planning.

Moreover, each hierarchical level has its own characteristics including the type of manager in charge of controlling the execution of the plan, the scope of the planning activity, the level of aggregation of the required information (and the form in which the information should be disaggregated when transferred to lower levels) and the time horizon of the decision. The lower one gets in the hierarchy, the narrower is the scope of the plan, the lower is the management level involved, the more detailed is the information needed, and the shorter is the planning time horizon. Each level of planning has its own objectives and constraints in which decisions have to be made. It is only natural, therefore, that a system designed to support the overall planning process should correspond to the hierarchical structure of the organization.

Finally, as Emery (1969) points out, when a high-level plan significantly restricts the options available at lower levels the plan becomes centralized. This can only be justified if centralization improves the overall performance of the organization, by recognizing broader objectives which cannot be perceived at lower organizational levels. This is particularly true when the degree of interaction existing among subunits is critical, as is usually the case when dealing with production and transportation decisions in a multiplant-multiproduct corporation.

In spite of the importance that this hierarchical approach has in production planning, very few integrated solutions to this problem have been reported. Most of the published efforts have concentrated on analysis of individual components of the overall problem. Although it is theoretically possible to develop iterative procedures which converge to an optimum final plan, by sequential adjustment of lower level actions (see Baumol/Fabian (1964)), this approach is not computationally feasible.

Prior to designing the hierarchical system we are about to describe, we considered the attractiveness of using other approaches, primarily those of Landon/Terjung (1971), Connors et al. (1972), Zangwill (1966), and Zoller (1971). We found, however, that those approaches were either difficult to implement or they were based on a given level of aggregation, ignoring the problems associated with detailed scheduling. We decided, therefore, to construct our own hierarchical system, making full use of the idiosyncracies of our particular problem. In taking this pragmatic approach we recognized there was a risk of arriving at planning and scheduling decisions that were not optimal. We are confident, however, that the reasons for isolating and linking manageable portions of the overall decision were sound enough to prevent major deviations from optimality and that we reached our goal of maintaining a simple design, relatively easy to implement. We will now describe the modeling effort to show why this is the case.

To decompose the overall problem we examined the extent to which the various planning and scheduling are coupled. If two sets of decisions were independent we could totally separate them in structuring the hierarchy of decisions.

Starting at the most detailed level, we found that Items sharing a major setup cost had to be scheduled jointly. If these Items are grouped into a Family, the production cost is substantially reduced relative to independent scheduling. Thus, the scheduling decisions for the Items in a Family are strongly coupled. On the other hand, the coupling between the schedules of Items in **different** Families is very weak. Items produced in one plant are produced with the same equipment and therefore compete for capacity, but otherwise are completely decoupled.

We also found that the scheduling decisions for a Family in one time period are strongly coupled to the scheduling decisions for the same Family in other time periods. This is a consequence of the need to accumulate seasonal stock in most Families of Items, due to the competition among Families for scarce capacity. If this were not so the Families could be scheduled independently and no seasonal stock would have to be accumulated. Furthermore, since we have batch production there is a potential coupling between the Family run length and the seasonal stock accumulation.

The latter coupling was removed by finding the optimum seasonal stock accumulation pattern for all Families simultaneously, under the assumption that the unit cost of production in accumulating seasonal stocks was independent of run length. In effect, we ignored run length economics in developing the Family seasonal stock accumulation pattern. However, we treated explicitly the coupling among Families in competing for capacity and the coupling among Family schedules in different time periods.

Ignoring run length economics in accumulating seasonal stocks is valid if the seasonal stock accumulation quantities do not influence the unit cost of production. This can be accomplished by neglecting the seasonal peak in calculating the economic run length for each Family and then extending the run length each time a batch is run to accumulate the seasonal stock needed without incurring an extra setup charge.

We found that the approaches which integrate production planning and economic lot sizes (see, for example, Manne (1958) and Dzielinski/Baker/Manne (1963)) are impractical because they lead to evaluation of a large number of production sequencing alternatives, present difficulties in aggregating and disaggregating information, and are computationally expensive for a problem of this size.

Families were aggregated into Types in performing the seasonal planning since this made the seasonal planning computations much simpler and also facilitated the development of the Family and Item schedules. Families which have the same seasonal pattern and the same production rate (measured by inventory investment produced per unit time) are indistinguishable from a seasonal planning point of view. If the cost to carry the inventory is the same for two Families in a Type, only the total accumulation for the two Families (or all Families in the Type) need be considered in developing the plan. The schedules of the Families in the Type are coupled only in that the total of the Family schedules must add up to the Type total as given by the seasonal plan.

Having several Families in which a given seasonal stock can be accumulated simplifies short term scheduling. It reduces the danger of too many Items running out at the same time and allows a general lengthening of runs relative to the constant demand economic run length, without inventory penalty.

Thus, the coupling between Family run lengths and the Family seasonal plans is removed by setting up a seasonal plan which assumes no incremental setups are incurred to accumulate seasonal stock and then accumulating the seasonal stock by extending runs which were scheduled to meet the current demand.

Finally, the schedule for a Family in one plant may be coupled to the schedules for the same Family in the other plants where the same Families can be produced. These schedules will be coupled only if it is desirable to produce in one plant to satisfy demand in another plant's territory and if transshipment is cheaper than seasonal stock accumulation as a means of dealing with the shortage. Since, in this case, it is always cheaper to carry the inventory than it is to ship the product, the plants are decoupled and may be scheduled independently. If there were a net **annual** shortage in one plant and an excess in another, the territories should be redefined or capacity adjusted by transfer of equipment.

The assignment of Families to plants for production is independent of the plant production schedules for the same reason. We found the most economic set of production locations for each Family and then used those locations, adjusting plant capacity as needed in the annual planning process. In this way we accumulate the seasonal stock without interplant shipments.

This decoupling process leads to the following hierarchy of decisions, as shown in FIGURE 1.

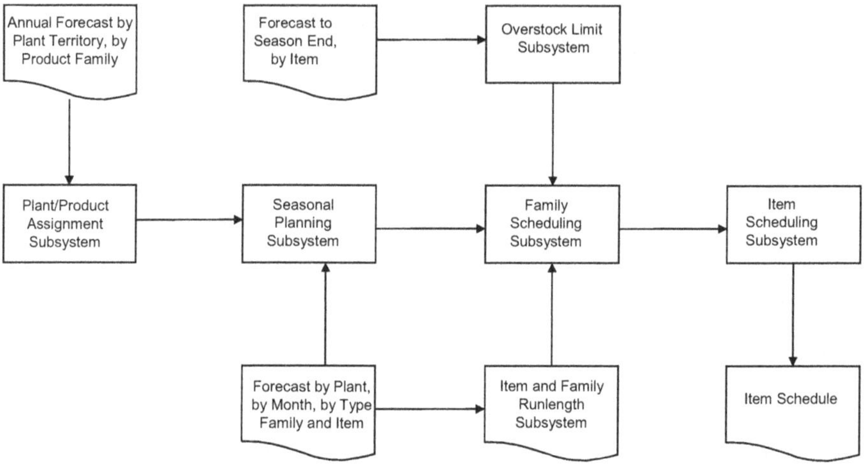

FIGURE 1: DECISION SEQUENCE IN PLANNING AND SCHEDULING

1. **Assignment of Items and Families to Plants.** The process starts with the forecasts of the annual demand in each plant territory. This information, as well as the additional capital investment to be incurred in each plant for the production of a given Family and the interplant transportation costs, is input to the Plant/Product Assignment Subsystem. This determines the plant locations at which each Family should be manufactured. This assignment is done annually.

2. **Seasonal Planning.** Once each Family has been assigned to a plant or plants, a monthly demand forecast for each product Type is computed for each plant. These forecasts, together with the available production capacity at each plant, are input to the Seasonal Planning Subsystem which defines a monthly production plan and a seasonal inventory accumulation strategy for each product Type at each plant location. The Seasonal Plan is updated monthly to take account of changes in actual inventory position and changes in forecast.

3. **Scheduling of Families.** The Item and Family Run Length Subsystem develops nominal minimum run quantities for all the Items and Families, and the Overstock Limit Subsystem imposes upper limits on the accumulation of seasonal stock in any one Item, based upon the risks of understock and overstock, and the associated costs at the end of each season. The run quantities and overstock limits are evaluated using standard procedures which will not be discussed in the present paper (see, for example, Groff/Muth (1972) and Magee/Boodman (1967)). The heart of the production planning and control system is the Family Scheduling Subsystem. This Subsystem uses the results of the Seasonal Planning, Item and Family Run Length and Overstock Limit Subsystems to determine the run quantities to be produced

for each Family in the coming month. This system is used monthly as soon as data are available from the preceding month and prepares a schedule for the following month.

4. **Scheduling of Items.** The Item Scheduling Subsystem allocates the Family production run quantity among the Items in the Family. Runout times are equalized, as this provides best service for the inventory investment, and Overstock Limits are observed. This system is used monthly, immediately following preparation of the Family Schedule.

It is clear that the proposed system requires several different kinds of forecasts. These were prepared using standard techniques discussed in Brown (1962), Chambers/Mullick/Smith (1971) and Draper/Smith (1966).

IV. PLANT/PRODUCT ASSIGNMENT SUBSYSTEM

When an Item Family can be manufactured at more than one plant, one must first decide where that product should be assigned for manufacture. The analysis must balance the cost of interplant freight and handling against the incremental capital investment cost required to set up the proper facilities, including tools, for the manufacture of the product in question. This decision is most critical for a new product, when it is necessary to determine the initial capital investments to be made, but it is also important for existing products because the variable manufacturing costs differ for each plant location and are subject to change. Furthermore, demand patterns may change, making it desirable to transfer production facilities from one plant to another.

It would have been desirable to develop a multiperiod, multiproduct model to take seasonality into account in the allocation of Item Families to the various plants and to impose capacity constraints on all the products. Such a model, which is very easy to formulate, gives rise to a very large mixed-integer program that cannot be solved within the existing state-of-the art. A capacity-free model offers several advantages. First, it provides the "ideal" distribution of plant capacity for the whole manufacturing process and, thus, it constitutes an effective tool for long-range capacity expansion policies. Second, it permits quantification of the desirability of transferring equipment from one plant, where it currently stands, to another plant, which represents the optimum location for processing the product. Finally, a capacity-free model allows each Item Family to be treated independently, since the only interaction among Items is their competition for scarce capacity.

The following notation will be used to describe the Plant/Product Assignment model in mathematical terms.

C_m = Incremental capital investment required to produce the Item Family at plant location m.

g_{mn} = cost of making one unit of the Item Family at plant m and transferring it to plant n for delivery to the final customer. If $m = n$, the transfer cost is zero.

F_m = forecast annual demand in plant territory m.

x_m = 1, if the Family is assigned to plant m;
0, otherwise

y_{mn} = production at plant m to fill demand in plant n territory

The model has as its objective the minimization of the total cost of capital, production and transportation.

$$\text{Cost} = \Sigma_m C_m x_m + \Sigma_m \Sigma_n g_{mn} x_m y_{mn}$$

subject to constraints:

$$\Sigma_m x_m y_{mn} = F_n \quad \text{for all } m, n$$

$$y_{mn} \geq 0$$

$$x_m = 0,1$$

If the number of plants is small, it is possible to enumerate all the alternatives given by the possible values of the variables x_m and proceed to solve a sequence of trivial linear programming problems that can be computed manually.

If there is a large number of plants to consider, the computational methods suggested by Manne (1964) have proven to be quite successful. After all Item Families have been processed, we can verify whether capacity constraints for each plant have been violated or not. If the aggregate solution is not feasible, reassignments should be made based on the reassignment cost penalties (i.e., increase in cost as a consequence of changing from the best solution to a second or third alternative) and the capacity constraints at each plant. A procedure for doing this has been suggested by Hax (1972). At this point, the model can be expanded into a multiperiod model to include seasonalities. In the application described here, this step was not required; the seasonalities were explicitly handled in the Seasonal Planning Subsystem.

V. SEASONAL PLANNING SUBSYSTEM

The Seasonal Planning Subsystem determines the production and seasonal stock accumulation for the product Types to be manufactured at each plant so that production and inventory costs are minimized, within the fixed constraints of the existing production facilities and the forecast demand requirements. Inventories are the logical way to smooth the production requirements created by the seasonal and fluctuating demand of the various product types, since it is uneconomical, and at times unfeasible, under these conditions to match perfectly the demand and production patterns. We used a linear programming model to accumulate seasonal stock, minimizing the total cost of inventory and production. Similar models have been developed by Bowman (1956), and Hanssmann/Hess (1960).

Planning Horizon and Number of Time Periods.

The total time horizon that should be considered in planning seasonal stock accumulation must be long enough to cover at least one seasonal cycle for all the products involved. On the other hand, it should be short enough to make the model computationally feasible and to allow demand forecasts to be reasonably accurate. In the problem under consideration, a time horizon of fifteen months met these two criteria.

The fifteen-month time horizon was divided into seven time periods; the first three are one month each, and the remaining four are one quarter (i.e., three months) each. The model was updated monthly for operational purposes; therefore, the aggregation of time periods did not lose any significant information, but it reduced substantially the computational time required to solve the model.

Product Aggregation.

The Seasonal Planning Subsystem was concerned exclusively with planning production of product Types. In a given product Type, we included all Items having a similar seasonal demand pattern and a similar inventory holding cost per hour of production. Considerations affecting individual run lengths and overstock limits for each Item were taken into account by other subsystems, as explained previously. These subsystems are described in later sections.

Regular and Overtime Production Capacities.

It is important to make a distinction between regular and overtime production because of the increase in labor cost when working overtime. The model properly balanced the tradeoff existing between the overtime cost penalty and the inventory carrying cost to determine the optimum production plan.

Demand Forecast.

A basic input to the subsystem was the demand forecast for each product Type during each time period covered by the planning time horizon. First, we replanned production monthly, updating the model once a month, in order to find the revised optimal production patterns as newer and more accurate demand forecasts were obtained. Second, by discounting the cost elements considered in the model, we gave more weight to the data describing earlier time periods, presumably known more accurately, than to the data describing more distant time periods. Third, the inventory at the end of each time period was required to be at least equal to the safety stock of the product Type. This safety stock is the sum of safety stocks of the Items belonging to the product Type (see Groff/Muth (1972)).

Cost Components.

The objective of the model is the minimization of the total cost involved in accumulating seasonal stock. We included the cost of overtime production and the cost of holding inventory for each product Type. Regular production labor costs had to be absorbed by the firm, so they were treated as fixed costs. Backorder costs and hiring and firing costs can be easily introduced in this formulation (see Hanssmann/Hess (1960)). The backorder problem was handled by a service policy statement and work force leveling was handled outside the model also, as described in the problem statement. Thus, these cost elements were not incorporated in the model.

Mathematical Formulation of the Seasonal Planning Model.

To describe the Seasonal Planning model (ignoring discounting and aggregated technological constraints) in precise mathematical terms, we will adopt the following notation:

R_{it} = hours of regular production of Type i to be scheduled during time period t

O_{it} = hours of overtime production of Type i to be scheduled during time period t

$(R)_t$ = total hours of regular production available during time period t

$(O)_t$ = total hours of overtime production available during time period t

I_{it} = inventory of Type i on hand at the end of time period t (units)

r_i = production rate for Type i (units/hr)

D_{it} = forecast demand of Type i during time period t (units)

$(CO)_{it}$ = cost of overtime production of Type i during time period t (\$/hr)

$(CI)_{it}$ = inventory holding cost for Type i during time period t (\$/unit-period)

SS_{it} = safety stock required for Type i at end of time period t (units).

Using this notation, the Seasonal Planning model has as its objective the minimization of the total of regular and overtime production cost and inventory holding cost given by

$$\text{Total Cost} = \Sigma_i \Sigma_t (CO)_{it} O_{it} + \Sigma_i \Sigma_t (CI)_{it} I_{it}$$

We find the best seasonal stock accumulation plan by minimizing this cost expression, subject to the following constraints:

$$x_i R_{it} \leq (R)_t \qquad \text{for all } t$$

$$x_i O_{it} \leq (O)_t \qquad \text{for all } t$$

$$r_i (R_{it} + O_{it}) - I_{it} + I_{i,t-1} = D_{it} \quad \text{for all } i, t$$

$$R_{it} \geq 0 \quad \text{for all } i, t$$

$$O_{it} \geq 0 \quad \text{for all } i, t$$

$$I_{it} \geq SS_{it} \quad \text{for all } i, t$$

The last constraint introduces lower bounds to the ending inventory at each time period to incorporate safety stock requirements.

VI. ITEM FAMILY SCHEDULING SUBSYSTEM

The Seasonal Planning Subsystem determines the production hours to be used for each product Type. To set the Family run quantities correctly and preserve the optimality of the seasonal stock accumulation plan, the Family Scheduling Subsystem must schedule just enough production in the Families in the product Type to use all the time available to the Type. It must also accumulate the required seasonal stock without incurring setup costs just for that purpose, and insure that customer service standards are met. Finally, the Family Scheduling Subsystem must avoid stock accumulations which exceed individual Item overstock limits. Within these constraints and because of the way in which product Types are defined, it is immaterial which Families belonging to the product Type we choose to run.

Accordingly, we use a system which, **first**, determines which Families in the Type must be run in the scheduling interval in order to meet Item service requirements; **second**, sets initial Family run quantities so as to minimize cycle inventory and changeover costs; **third**, adjusts the run quantities of the Families scheduled to use all the production time available to the Type, while, **fourth**, observing Item overstock limits.

Two general methods are available for accumulating the seasonal stock needed in each product Type. We can increase the run length of each Family which must be run in order to meet service requirements, or we can include more Families, going "deeper into the run-out list," making minimum or nominal runs of each Family. We chose the first method since this allows us to save on setup costs, gaining longer runs with no inventory penalty. Further, we selected an allocation logic which balances inventories relative to overstock limits, to minimize the problems of end-of-season carryover stocks.

A Family must be produced in the schedule interval if the available stock A_{kj} of any Item k in Family j, is less than forecasted demand of the Item during the schedule interval plus the desired amount of safety stock S_{kj}. S_{kj} is established using standard safety stock methods. The minimum or base run length Q_j is set for each Family using standard methods for items grouped for production. The overstock limits, L_{kj}, are set using standard newsboy methods.[1] The cost of understock is equal to the changeover cost associated with having to make an extra run between now and the end of the season; the cost of overstock is equal to the cost of carrying stock over to the next season. Using these together with the distribution of uncertainty in Item forecast to the end of the season yields the desired overstock limit L_{kj}.

To establish the Family schedule we proceed as follows:

Define the runout time RT_j for each Family as the time at which the available inventory is expected to reach the safety stock level:

$$RT_j = \min_k \left[(A_{kj} - S_{kj}) / F_{kj}\right]$$

where

A_{kj} = available stock of the *k-th* Item in Family *j*

S_{kj} = Item Safety Stock level required at the end of the schedule period to meet the service requirement

F_{kj} = forecast usage rate of the *k-th* Item.

In this case the schedule period is one month and the available stock is measured at the beginning of the prior month. Thus, the safety stock must cover the forecast errors during a two month interval. All those Families with RTj less than the time until the end of the period being scheduled must be put into the schedule for that period. In the remainder of this subsystem description the index j will be restricted to the set of Families in the current schedule.

Define a trial Family run quantity

$$RQ_j = \Sigma_k RQ_{kj} = \Sigma_k \min\left[Q_{kj}, (L_{kj} - A_{kj})\right]$$

as the sum of the minimum runs for each Item in the Family.

We now sum the trial run quantities for all the Families in the Type and compare the sum with the total production available to the Families in the Type, r_i $(R_i + O_i)$ where R_i and O_i are the regular and overtime hours allocated to Type *i* and r_i is the production rate in units per hour.

If $\Sigma_j RQ_j < r_i (R_i + O_i)$, schedule RQ_j^* for each Family were

[1] For these standard methods consult any good inventory control text such as Groff/Muth (1972) or Magee/Boodman (1967).

$$(1)\ RQ_j^* = max\{\Sigma_k (L_{kj} - A_{kj}), RQ_j + [r_i (R_i + O_i) - \Sigma_j RQ_j)]\ \Sigma_k (L_{kj} - A_{kj}) / \Sigma_j \Sigma_k (L_{kj} - A_{kj})\}$$

The seasonal stock accumulation amount for the *j-th* Family is the difference between RQ_j and RQ^*. That cannot exceed the amount required to reach the overstock limit for each Item in the Family, $\Sigma_j (L_{kj} - A_{kj})$. If overstock limits do not interfere (and they usually do not), the amount of production required in the Type, $r_i (R_i + O_i)$ in excess of the base run quantities ΣRQ_j, is allocated among the Families so that after allocation all Families will have the same fractions of their overstock limits, keeping inventories in good balance.

Notice that when the overstock limit does not interfere, RQ_j^* will be determined by the second term inside the bracket in equation (1), then

$$\begin{aligned} \Sigma_j RQ_j^* &= \Sigma_j RQ_j + [r_i (R_i + O_i) - \Sigma_j RQ_j]\ \Sigma_j \Sigma_k (L_{kj} - A_{kj}) / \Sigma_j \Sigma_k (L_{kj} - A_{kj}) \\ &= r_i (R_i + O_i) \end{aligned}$$

as required.

If $\Sigma_j RQ_j \geq r_i (R_i + O_i)$, schedule RQ^* for each Family where

$$\mathrm{R}\ RQ_j^* = r_i (R_i + O_i)\ RQ_j / \Sigma_j RQ_j$$

This expression adjusts run quantities downward if the total production time allocated to the Type is insufficient to produce base run quantities of all the Families which need to be run in the schedule interval. Here we also have

$$\Sigma_j RQ_j^* = r_i (R_i + O_i)\ \Sigma_j RQ_j / \Sigma_j RQ_j = r_i (R_i + O_i)$$

When the overstock limits do interfere, it may be impossible to produce all the assigned quantity $r_i (R_i + O_i)$ with those Families belonging to Type *i* which have been put into the current schedule (because their runout times were less than the time to the end of the schedule period). In this case, one should go deeper into the runout list and add additional Families belonging to the product Type. The new Families should be run up to their maximum allowable quantities, given by $\Sigma_k (L_{kj} - A_{kj})$, until the product Type production requirements are met.

No seasonal stock is being accumulated. This simply reflects the realities of the end game as the peak season draws to a close. The Family scheduling logic is shown in flow chart form in Appendix A.

VII. ITEM SCHEDULING SUBSYSTEM

The Item Scheduling Subsystem determines the production quantities for each Item so that the total of the quantities equals the Family schedules. In doing this we must continue to observe the overstock limits and attempt to maximize the customer service which can be obtained with the stock which will be produced. We do this by equalizing the expected runout times for the Items in the Family. This maximizes the expected time until the first Item in the Family runs out. Such a runout requires scheduling the Family again and therefore should be deferred as long as possible, within the constraints of the overstock limits and the total Family production quantity available.

The Item run quantity which accomplishes this is

(2) $RQ_{kj}^{*} = S_{kj} - A_{kj} + F_{kj}\,[RQ_j^{*} + \Sigma_k (A_{kj} - S_{kj})] / \Sigma_k F_{kj}$

where

RQ_{kj}^{*} = desired run quantity for *k-th* Item belonging to Family *j*

F_{kj} = forecast demand rate for *k-th* Item

RQ_j^{*} = $\Sigma_k RQ_{kj}^{*}$ = Family run quantity, from the Family Scheduling System

A_{kj} = available inventory of the *k-th* Item

S_{kj} = desired safety stock of the *k-th* Item at the end of the schedule interval

Notice that the new runout time for Item *k* will be

$$RT_k = (RQ_j^{*} + A_{kj} - S_{kj}) / F_{kj}$$

and, by expression (2), this is equal to

$$RT_k = [RQ_j^{*} + \Sigma_k (A_{kj} - S_{kj})] / \Sigma_k F_{kj}$$

and this is constant for every Item *k*. This equalizes the expected runout time for all the Items in the Family. Moreover, adding each side of equation (2) with respect to *k* gives us:

$$\Sigma_k RQ_{kj}^{*} = RQ_j^{*}$$

and, therefore, guarantees that the total amount scheduled for the Family, RQ_j^{*} has been allocated among the Items belonging to that Family.

The resulting run quantities must be tested for negativity and against the overstock limits for each Item. If the Item run quantity does not lie between these limits it is set to zero or to the overstock limit, as appropriate. The normalizing constant, $[RQ_j^{*} + \Sigma_k (A_{kj} - S_{kj})] / \Sigma_k F_{kj}$, is then increased (by eliminating the Item from the summation) for the remaining Items until the total Family run quantity has been scheduled. A logical flow chart of the Item Scheduling System is shown in Appendix B.

We did not allocate the Family production run among Items in the same way we allocated the Type production time among Families. The Type allocation should build a balanced seasonal stock and avoid carryover stocks. The Family allocation, within the constraint of a specified seasonal stock accumulation, should defer the next requirement to run the Family as long as possible. For this reason we allocate the Family production run among Items so as to equalize expected Item runout times.

VIII. IMPLEMENTATION

We will divide the implementation activities into two parts – technical and tactical. The technical part refers to the transformation of the system logic into computer programs and input and output systems and procedures. The tactical part is concerned with the development of management familiarity with the system and an effective reduction of the system to practice.

Technical Parts

Surprisingly little difficulty was encountered in the technical parts of the system. We believe this is the result of what some would call a profligate use of computer test time and programmer time. The major computer systems are the Seasonal Planning Subsystem, the Family and Item Run Length Subsystem, and the Monthly Family and Item Scheduling Subsystem. We first set up the LP for seasonal planning by hand and used the OMEGA code available from University Computing for use on their Univac 1108. In all but the earliest versions, we used a matrix generator code which we wrote for the IBM 360/40 at Arthur D. Little, Inc. We transmitted the matrix to UCC, where the LP was run as a remote entry batch job on the Univac 1108.

The client then programmed the matrix generator for his own IBM 1130. He also ran the LP on the 1130 using the IBM Mathematical Programming System (MPS). This was initially intended to be a pilot version. The plan was to convert it to MPS on the client's 360/40 or use a service bureau. At present, they plan to use the 1130 indefinitely for this job. It provides a great deal of flexibility in use, allows rapid turnaround time and is not expensive, if figured at marginal cost. Most importantly, the programmer analyst is very involved in the problem and is close enough to the computer to resolve his own difficulties.

The Family and Item Run Length Subsystem was programmed three times. It was first programmed on a time-shared console by us. This provided for "conversational" mode development of the logic. It was then programmed on the 1130 by the client. This provided pilot testing and some first-order changes. In fact, there was one major reprogramming of the system after some significant problems were uncovered. So, in a sense, there were two 1130 versions. Finally, it was programmed in a "final" version on the 360/40.

The scheduling system has been programmed only once, on the 1130.

This may sound like a lot of programming and reprogramming. It has not, however, been an expensive strategy. By knowing that initial versions would be modified and reprogrammed, we could then take a more casual view of the requirements and specifications and could get that job done quickly and cheaply. We could then see which aspects of the system needed most change and which changes were most important. This evaluation could be made with results in hand, so the discussions about program specification and design were concrete rather than abstract.

All in all, we believe that we saved programming time and cost by taking this iterative approach to the problem. We could, by running pilot versions of the logic, pave the way for the client and make his work not only easier but more certain to succeed. When he programmed the problem, he could concentrate on system problems, not logic or algorithms, since they had been worked out earlier.

The bulk of the implementation effort went into development of the data base. As is usually the case in firms with primarily manual information and control systems, the existing records contained many errors – obsolete part numbers, incorrect production rates, old costs, etc. Development of a system which would accurately maintain the data file and getting the file cleaned up initially was the major barrier to implementation.

Cleaning up this file had major side benefits. The cost files are now accurate and a better base for planning. Capacity estimation is also much more accurate. Perhaps most important is the development of a common language for production and distribution, essentially eliminating production and shipping errors as a consequence of incorrect part numbers.

Tactical Parts

The tactical implementation of a system of this type is primarily a question of developing management familiarity with the system – what it does and why, and how to make it achieve a desirable result.

A primary problem has been the rigidity of the system, or put differently, the discipline that such a system imposes. In particular, the seasonal planning system will not accept inconsistent sets of requirements and capabilities. If the time available to complete the required production is insufficient, the system will not produce a "best efforts" result; it just produces a printout which says there is no feasible solution.

This kind of system characteristic can be a problem to a manager who is accustomed to "dealing with those problems when we get to them." Such a manager wants to avoid deciding which of two (or twenty) competing items will be produced with scarce production capacity, until faced with the fact of having both items run out. The manager hopes this will not happen and that he will have made an economic choice by having less than the expected required capacity available.

With the new system, the manager must choose a total production quantity he is prepared to accept; the system will then develop schedules such that the risk of simultaneous runout of two or more items is kept manageably low. This does not require the kind of long-term forecast accuracy or willingness to make blind commitments as may appear at first. The planning model is updated monthly in a routine way and can be updated any time a change large enough to justify such updating has occurred. Each such updating is capable of accepting new production requirements statements and new capacities, as well as reflecting what has happened thus far, via beginning inventories.

A second problem has been to develop an understanding of the difference between a forecast and a plan. For most manufacturers, a forecast of demand is what they plan to make, even though they recognize the forecast is uncertain. Often a seasonal stock accumulation plan is just a level production plan, by item, to reach the annual forecast.

In an explicit planning system of the sort described here, the plan at the beginning of the season may differ from the forecast because of capacity limitations. It certainly does not lead to a level plan by item. Thus, a manager must develop a new working style in using this planning system.

A third problem arises because of the need to decide to accumulate seasonal stocks before forecast uncertainties are resolved. Once the amount to be produced has been established, the seasonal planning system will find the most economic production pattern. It is not easy, however, to decide how much to produce, given forecast uncertainties. Although it is possible to incorporate costs of lost sales in such a model in order to treat this problem explicitly, this has not been done – primarily to avoid the additional complexity, but also because our understanding of these costs does not justify such a quantitative treatment.

The seasonal planning approach used here deals with these uncertainties in a heuristic way, by replanning each month with an updated forecast. While this produces a well-behaved production plan, it is unsettling to a manager who sees his plan greater than the forecast one month and less the following month. As was pointed out above many managers prefer to adjust the plan to match the forecast and make no attempt to deal with uncertainty until faced with the actual requirement. When requirements are a long way off in time, it seems as though an excess of such commitments could be accepted since some of them will certainly not materialize. It has been difficult for management to accept the fact that production resources must be committed to these items long in advance of need, because of seasonal planning; they have, therefore, been slow in providing the basic commitment decisions.

Costs and Benefits

This system required the efforts of a project director, about half time, for a period of about two years. He had a full-time system analyst and about half-time help from a production manager assigned to the project. The total cost of the development, including consultants' fees was $150–200,000.

We cannot say yet how large the benefits will be. The primary benefit will be smoother production with fewer emergency interrupts. This will save money both in setup costs and in improved service. A secondary saving will be reduction in inventory carrying costs. These together are expected to amount to more than $200,000 each year in each of the plants.

Acknowledgement

The project director for the manufacturing firm and his colleagues provided invaluable assistance in describing the nature of the operation to us and in reviewing the system as it developed. Unfortunately, they must remain anonymous. D.J. Smith and A.S. Freedman of Arthur D. Little, Inc., were responsible for the development of the Plant/Product Assignment model and the Family and Item Run Length model, respectively.

REFERENCES

Baumol/Fabian (1964): Baumol, W.J. and T. Fabian: Decomposition, Pricing for Decentralization and External Economies. In: *Management Science*, Vol. 11, No. 1, September 1964.

Bowman (1956): Bowman, W.H.: Production Scheduling by the Transportation Method of Linear Programming. In: *Operations Research*, Vol. 4, No. 1, February 1956.

Brown (1962): Brown, R.G.: *Smoothing, Forecasting and Prediction of Discrete Time Series*, Prentice-Hall, 1962.

Chambers/Mullick/Smith (1971): Chambers, J.C., S.K. Mullick, and D.D. Smith: How to Choose the Right Forecasting Technique, *Harvard Business Review*, July-August 1971.

Connors et al. (1972): Connors, M.M.; C. Coray; C.J. Cuccaro, W.K. Green, D.W. Low, and H.M. Markowitz: The Distribution System Simulator. In: *Management Science*, Vol. 18, No. 8, April 1972.

Draper/Smith (1966): Draper, N.R., and H. Smith: *Applied Regression Analysis*, John Wiley and Sons, 1966.

Dzelinski/Baker/Manne (1963): Dzielinski, B.P., C T. Baker, and A. S. Manne: Simulation Tests of Lot Size Programming. In: *Management Science*, Vol. 9, No. 2, January 1963.

Emery (1969): Emery, J.C.: *Organizational Planning and Control Systems*, Macmillan, 1969.

Groff/Muth (1972): Groff, G.K., and J.F. Muth: *Operations Management-Analysis for Decisions*, D. Irwin, 1972.

Hanssmann/Hess (1960): Hanssmann, F. and S.W. Hess: A Linear Programming Approach to Production and Employment Scheduling. In: *Management Technology*, Vol. 1, January 1960.

Hax (1972): Hax, A.C.: *Integration of Aggregate and Detailed Scheduling via Linear Programming*, Forthcoming. Also, Aluminum Products International (A) and (B), Harvard Business School, 1972.

Iglehart (1967): Iglehart, D.L.: Recent Results in Inventory Theory. In: *Journal of Industrial Engineering*, Vol. 17, January 1967.

Landon/Terjung (1971): Landon, L.S., and R.C. Terjung: An Efficient Algorithm for Multi-Item Scheduling. In: *Operations Research*, Vol. 19, No. 4, July-August 1971.

Magee/Boodman (1967): Magee, J.H., and D.M. Boodman: *Production Planning and Inventory Control*, McGraw-Hill, 1967.

Manne (1958): Manne, A.S.: Programming of Economic Lot Sizes. In: *Management Science*, Vol. 4, No. 2, January 1958.

Manne (1964): Manne, A.S.: Plant Location Under Economies-of-Scale – Decentralization and Computation. In: *Management Science*, Vol. II, No. 2, November 1964.

Zangwill (1966): Zangwill, W.I.: A Deterministic Multiproduct, Multifacility Production and Inventory Model. In: *Operations Research*, Vol. 14, No. 3, May-June 1966.

Zoller (1971): Zoller, K.: Optimal Disaggregation of Aggregate Production Plans. In: *Management Science*, Vol. 17, No. 8, April 1971.

APPENDIX A

Logical Flow Chart for Family Scheduling Subsystem

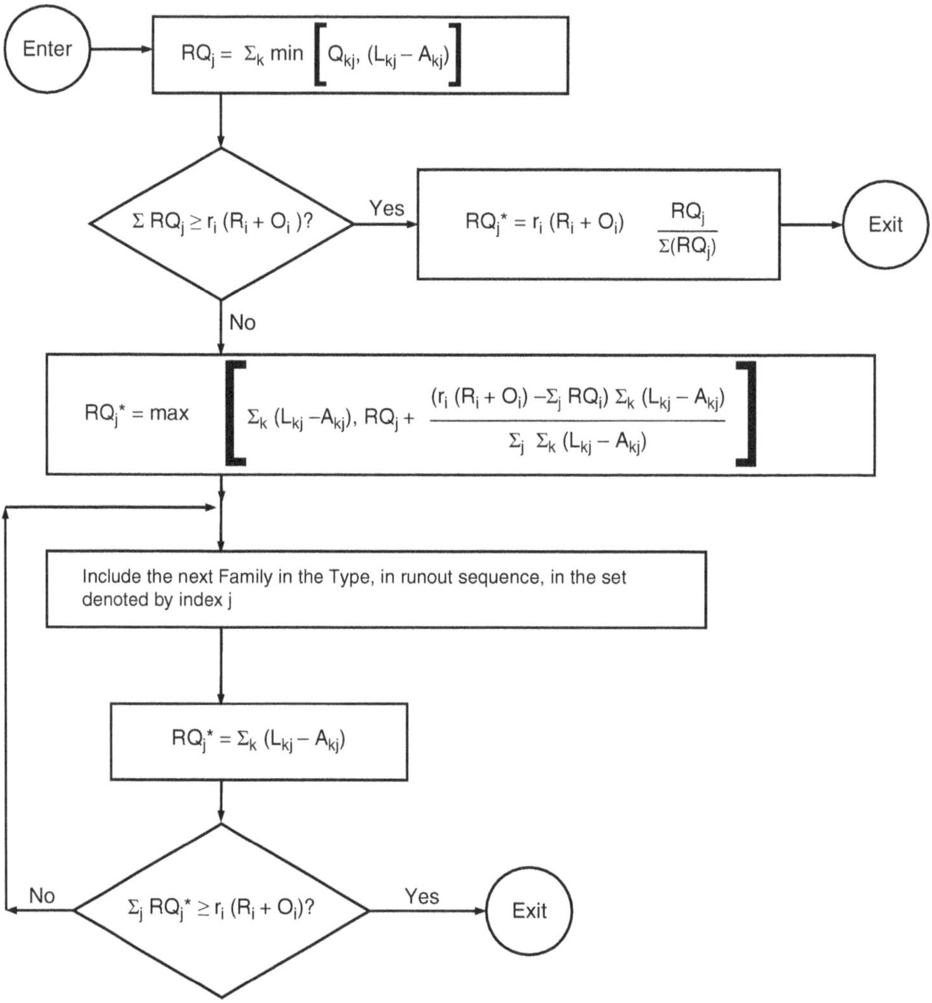

Symbols are as defined in Sections V and VI, initially the index j is limited to those Families in the Type which must be run in the period to meet service requirements.

APPENDIX B

Logical Flow Chart for Item Scheduling Subsystem

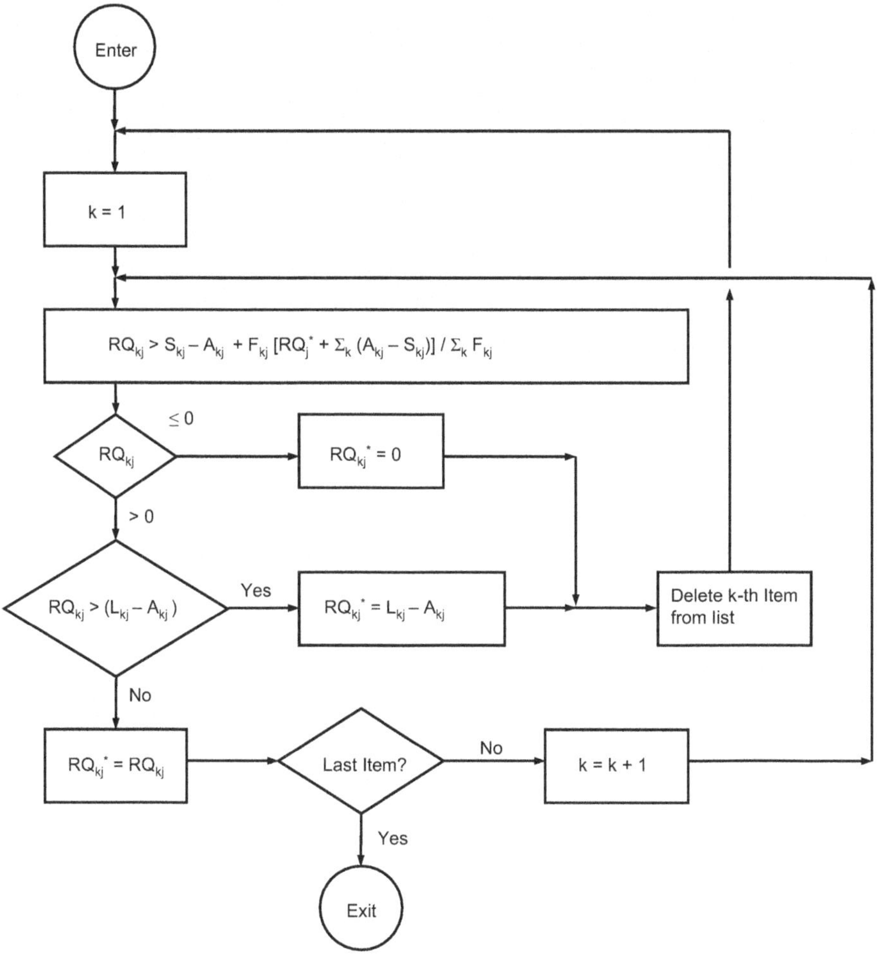

Initially the index k runs over all the Items in the Family. As Items are deleted as a consequence of having their run quantities set to zero or to the overstock limit, the list of Items identified by k is modified.

Theodore Levitt (1972)

PRODUCTION-LINE APPROACH TO SERVICE

In tracing the roots of the Science of Logistics in this collection of precursors and classical contributions to field, a rich array of terminological, conceptual, and methodological innovations has been uncovered so far. They provide answers to questions which concern the field's identity as a science. If one reviews these answers and questions, it becomes apparent, that most of them deal with "networks and flows of objects through time and space", aiming to "supply recommendations for action on the design, implementation and mobilization of such flows and networks" at a rather abstract, strategic level: How is it that logistics creates utilities and value in an economy? What is the specific perspective which distinguishes logistics from other fields of science? What do we know about appropriate architectures of logistical systems?

But the everyday practice of logistics is characterized at least as much by the more mundane operational issues of how to run logistical systems and supply chains efficiently and financially successful, and how to continuously rationalize and improve them.

Among the scholars who delivered early and visionary contributions to this aspect of the Science of Logistics Theodore Levitt (1925–2006) is most prominent, a much quoted professor of Marketing at Harvard Business School. In his 1972 article *Production-line Approach to Service* he took up the question why the production of services traditionally seems to have been much less efficient than industrial production. By using McDonald's and some other pioneers of innovative marketing and operations practices in the service industries as prototypical examples, Levitt demonstrated how service processes can be executed in large numbers in a repetitive way like industrial processes – based on highly standardized products and processes, supported by automated technology, systematic quality controls and professionalized marketing.

A "production-line approach" to logistics services and – more generally the "industrialization" of logistics services (cf. Levitt's follow-up paper of 1976) – is an obvious, powerful approach to address the issues of efficiency and quality in the logistics industry. Levitt's work is the starting point for other influential contributions to successful management at more operational levels – with high relevance to logistics. Other, later work which followed the course set by Levitt and helped to expand the logisticians' toolkits for everyday management is Cooper/Kaplan's (1988) contribution on "activity-based costing", Camp's (1989) on "benchmarking", and Hammer's (1990) on "business process reengineering".

P. Klaus and S. Müller (eds.), *The Roots of Logistics*, DOI 10.1007/978-3-642-27922-5_20,

Reference:

Theodore Levitt: Production-line Approach to Service. In: *Harvard Business Review*, Vol. 50, October-November 1972, pp. 41–52.

Further reading:

Camp, Robert C.: *Benchmarking: The Search for Industry Best Practices that Lead to Superior Performance*. Milwaukee/WI: ASQC Quality Press, 1989.

Cooper, R. and Kaplan, RS: Measure costs right: make the right decisions, in: *Harvard Business Review*, September-October 1988, pp. 96–103.

Hammer, Michael: Don't Automate, Obliterate! In: *Harvard Business Review*, July-August 1990, pp. 104–112.

Levitt, Theodore: The Industrialization of Service. In: *Harvard Business Review*, September-October 1976, pp.63–71.

Theodore Levitt: Production-line Approach to Service

Once service "in the field" receives the same attention as products "in the factory" a lot of new opportunities become possible

FOREWORD

We think about service in humanistic terms; we think about manufacturing in technocratic terms. This, according to the author, is why manufacturing industries are forward-looking and efficient while service industries and customer service are, by comparison, primitive and inefficient. He argues that if companies stop thinking of service as servitude and personal ministration, they will be able to effect drastic improvements in its quality and efficiency. Then he shows companies how to take a manufacturing approach to this activity, one that substitutes "technology and systems for people and serendipity."

Mr. Levitt is Professor of Business Administration at the Harvard Business School, and a familiar author to HBR readers. His articles have won three McKinsey awards and a John Hancock Award for Excellence. His most recent book is *The Marketing Mode* (New York, McGraw-Hill, 1969).

The service sector of the economy is growing in size but shrinking in quality. So say a lot of people. Purveyors of service, for their part, think that they and their problems are fundamentally different from other businesses and their problems. They feel that service is people-intensive, while the rest of the economy is capital-intensive. But these distinctions are largely spurious. There are no such things as service industries. There are only industries whose service components are greater or less than those of other industries. Everybody is in service.

Often the less there seems, the more there is. The more technologically sophisticated the generic product (e.g., cars and computers], the more dependent are its sales on the quality and availability of its accompanying customer services (e.g., display rooms, delivery, repairs and maintenance, application aids, operator training, installation advice, warranty fulfillment). In this sense, General Motors is probably more service-intensive than manufacturing-intensive. Without its services its sales would shrivel.

Thus the service sector of the economy is not merely comprised of the so-called service industries, such as banking, airlines, and maintenance. It includes the entire abundance of product-related services supplied by manufacturers and the sales-related services supplied by retailers. Yet we confuse things to our detriment by an outdated taxonomy. For example:

- The First National City Bank (Citibank) is one of the biggest worldwide banks. It has about 37,000 employees, over half of whom deal directly with the public, either selling them things (mostly money and deposit services) or helping them with things they have already bought (cashing checks, taking additional deposits, writing letters of credit, opening lockboxes, managing corporate cash). Most of the other employees work back in what is called "the factory" – a massive congeries of people, paper, and computers that processes, records, validates, and scrutinizes everything the first group has done. All the corporate taxonomists, including the U.S. Department of the Census, classify Citibank as a service company.
- IBM is the biggest worldwide manufacturer of computers. It has about 270,000 employees, over half of whom deal directly with the public, either selling them things (mostly machines) or helping them with the things they have already bought (installing and repairing machines, writing computer programs, training customers). Most of the other employees work back in the factory – a massive congeries of wires, microminiature electronic components, engineers, and assemblers. All the corporate taxonomists, including the U.S. Department of the Census, classify IBM as a manufacturing company.

Something is wrong, and not just in the Bureau of the Census. The industrial world has changed more rapidly than our taxonomies. If only taxonomy were involved, the consequences of our contradictory classifications would be trivial. After all, man lives perfectly well with his contradictions: his simultaneous faith, for instance, in both God and science; his attachment to facts and logic when making important business decisions, but reliance on feelings and emotion when making far more important life decisions, like marriage.

I hope to show in this article that our contradictory notions about service may have malignant consequences. Not until we clarify the contradictions will companies begin to solve problems that now seem so intractible. In order to do so, they must think of themselves as performing manufacturing functions when it comes to their so-called "service" activities. Only then will they begin to make some significant progress in improving the quality and efficiency of service in the modern economy.

FIELD VERSUS FACTORY

People think of service as quite different from manufacturing. Service is presumed to be performed by individuals for other individuals, generally on a one-to-one basis. Manufacturing is presumed to be performed by machines, generally tended by large clusters of individuals whose sizes and configurations are themselves dictated by the machines' requirements. Service (whether customer service or the services of service industries) is performed "out there in the field" by distant and loosely supervised people working under highly variable, and often volatile, conditions. Manufacturing occurs "here in the factory" under highly centralized, carefully organized, tightly controlled, and elaborately engineered conditions.

People assume, and rightly so, that these differences largely explain why products produced in the factory are generally more uniform in features and quality than the services produced (e.g., life insurance policies, machine repairs) or delivered (e.g., spare parts, milk) in the field. One cannot as easily control one's agents or their performance out there in the field. Besides, different customers want different things. The result is that service and service industries, in comparison with manufacturing industries, are widely and correctly viewed as being primitive, sluggish, and inefficient.

Yet it is doubtful that things need be all that bad. Once conditions in the field get the same kind of attention that conditions inside the factory generally get, a lot of new opportunities become possible. But first management will have to revise its thinking about what service is and what it implies.

Limits of servitude

The trouble with thinking of oneself as providing services – either in the service industries or in the customer-service sectors of manufacturing and retailing companies – is that one almost inescapably embraces ancient, pre-industrial modes of thinking. Worse still, one gets caught up in rigid attitudes that can have a profoundly paralyzing effect on even the most resolute of rationalists.

The concept of "service" evokes, from the opaque recesses of the mind, time-worn images of personal ministration and attendance. It refers generally to deeds one individual performs personally for another. It carries historical connotations of charity, gallantry, and selflessness, or of obedience, subordination, and subjugation. In these contexts, people serve because they want to (as in the priestly and political professions) or they serve because they are compelled to (as in slavery and such occupations of attendance as waiter, maid, bellboy, cleaning lady).

In the higher-status service occupations, such as in the church and the army, one customarily behaves ritualistically, not rationally. In the lower-status service occupations, one simply obeys. In neither is independent thinking presumed to be a requisite of holding a job. The most that can therefore be expected from service improvements is that, like Avis, a person will try harder. He will just exert more animal effort to do better what he is already doing.

So it was in ancient times, and so it is today. The only difference is that where ancient masters invoked the will of God or the whip of the foreman to spur performance, modern industry uses training programs and motivation sessions. We have not in all these years come very far in either our methods or our results. In short, service thinks humanistically, and that explains its failures.

Promise of manufacturing

Now consider manufacturing. Here the orientation is toward the efficient production of results, not toward attendance on others. Relationships are strictly businesslike, devoid of invidious connotations of rank or self.

When we think about how to improve manufacturing, we seldom focus on ways to improve our personal performance of present tasks; rather, it is axiomatic that we try to find entirely new ways of performing present tasks and, better yet, of actually changing the tasks themselves. We do not think of greater exertion of our animal energies (working physically harder, as the slave), of greater expansion of our commitment (being more devout or loyal, as the priest), or of greater assertion of our dependence (being more obsequious, as the butler).

Instead, we apply the greater exertion of our minds to learn how to look at a problem differently. More particularly, we ask what kinds of tools, old or new, and what kinds of skills, processes, organizational rearrangements, incentives, controls, and audits might be enlisted to greatly improve the intended outcomes. In short, manufacturing thinks technocratically, and that explains its successes.

Manufacturing looks for solutions inside the very tasks to be done. The solution to building a low-priced automobile, for example, derives largely from the nature and composition of the automobile itself. (If the automobile were not an assembly of parts, it could not be manufactured on an assembly line.) By contrast, service looks for solutions in the performer of the task. This is the paralyzing legacy of our inherited attitudes: the solution to improved service is viewed as being dependent on improvements in the skills and attitudes of the performers of that service.

While it may pain and offend us to say so, thinking in humanistic rather than technocratic terms ensures that the service sector of the modern economy will be forever inefficient and that our satisfactions will be forever marginal. We see service as invariably and undeviatingly personal, as something performed by individuals directly for other individuals.

This humanistic conception of service diverts us from seeking alternatives to the use of people, especially to large, organized groups of people. It does not allow us to reach out for new solutions and new definitions. It obstructs us from redesigning the tasks themselves, – from creating new tools, processes, and organizations, – and, perhaps, even from eliminating the conditions that created the problems.

In sum, to improve the quality and efficiency of service, companies must apply the kind of technocratic thinking which in other fields has replaced the high-cost and erratic elegance of the artisan with the low-cost, predictable munificence of the manufacturer.

THE TECHNOCRATIC HAMBURGER

Nowhere in the entire service sector are the possibilities of the manufacturing mode of thinking better illustrated than in fast-food franchising. Nowhere have manufacturing methods been employed more effectively to control the operation of distant and independent agents. Nowhere is "service" better.

Few of today's successful new commercial ventures have antecedents that are more humble and less glamorous than the hamburger. Yet the thriving nationwide chain of hamburger stands called "McDonald's" is a supreme example of the application of manufacturing and technological brilliance to problems that must ultimately be viewed as marketing problems. From 1961 to 1970, McDonald's sales rose from approximately $54 million to $587 million. During this remarkable ascent, the White Tower chain, whose name had theretofore been practically synonymous throughout the land with low-priced, quick-service hamburgers, practically vanished.

The explanation of McDonald's thundering success is not a purely fiscal one – i.e., the argument that it is financed by independent local entrepreneurs who bring to their operations a quality of commitment and energy not commonly found among hired workers. Nor is it a purely geographical one – i.e., the argument that each outlet draws its patronage from a relatively small geographic ring of customers, thus enabling the number of outlets easily and quickly to multiply. The relevant explanation must deal with the central question of why each separate McDonald's outlet is so predictably successful, why each is so certain to attract many repeat customers.

Entrepreneurial financing and careful site selection do help. But most important is the carefully controlled execution of each outlet's central function – the rapid delivery of a uniform, high-quality mix of prepared foods in an environment of obvious cleanliness, order, and cheerful courtesy. The systematic substitution of equipment for people, combined with the carefully planned use and positioning of technology, enables McDonald's to attract and hold patronage in proportions no predecessor or imitator has managed to duplicate. Consider the remarkable ingenuity of the system, which is worth examining in some detail:

To start with the obvious, raw hamburger patties are carefully prepacked and premeasured, which leaves neither the franchisee nor his employees any discretion as to size, quality, or raw-material consistency. This kind of attention is given to all McDonald's products. Storage and preparation space and related facilities are expressly designed for, and limited to, the predetermined mix of products. There is no space for any foods, beverages, or services that were not designed into the system at the outset. There is not even a sandwich knife or, in fact, a decent place to keep one. Thus the owner, has no discretion regarding what he can sell – not because of any contractual limitations, but because of facilities limitations. And the employees have virtually no discretion regarding how to prepare and serve things.

Discretion is the enemy of order, standardization, and quality. On an automobile assembly line, for example, a worker who has discretion and latitude might possibly produce a more personalized car, but one that is highly unpredictable. The elaborate care with which an automobile is designed and an assembly line is structured and controlled is what produces quality cars at low prices, and with surprising reliability considering the sheer volume of the output. The same is true at McDonald's, which produces food under highly automated and controlled conditions.

French-fried automation

While in Detroit the significance of the technological process lies in production, at McDonald's it lies in marketing. A carefully planned design is built into the elaborate technology of the food-service system in such a fashion as to make it a significant marketing device. This fact is impressively illustrated by McDonald's handling of that uniquely plebeian American delicacy, french-fried potatoes.

French fries become quickly soggy and unappetizing; to be good, they must be freshly made just before serving. Like other fast-food establishments, McDonald's provides its outlets with precut, partially cooked frozen potatoes that can be quickly finished in an on-premises, deep-fry facility. The McDonald's fryer is neither so large that it produces too many french fries at one time (thus allowing them to become soggy) nor so small that it requires frequent and costly frying.

The fryer is emptied onto a wide, flat tray adjacent to the service counter. This location is crucial. Since the McDonald's practice is to create an impression of abundance and generosity by slightly overfilling each bag of french fries, the tray's location next to the service counter prevents the spillage from an overfilled bag from reaching the floor. Spillage creates not only danger underfoot but also an unattractive appearance that causes the employees to become accustomed to an unclean environment. Once a store is unclean in one particular, standards fall very rapidly and the store becomes unclean and the food unappetizing in general.

While McDonald's aims for an impression of abundance, excessive overfilling can be very costly for a company that annually buys potatoes almost by the trainload. A systematic bias that puts into each bag of french fries a half ounce more than is intended can have visible effects on the company's annual earnings. Further, excessive time spent at the tray by each employee can create a cumulative service bottleneck at the counter.

McDonald's has therefore developed a special wide-mouthed scoop with a narrow funnel in its handle. The counter employee picks up the scoop and inserts the handle end into a wall clip containing the bags, One bag adheres to the handle. In a continuous movement the scoop descends into the potatoes, fills the bag to the exact proportions its designers intended, and is lifted, scoop facing the ceiling, so that the potatoes funnel through the handle into the attached bag, which is automatically disengaged from the handle by the weight of the contents. The bag comes to a steady, nonwobbling rest on its flat bottom.

Nothing can go wrong – the employee never soils his hands, the floor remains clean, dry, and safe, and the quantity is controlled. Best of all, the customer gets a visibly generous portion with great speed, the employee remains efficient and cheerful, and the general impression is one of extravagantly good service.

Mechanized marketing

Consider the other aspects of McDonald's technological approach to marketing. The tissue paper used to wrap each hamburger is color-coded to denote the mix of condiments. Heated reservoirs hold pre-prepared hamburgers for rush demand. Frying surfaces have spatter guards to prevent soiling of the cooks' uniforms. Nothing is left to chance or the employees' discretion.

The entire system is engineered and executed according to a tight technological discipline that ensures fast, clean, reliable service in an atmosphere that gives the modestly paid employees a sense of pride and dignity. In spite of the crunch of eager customers, no employee looks or acts harassed, and therefore no harassment is communicated to the customers.

But McDonald's goes even further. Customers may be discouraged from entering if the building looks unappealing from the outside; hence considerable care goes into the design and appearance of the structure itself.

Some things, however, the architect cannot control, especially at an establishment where people generally eat in their parked cars and are likely to drop hamburger wrappings and empty beverage cartons on the ground. McDonald's has anticipated the requirement: its blacktop parking facilities are dotted like a checkerboard with numerous large, highly visible trash cans. It is impossible to ignore their purpose. Even the most indifferent customer would be struck with guilt if he simply dropped his refuse on the ground. But, just in case he drops it anyway, the larger McDonald's outlets have motorized sweepers for quick and easy cleanup.

What is important to understand about this remarkably successful organization is not only that it has created a highly sophisticated piece of technology, but also that it has done this by applying a manufacturing style of thinking to a people-intensive service situation. If machinery is to be viewed as a piece of equipment with the capability of producing a predictably standardized, customer-satisfying output while minimizing the operating discretion of its attendant, that is what a McDonald's retail outlet is. It is a machine that produces, with the help of totally unskilled machine tenders, a highly polished product. Through painstaking attention to total design and facilities planning, everything is built integrally into the machine itself, into the technology of the system. The only choice available to the attendant is to operate it exactly as the designers intended.

TOOLING UP FOR SERVICE

Although most people are not aware of it, there are many illustrations of manufacturing solutions to people-intensive service problems. For example:

- Mutual funds substitute one sales call for many; one consultation for dozens; one piece of paper for thousands; and one reasonably informed customer choice for numerous, confused, and often poor choices.

- Credit cards that are used for making bank loans substitute a single credit decision (issuing the card in the first place) for the many elaborate, costly, people-intensive activities and decisions that bank borrowing generally entails.
- Supermarkets substitute fast and efficient self-service for the slow, inefficient, and often erratic clerks of the traditional service store.

In each of these examples a technological device or a manufacturing type of process has replaced what had been resolutely thought of as an irrevocably people-requiring service. Similar devices or processes can be used to modify and alleviate the customer-repelling abrasions of other people-intensive service conditions.

Consider the airlines. This industry is highly unusual. It is exceedingly capital-intensive in the creation of the facilitating product (the airplane), but it is extremely people-intensive in the delivery of the product (travel arrangements and the customer's flight experience). The possibilities for revenue production that a $20-million airplane represents are quickly vitiated by a surly or uncooperative reservations clerk. The potentials of repeat business that the chef so carefully builds into his meals can be destroyed by a dour or sloppy stewardess.

In fact, stewardesses have a particularly difficult job. A hundred passengers, having paid for reasonable service, understandably expect to be treated with some care. While three young ladies are there to serve them, a number of these passengers must inevitably get their drinks and meals later than others. Most experienced travelers are understanding and tolerant of the rushed stewardesses' problems, but a few usually harass them. The pressure and abuse can easily show in the stewardesses' personal appearance and behavior, and are likely to result in nearly all passengers being reciprocally mistreated. This is human. Besides, the ladies may have been on their feet all day, or may have slept only a few hours the night before.

"More and better training" is not likely to help things very much. When the pressure is on, service deteriorates. And so does a stewardess's cheerful manner and appearance, no matter how well schooled she is in personal care and keeping her cool or how attractively her clothes are designed.

But it might help to put mirrors in the airplane galley, so that each time a stewardess goes in she sees herself. There is some reason to expect that she'll look into the mirror each time she passes it, and that she'll straighten her hair, eliminate that lipstick smudge, put on a more cheerful face. Improvement will be instantaneous. No training needed.

Here is another possibility: the stewardess makes a quick trip down the aisle, passing out rum-flavored bonbons and explaining, "For those who can't wait till we get the ice out." This breaks the tension, produces an air of cheerfulness, acknowledges the passengers' eagerness for quick service, and says that the ladies are trying their hurried best. Further, it brings the stewardess into friendly personal contact with the passenger and reduces the likelihood of her being pressured and abused. She, in turn, is less likely to irritate other passengers.

From the manufacturing point of view, these two modest proposals represent the substitution of tools (or, as I prefer, technology) for motivation. Mirrors are a tool for getting self-motivated, automatic results in the stewardesses' appearance and personal behavior. Bonbons are a tool for creating a benign interpersonal ambience that reduces both the likelihood of customer irritation and the reciprocal and contagious stewardess irritation of others. They are small measures, but so is a company president's plant tour.

In each case there is considerable presumption of solid benefits. Yet to get these benefits one must think, as the factory engineer thinks, about what the problems are and what the desired output is; about how to redesign the process and how to install new tools that do the job more automatically; and, whenever people are involved, about how to "control" their personal behavior and channel their choices.

Hard & soft technologies

There are numerous examples of strictly "hard" technologies (i.e., pieces of equipment) which are used as substitutes for people – coffee vending machines for waitresses, automatic check-cashing machines for bank tellers, self-operated travel-insurance-policy machines for clerks. Although these devices represent a manufacturing approach to service, and while their principles can be extended to other fields, even greater promise lies in the application of "soft" technologies (i.e., technological systems). McDonald's is an example of a soft technology. So are mutual funds. Other examples are all around us, if we just think of them in the right way. Take the life insurance industry:

A life insurance salesman is said to be in a service industry. Yet what does he really do? He researches the prospect's needs by talking with him, designs several policy models for him, and "consumer-use tests" these models by seeking his reactions. Then he redesigns the final model and delivers it for sale to the customer. This is the ultimate example of manufacturing in the field. The factory is in the customer's living room, and the producer is the insurance agent, whom we incorrectly think of as being largely a salesman. Once we think of him as a manufacturer, however, we begin to think of how best to design and manufacture the product rather than how best to sell it.

The agent, for example, could be provided with a booklet of overlay sheets showing the insurance plans of people who are similar to the customer. This gives the customer a more credible and informed basis for making a choice. In time, the agent could be further supported by similar information stored in telephone-access computers.

In short, we begin to think of building a system that will allow the agent to produce his product efficiently and effectively by serving the customer's needs instead of performing a manipulative selling job.

Manufacturers outside the factory

The type of thinking just described applies not only to service industries but also to manufacturing industries. When the computer hardware manufacturer provides installation and maintenance services, debugging dry-runs, software programs, and operator training as part of his hardware sales program, he acknowledges that his "product" consists of considerably more than what he made in the factory. What is done in the field is just as important to the customer as the manufactured equipment itself. Indeed, without these services there would generally be no sale.

The problem in so many cases is that customer service is not viewed by manufacturers as an integral part of what the customer buys, but as something peripheral to landing the sale. However, when it is explicitly accepted as integral to the product itself and, as a consequence, gets the same kind of dedicated attention as the manufacture of the hardware gets, the results can be spectacular. For example:

- In the greeting card industry, some manufacturer-provided retail display cases have built-in inventory replenishment and reordering features. In effect, these features replace a company salesman with the willing efforts of department managers or store owners. The motivation of the latter to reorder is created by the visible imminence of stockouts, which is achieved with a special color-coded card that shows up as the stock gets low. Order numbers and envelopes are included for reordering. In earlier days a salesman had to call, take inventory, arrange the stock; and write orders. Stock-outs were common.

  The old process was called customer service and selling. The new process has no name, and probably has never been viewed as constituting a technological substitute for people. But it is. An efficient, automatic, capital-intensive system, supplemented occasionally by people, has replaced an inefficient and unreliable people-intensive system.

- In a more complex situation, the A.O. Smith Company has introduced the same kind of preplanning, routinizing, people-conserving activity. This company makes, among other things, grain storage silos that must be locally sold, installed, serviced, and financed. There are numerous types of silos with a great variety of accessories for loading, withdrawing, and automatically mixing livestock feed. The selling is carried out by local distributor-erectors and is a lengthy, difficult, sophisticated operation.

  Instead of depending solely on the effective training of distributors, who are spread widely in isolated places, A.O. Smith has developed a series of sophisticated, colorful, and interchangeable design-module planning books. These can be easily employed by a distributor to help a farmer decide what he may need, its cost, and its financing requirements. Easy-to-read tables, broken down by the size of farm, numbers and types of animals, and purpose of animals (cattle for meat or cows for milk), show recommended combinations of silo sizes and equipment for maximum effectiveness.

  The system is so thorough, so easy to use and understand, and so effective in its selling capability that distributors use it with great eagerness. As a consequence, A.O. Smith, while sitting in Milwaukee, in effect controls every sales presentation made by every one of its far-flung distributors. Instead of constantly sending costly company representatives out to retrain, cajole, wine-and-dine, and possibly antagonize distributors, the supplier sends out a tool that distributors want to utilize in their own self-interest.

Product-line pragmatics

Thinking of service as an integral part of what is sold can also result in alteration of the product itself – and with dramatic results. In 1961, the Building Controls and Components Group of Honeywell, Inc., the nation's largest producer of heating and air conditioning thermostats and control devices, did a major part of its business in replacement controls (the aftermarket). These were sold through heating and air conditioning distributors, who then supplied plumbers and other installation and repair specialists.

At that time, Honeywell's product line consisted of nearly 18,000 separate catalog parts and pieces. The company had nearly 5,000 distributor accounts, none of which could carry a full line of these items economically, and therefore it maintained nearly 100 fully stocked field warehouses that offered immediate delivery to distributors. The result was that, in a large proportion of cases, distributors sold parts to plumbers that they did not themselves have in stock. They either sent plumbers to nearby Honeywell warehouses for pickup or picked up parts themselves and delivered them directly to the plumbers. The costs to Honeywell of carrying these inventories were enormous, but were considered a normal expense of doing business.

Then Honeywell made a daring move – it announced its new Tradeline Policy. It would close all warehouses. All parts would have to be stocked by the distributors. The original equipment, however, had been redesigned into 300 standard, interchangeable parts. These were interchangeable not only for most Honeywell controls, but also for those of its major competitors. Moreover, each package was clearly imprinted to show exactly what Honeywell and competing products were repairable with the contents.

By closing its own warehouses, Honeywell obviously shifted the inventory-carrying costs to its distributors. But instead of imposing new burdens on them, the new product lines, with their interchangeability features, enabled the distributors to carry substantially lower inventories, particularly by cutting down the need for competitive product lines which the distributors could nonetheless continue to service. Thus they were able to offer faster service at lower costs to their customers than before.

But not all distributors were easily persuaded of this possibility, and some dropped the line. Those who were persuaded ultimately proved their wisdom by the enormous expansion of their sales. Honeywell's replacement market share almost doubled, and its original equipment share rose by nearly 50%. Whereas previously nearly 90% of Honeywell's replacement sales were scattered among 4,000 distributors, within ten years after Tradeline's introduction the same proportion (of a doubled volume) was concentrated among only about 900 distributors. Honeywell's cost of servicing these fewer customers was substantially less, its trade inventory carrying costs were cut to zero, and the quality of its distributor services was so substantially improved that only 900 of its distributors captured a larger national market share than did the nearly 4,000 less efficient and more costly distributors.

Again, we see a people-intensive marketing problem being solved by the careful and scrupulous application of manufacturing attitudes. Motivation, hard work, personalization, training, and merchandising incentives were replaced with systematic programming, comprehensive planning, attention to detail, and particularly with imaginative concern for the problems and needs of customers (in this case, the company's distributors).

STOPGAPS: COMPLEXITY...

Exaggeration is not without its merits, especially in love and war. But in business one guards against it with zeal, especially when one tries to persuade oneself. The judicious application of the manufacturing mentality may help the service industries and the customer-service activities of others. Yet this does not necessarily mean the more technology, the better.

Entrepreneurial roadsides are littered with the wrecks of efforts to install Cadillac technologies for people who cannot yet handle the Model T. This point is illustrated by the failure of two exceedingly well-financed joint ventures of highly successful technology companies. These joint ventures attempted to provide computerized medical diagnostic services for doctors and hospitals. The companies developed console hook-ups to central diagnostic computers, so that everybody could stop sending off samples to pathology laboratories and agonizingly poring through medical texts to diagnose the patients' symptoms.

The ventures failed because of hospital and doctor resistance, not for want of superior or reliable products. The customer was compelled suddenly to make an enormous change in his accustomed way of doing things, and to employ a strange and somewhat formidable piece of equipment that required special training in its use and in the interpretation of its output.

Interactive teaching machines are meeting a similar fate. The learning results they achieve are uniformly spectacular. The need for improved learning is a visible reality. The demand for greater individualization of teaching is widespread. But the equipment has not sold because technologists have created systems employing equipment that is at the cutting edge of technological progress. The teachers and school systems that must use them are far behind, and already feel badly bruised by their failure to comprehend even simple new technologies. For them, the new Cadillac technologies do not solve problems. They create problems.

... & compromise

On the other hand, failure to exploit technological possibilities can be equally destructive. When a major petroleum company with nearly 30,000 retail outlets in the United States was persuaded to pioneer a revolutionary automobile repair and servicing system, compromises of the original plan ensured the system's failure.

The theory was to build a gigantic service and repair system that could handle heavy volumes of continuous activity by using specialized diagnostic and repair equipment. With this equipment (rather than a harried and overworked man at a gas station) pinpointing the exact problems, cars could be shuttled off to specific stations in the repair center. Experts would work only on one kind of problem and section of a car, with newly designed, fast-action tools. Oil changes would me made in assembly-line fashion by low-paid workers, electrical work would be performed by high-paid technicians doing only that, and a post-diagnostic checkup would be made to guarantee success.

Since profitability would require high volume, the center would have to draw on a vast population area. To facilitate this, the original proposal called for a specially constructed building at a center-city, old warehouse location – the land would be cheaper, the building would be equally accessible throughout the entire metropolitan area, the service center's technological elegance and see-through windows for customers would offset any run-down neighborhood disadvantages, and volume business would come from planned customer decisions rather than random off-street traffic.

The original concept also called for overnight pickup and delivery service; thus a car could be repaired at night while its owner slept, rather than during the day when he would need it. And because the required promotion of this service would tend to alienate the company's franchised service station dealers, perhaps driving them into the hands of competitors, it was recommended that the first center be installed in a major city where the company had no stations.

This sounds like an excellent manufacturing approach to a service situation; but the company made three fatal compromises:

1) It decided to place the center in a costly, high-traffic suburban location, on the grounds that "if the experiment fails, at least the building will be in a location that has an alternative use." The results were an awkward location, a land-acquisition cost five times higher than the original center-city location, and, therefore, a vastly inflated break-even point for the service center.

2) It decided not to offer overnight service, on the grounds that "we'd better crawl before we walk. And besides, we don't think people will leave their cars overnight in a strange and distant garage." The fact that the results would be guaranteed by a reputable, nationally known petroleum company operating an obviously sophisticated new type of consumer service facility was not persuasive to the corporate decision makers.

3) It decided to put the first center in a city occupied by its own franchised dealers, on the grounds that "we know it better." To offset the problem of not being able to advertise aggressively for business, the company offered its dealers a commission to send their repair jobs to the center. The dealers did this, but only with jobs they could not, or did not want to, do themselves. As a result, the traffic at the big, expensive center was miserably low.

Companies that take a manufacturing approach to service problems are likely to fail if (a) they compromise technological possibilities at the conception and design stage, or (b) they allow technological complexity to contaminate the operating stage. The substitution of technology and systems for people and serendipity is complex in its conception and design; only in its *operation*, as at McDonald's, is it simple.

It is the simplicity of mutual funds that, after all, accounts for their success. But the concept is in fact much more complex than that of selling individual stocks through a single customer-man sitting at a desk. Mutual funds are the financial community's equivalent of McDonald's. They are a piece of technology that not only simplifies life for both the seller and the buyer but also creates many more buyers and makes production more profitable.

Mass merchandising is similar. It substitutes a wide selection and fast, efficient self-service for a narrow selection and slow, incompetent salesclerk service. The mass merchandising retail store (e.g., general merchandise supermarket) is a new technology, incorporating into retailing precisely the thinking that goes into the assembly line, except that the customer does his own assembling.

WHY THINGS GO WRONG

The significance of all this is that a "product" is quite different from what it is generally assumed to be. When asked once what he did, Charles Revson, head of Revlon, Inc., made the now well-known reply, "In the factory we make cosmetics, in the store we sell hope." He defined the product in terms of what the consumer wanted, not in terms of what the manufacturer made. McDonald's obviously does the same – not just hamburgers but also speed, cleanliness, reassurance, cheerfulness, and predictable consistency. Honeywell defined it not in terms of replacement parts but, rather, in terms of those needs of its distributors which, if met, would result in substantially larger proportions of patronage for Honeywell. Thus a product is not something people buy, but a tool they use – a tool to solve their problems or to achieve their intentions.

So many things go wrong because companies fail to adequately define what they sell. Companies in so-called service industries generally think of themselves as offering services rather than manufacturing products; hence they fail to think and act as comprehensively as do manufacturing companies concerned with the efficient, low-cost production of customer-satisfying products.

Moreover, manufacturing companies themselves do not generally think of customer service as an integral part of their products. It is an afterthought to be handled by the marketing department.

The marketing department, in turn, thinks of itself as providing customer services. There is a hidden and unintentional implication of giving something away for free. One is doing something extra as a favor. When this is the underlying communication to one's own organization, the result is about what one would expect – casual, discretionary attitudes and little attention to detail, and certainly no attention to the possibilities of substituting systems and preplanning for people and pure effort. Hence products are designed that cannot be easily installed, repaired, or modified.

(Motorola's "works in a box" television set, which has been promoted so successfully on the basis of its easy replacement and repairability, is an outstanding example of the sales-getting potential of proper care in design and manufacturing.)

Chill winds from ice cream

An excellent example of the confusion between what a company "makes" and what a customer "buys" is provided by a producer of private-label ice cream products for supermarket chains. Since supermarkets need to create low-price impressions in order to attract and hold customers, selling successfully to them means getting down to rock-bottom prices. The company (call it the Edwards Company) became extraordinarily good at producing a wide line of ice cream products at rock-bottom costs. It grew rapidly while others went bankrupt. It covered ten states with direct deliveries to stores out of its factory and factory warehouse, but continued growth eventually required establishing plant, distribution, and marketing centers elsewhere. The result was disaster, even though the company manufactured just as efficiently in the new locations as it did in the old.

Under the direct and constant supervision of the president in the original Edwards location, an exceedingly efficient telephone ordering and delivery system had been working to meet the supermarkets' rather stringent requirements. Because of limited storage and display space, they required several-times-a-week delivery at specified, uncrowded store hours. To make up for low volume in slow periods, they needed regular specials as well as holiday and summer specials. Over time, these needs had become so automatically but efficiently supplied from the original Edwards factory location that this delivery service became routinized and therefore taken for granted.

In building the new plant, the president and his compact management team focused on getting manufacturing costs down to rock bottom. After all, that is what made the sale – low prices. Not being very conscious of the fact that they had created in the original location an enormously customer-satisfying, efficient, automatic ordering and delivery system, they did not know exactly what to look for in evaluating how well they were working out these "service" details at the new plant, distribution, and marketing centers.

In short, they did not know what their product really was (why Edwards had become so successful) and they failed to expand Edwards' success. Service was not considered an integral part of the company's product. It was viewed merely as "something else" you do in the business. Accordingly, service received inadequate attention, and that became the cause of the Edwards Company's failure.

CONCLUSION

Rarely is customer service discretionary. It is a requisite of getting and holding business, just like the generic product itself. Moreover, if customer service is consciously treated as "manufacturing in the field," it will get the same kind of detailed attention that manufacturing gets. It will be carefully planned, controlled, automated where possible, audited for quality control, and regularly reviewed for performance improvement and customer reaction. More important, the same kinds of technological, labor-saving, and systems approaches that now thrive in manufacturing operations will begin to get a chance to thrive in customer service and service industries.

Once service-industry executives and the creators of customer-service programs begin seriously to think of themselves as actually manufacturing a product, they will begin to think like product manufacturers. They will ask: What technologies and systems are employable here? How can things be designed so we can use machines instead of people, systems instead of serendipity? Instead of thinking about better and more training of their customer-service representatives, insurance agents, branch bank managers, or salesmen "out there," they will think about how to eliminate or supplement them.

If we continue to approach service as something done by individuals rather than by machines or systems, we will continue to suffer from two distortions in thinking:

a. Service will be viewed as something residual to the ultimate reality – to a tangible product, to a specific competence (like evaluating loans, writing insurance policies, giving medical aid, preparing on-premises foods). Hence it will have residual respectability, receive residual attention, and be left, somehow, for residual performers.

b. Service will be treated as purely a human task that must inevitably be diagnosed and performed by a single individual working alone with no help or, at best, with the rudimentary help of training and a variety of human-engineering motivators. It will never get the kind of manufacturing-type thinking that goes into tangible products.

Until we think of service in more positive and encompassing terms, until it is enthusiastically viewed as manufacturing in the field, receptive to the same kinds of technological approaches that are used in the factory, the results are likely to be just as costly and idiosyncratic as the results of the lonely journeyman carving things laboriously by hand at home.

H. Igor Ansoff (1975)

MANAGING SURPRISE AND DISCONTINUITY: STRATEGIC RESPONSE TO WEAK SIGNALS

The question of how to protect the functioning of logistics systems from risk, crises, extreme uncertainty or "strategic surprise" – as Igor Ansoff (1918–2002) called it – is another of the priorities of the field. The events of "9-11", natural catastrophes, and the general increase in the volatility of the global economy have been adding urgency recently.

Ansoff, the Russian-American Professor of Management, who is credited with founding the discipline of Strategic Management, in several contributions of the 1970s was among the first authors to suggest systematic analysis and responses to the issues of risk management – with obvious relevance to the issue of supply chain risk management.

In the article reprinted here he distinguishes between the concepts of "crisis preparedness" and "strategic preparedness". With the first group of concepts he is referring to the ability of organizations to deal with crises – which may be achieved through structural adaptability and robustness, as suggested in the contributions by Simon (1962) and Weick (1976) in this volume. Strategic preparedness, which is Ansoff's focus in the paper presented here, refers to an organization's ability to make the best uses of information that can be had about impending risks: the ability to hear "weak signals" and respond to them.

Reference:

H. Igor Ansoff: Managing Strategic Surprise by Response to Weak Signals. In: *California Management Review*, Vol. XVIII, No. 2, 1975, pp. 21–33.

P. Klaus and S. Müller (eds.), *The Roots of Logistics*, DOI 10.1007/978-3-642-27922-5_21,

H. Igor Ansoff: Managing Surprise and Discontinuity: Strategic Response to Weak Signals

I. CONCEPTUAL FRAMEWORK

1. The Problem

The paradox of strategic military surprise has been a familiar phenomenon throughout recorded human history. From the Trojan Horse to Pearl Harbour to the Kippur war, nations and armies have been confronted with sudden crises, in spite of ample information about enemy intentions.

The recent "petroleum crises" was a comparable event in the industrial world: large and important firms were suddenly confronted with a major discontinuity, although advanced forecasts of Arab action were not only publicly available, but on the day of the surprise, were to found on the desks of some of the surprised managers.

The petroleum crisis, because of its pervasive scope, highlighted the danger of strategic business surprises. But such surprises had overtaken numerous firms, one by one, from the early 1950s, enough of them to provide material for a "Fortune" book titled "Corporations in Crisis".

In the aftermath, it was argued that these corporations were caught unawares, because they lacked modern forecasting and planning systems. But in the 1970s a majority of the firms caught by the petroleum crisis had such systems. In the mid 1960s, the management of one of the world's largest conglomerates proudly displayed its planning and control. A week after public display, the same management made a red-faced admission of **two** multi-million dollar surprises: a major overrun in its office furniture division and another in its ship-building division.

The American automotive industry, a leader in modern planning and control, was certainly unprepared for the forceful Congressional position on automotive safety. And a bare four years later it was again "surprised" by the success of the small car.

Such events need little support from the voluminous literature on futurology to predict that discontinuities and surprises will occur with increasing frequency. If, as experience suggests, modern planning technology does not insure against surprises, the technology needs to be extended to provide such insurance. An exploration of such extension is the purpose of this paper.

2. The nature of Strategic Surprise

EXHIBIT 1 plots, against time, the growth of a firm, which can be measured by any one of common yardsticks, such as sales, profits, ROI. The middle line shows smooth extrapolation of past experience into the future. The two branching curves, a threat and an opportunity, show a significant departure, a discontinuity from the past. In principle, such discontinuities can be anticipated by available forecasting techniques. Given enough warning, the firm should be able to avert the threat, or seize the opportunity.

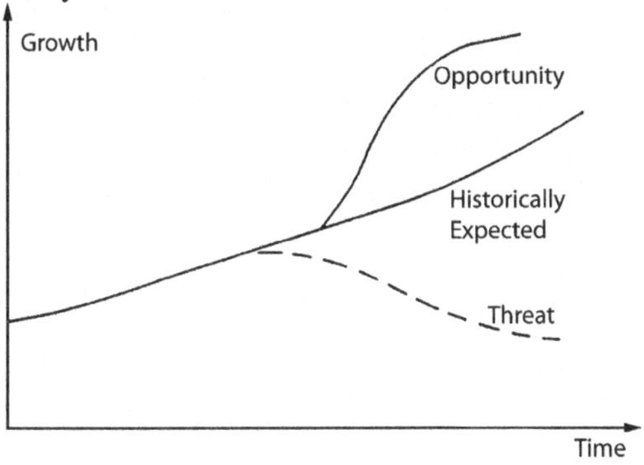

EXHIBIT 1: IMPACT OF THREAT/OPPORTUNITY

In fact, firms often fail to anticipate and suddenly discover that a fleeting opportunity has been missed. Typically, at the "clear moment of truth", neither the causes nor the possible responses are clear and the firm confronts an unfamiliar and often threatening event. We shall call such events **strategic surprises**: sudden, urgent, unfamiliar changes in the firm's perspective which threaten either a major profit reversal or loss of a major opportunity.

A firm, which wishes to prepare for strategic surprises, has two options:

(1) To develop a capability for effective **crisis management**: fast and efficient, **after the fact** responsiveness to sudden discontinuities. A useful prototype is a fire-fighting company. Unable to predict or control occurrence of fires, it prepares itself, through repeated practice, to respond quickly and effectively a whole range of different alarms.

(2) The second approach is to treat the problem **before the fact**, minimize the probability of strategic surprises: to prepare in such a way, that by the time it strikes, a strategic discontinuity has lost its suddenness, urgency and unfamiliarity.

Both approaches deserve management attention: before the fact strategic preparedness because it is the more efficient approach; crisis preparedness because, even the best advance efforts do not assure immunity from surprises. In a companion paper we have dealt with crisis preparedness.[1] In this paper we shall deal with before the fact strategic preparedness.[2]

3. Focus on Responsiveness

A common feature of strategic surprises is the failure of decision makers to act on information available in advance of the event. We need, therefore, to distinguish between the **available** and the **utilized** information. A way to do this is illustrated in EXHIBIT 2, which suggests three levels of information. The first level is the "common knowledge" available in the environment of the firm through word of mouth, in publications, in intelligence analyses, in forecasts.

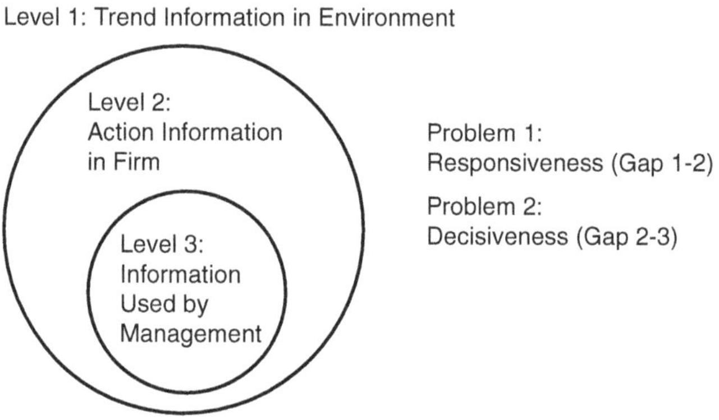

EXHIBIT 2: INFORMATION LEVELS IN THE "SURPRISE PROBLEM"

Level 2 is the information generally available within the firm. Part of it is in the same form as the outside information, but the bulk interprets this "raw" data into consequences for the firm: competitive analyses, threats and opportunities, sales and profits forecast, plans, program budgets. As the petroleum crisis has shown, typically, only historically familiar raw data finds its way into the interpreted consequences. Reports on unfamiliar discontinuities, if they find their way into the firm, remain in raw form, because the methods and approaches for converting them into action typically do not exist.

Thus, as EXHIBIT 2 suggests, there is a gap between the information available in the environment and actionable information available to managers. But there is also another gap between the latter and the information actually employed in managerial decision making and action. Again, it is the familiar, surprise-threatening information, that is typically rejected by managers (until the moment of crisis), as "too ab-

---

1 H. Igor Ansoff, J. Eppink, and H. Gomer: *Management of Surprise and Discontinuities: Increasing Managerial Responsiveness* (forthcoming).

2 Op. Cit. Ref. 1.

stract" unreliable, unrelated to past experience and therefore irrelevant to the problems of the firm.

Thus avoiding of surprise involves closing of two important **gaps**:

(1) The **responsiveness** gap between information available in the environment and action plans within the firm, and

(2) The **decisiveness** gap between the available plans and managerial use of plans.

Both need to be analyzed and reduced, if the firm is to avoid surprise. Since we have treated the latter in another paper[3] this paper focuses on the responsiveness gap.

4. Limitations of Environmental Information

There is now a well developed technology, called **strategic planning**, for converting environmental information about threats and opportunities into concrete action plans, programs and budgets. But to date strategic planning has had little success in dealing with surprises. One reason is the cumbersome, costly and time-consuming process by which most firms prepare their plans. A potential surprise which surfaces suddenly needs to be of crisis proportions to justify interrupting the annual planning cycle. A majority of strategic discontinuities, unless they conveniently match the planning calendar, must wait for attention until next cycle.

But lack of speed and responsiveness to environmental discontinuities is only one problem in treatment of surprises. Another, more fundamental, problem is inherent limitation in the available information.

To be useful in strategic planning, information must satisfy two conditions. First, it must be forecastable far enough into the future to permit timely preparation and response to the threat/opportunity.

Second, the content of the forecast must be adequate for estimating the impact on the firm, for identifying specific responses, and for estimating the potential profit impact of these responses.

In both strategic planning literature and practice, an implicit assumption is made that both conditions can be satisfied; that the forecaster can meet the needs of the planner. This assumption turns out to be true when planning is concerned either with simple extrapolation of the past, or with "logical", incremental developments of historical trends. But when a potential surprise originates in an alien technology, or with a previously unknown competitor, or with a new political coalition, or with a new economic phenomenon – in all of these cases both the timeliness and the content requirements frequently cannot be met. The firm **can** have a longer time perspective, if it is willing to put up with increasing vagueness of content; or the firm can wait for originally vague information to become specific.

[3] Op. Cit. Ref. 1.

In the past waiting has been the typical behavior. If by the time the information became specific, there was still enough time left for an orderly response, nothing was lost. But frequently the information arrived too late and the firm faced a crisis.

The ability of the firm to effect a timely response depends on two variables: the external rapidity with which the threat/opportunity rises and decays, and the amount of time needed by the firm to plan and effect the response. Since the 1950s these two variables have been on a collision course: the rate of environmental change has accelerated, and the firm's response has been made slower by growing size and complexity.

Thus an apparent paradox: if the firm waits for information it will be increasingly surprised by a crisis, if it accepts vague information, the content will not be specific enough to estimate impact, alternatives and consequences. A way out of this paradox is to change the planning approach: instead of deriving strategic information needs from **a priori** decision on what responses are **desirable**; to determine what responses are **feasible** in the light of available information. This implies that, early in the life of a threat, when the information is vague and its future course is unclear, the responses will be correspondingly unfocussed, aimed at increasing flexibility. As the information becomes more precise, so will the firm's response, terminating eventually in a direct attack on the threat or an opportunity.

We might call this graduated response "amplification and response to weak signals"[4] in contrast to the conventional strategic planning which depends on the existence of "strong signals". It is our purpose in this paper to develop a practical planning approach to such response. Our first task is to explore the range of "weak signals" which can be typically expected from a strategic discontinuity.

5. States of Ignorance

EXHIBIT 3 illustrates the threat information typically required in strategic planning. The three curves in the left part of the exhibit trace the alternative impact of a threat on the profits of the firm, the right part of the exhibit shows three possible responses to one of the impact curves. Without a detailed discussion of the mathematical notation, the multiplicity of probable paths gives the impression of information which is imperfect and uncertain. But a closer look at EXHIBIT 3 shows that while uncertain, this is very **content rich** information: the threat is well enough understood to compute the possible profit consequences, the responses are well enough developed to estimate both their costs and their counter effect on the threat. It is reasonable to expect this much knowledge from a threat/opportunity which arises from a familiar prior experience. This will be the case when a competitor introduces a new marketing strategy, a new product, a new pricing strategy. But when the threat/opportunity is discontinuous (such as impact of laser technology on land sur-

---

[4] I owe this happy phrase to W.W. Bryant, Manager of TEO Central, in the Philips Company.

veying, or of large-scale integration on electronic components), then, in the early stages, the nature, the impact and the possible responses are unclear. Frequently it is not even clear whether the discontinuity will develop into a threat or an opportunity.

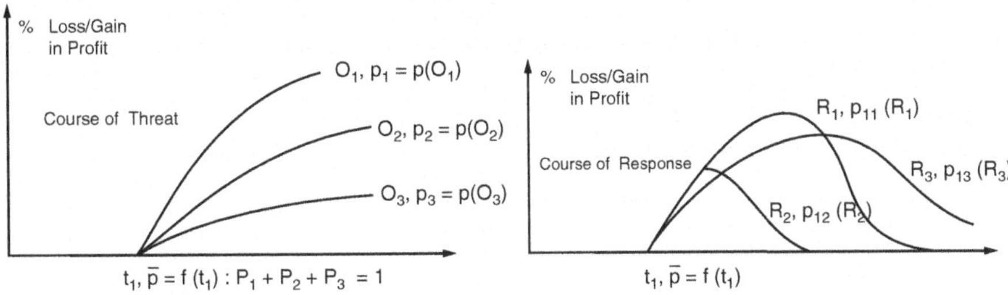

EXHIBIT 3: EXAMPLE OF INFORMATION REQUIRED FOR THREAT ANALYSIS IN STRATEGIC PLANNING

Thus the information contained in EXHIBIT 3 while desirable, is not to be arbitrarily expected from unfamiliar discontinuities. Initially, we must be prepared for very vague information which will progressively develop and improve with time. We shall characterize such progression by the successive **states of ignorance**. These are illustrated in EXHIBIT 4 (on the next page), where stage number five, the most information-rich state, contains exactly the same information as EXHIBIT 3. As entries in the left-hand column indicate, enough is known to compute both the probable profit impact of the discontinuity and the profit impact of the response.

At the other extreme, in column (1), is the vaguest state of information which can be of use to management. As the "NO's" show, all that is known is that **some** threats and opportunities will undoubtedly arise, but their shape and nature and source are not yet known. In today's "political and economic fog of uncertainty"[5] many firms find themselves in such a state of ignorance. Having witnessed and experienced shocks of change in recent past, managers are convinced that new discontinuities are coming, but they cannot identify the source.

States of ignorance on level two **improves** matters somewhat. For example, in the early 1940s, it was generally recognized by physicists that solid-state physics was pregnant with potential for the electronics industry. But the invention of the specific discontinuity, the transistor, was still several years off. The source of the threat was clear, but not the threat itself.

When the transistor was invented by **Shockley** and his team, the ignorance was reduced to State three, but at the outset, the ramifications of the inventions were unclear, nor the defensive and the aggressive responses which different firms were eventually to make.

[5] My thanks to Mr. Leslie Smith, Chairman of the BOC Limited, London.

When the firms developed and made the initial responses, and ignorance was reduced to level four, the eventual investments and the profits were not yet visible. Pioneering firms were investing boldly into the new technology with little experience to guide them, in high hopes that their entrepreneurial risk taking will pay off.

State five was not reached until knowledge of crystal yields and of manufacturing process costs was sufficient to make reasonable predictions of the ultimate technology and its profitability.[6] But, by then, the leaders were entrenched and those who originally held back had to pay a high cost of entry into the industry.

States of Ignorance / Info Content	(1) Sense of Threat/ Opportunity	(2) Source of Threat/ Opportunity	(3) T/O Concrete	(4) Response Concrete	(5) Outcome Concrete
Conviction that discontinuities are impending	YES	YES	YES	YES	YES
Area or organization is identified which is the source of discontinuity	NO	YES	YES	YES	YES
Characteristics of threat, nature of impact, general gravity of impact, timing of impact	NO	NO	YES	YES	YES
Response identified: timing, actions, programs, budgets	NO	NO	NO	YES	YES
Profit impact and consequences of responses are computable	NO	NO	NO	NO	YES

EXHIBIT 4: STATES OF IGNORANCE UNDER DISCONTINUITY

[6] We should note in passing that the information in **each state** of ignorance may be certain, uncertain, or risky in the sense of definitions commonly used in statistical decision theory. The focus in the table is on illustrating the variability of **content** and not in the **state of uncertainty**. The dimension of uncertainty can be easily added orthogonally to the table, thus creating a cube of possible states of information. In this cube, the states of information, treated in statistical decision theory would be included in slice number five.

As indicated by the growing number of YES's in EXHIBIT 4, ignorance is reduced and information is enriched as a threat/opportunity evolves from state one to state five. In the history of a given threat/opportunity, this evolution takes place over time. The speed of the evolution limits the time available to the firm for mounting a response. A useful measure is the time remaining before the impact of the discontinuity on the firm passes a critical profit benchmark. For a threat this is the level of loss at which the survival of the firm is threatened; for an opportunity this is the point beyond which it is too late: the cost of climbing on the bandwagon can no longer be recovered through profits.

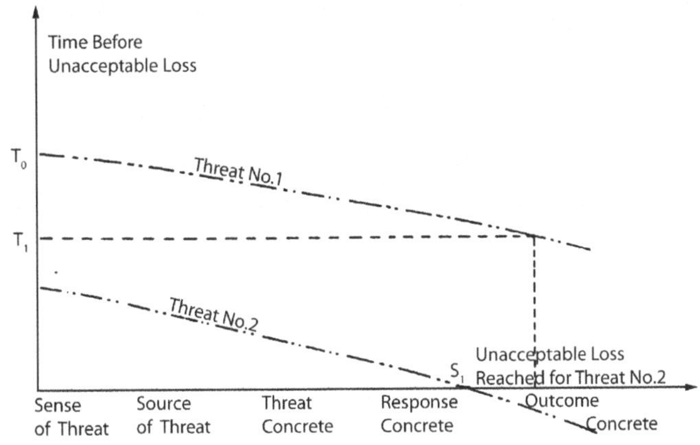

EXHIBIT 5: EVOLUTION OF THREAT

Two hypothetical threat histories are illustrated in EXHIBIT 5. Threat No. 1, appears to develop favorably to the firm. State five information, adequate for strategic planning, becomes available T_1 in advance of the unacceptable threat level, allowing time for planning and response. Threat Nr. 2, on the other hand is a potential source of surprise. It arises late, leaving little time before crisis, the crisis level is reached at State S_1, before the information is concrete enough for strategic planning.

It is clear that Threat No. 2 is a trouble maker; the quality of the Threat No. 1 is less clear. The answer depends on whether the time T_1 is adequate to prepare a strategic plan and to execute it in a timely manner. This depends on how well the firm uses the period since the sense of threat was felt, T_0 before the critical point.[7] A firm which requires state five information before it responds, will at best, monitor the evolution of the threat. A firm prepared to respond to partial information can use the period T_0–T_1 to enhance the responsiveness of the firm in a way so that the residual time T_1 **becomes adequate**! It is to this possibility that we now turn.

---

[7] Beyond this point content of threat and response are well understood. But information will continue to get better, thanks to progressive reduction of uncertainty.

6. Alternative Response Strategies

Just as we have expanded the states of information to include poorer knowledge, we need to enlarge the repertoire of responses to permit weaker responses. Such repertoire is shown in EXHIBIT 6, where management options are subdivided into two groups (1) responses which change the firm's relationship with the **environment** and (2) responses which change the **internal dynamics and structure** of the firm. For each groups there are three categories of response strategies:

(i) those that enhance the firm's awareness of its outer and inner worlds;

(ii) those that increase the firm's flexibility for response to vague threat/opportunities and

(iii) those that pin-point the threat and respond directly to it.

Thus the entries in the table provide a total of six response strategies.

Response Strategies / Domain of Response	Direct Response	Flexibility	Awareness
Relationship to Environment	External action (strategic planning and implementation)	External flexibility	Environmental awareness
Internal Configuration	Internal readiness (contingency planning)	Internal flexibility	Self-awareness

EXHIBIT 6: ALTERNATIVE RESPONSE STRATEGIES

The **External Action** strategy, as its name implies, mounts a direct counteraction against identified threats or opportunities. It proceeds through selection of the attack strategy, translation into plans and programs and implementation of the latter. The end result is a threat averted, or an opportunity captured, an enhanced potential for future profits. Preparation for direct action is the object of strategic planning. The range of specific actions available under the direct action strategy can be divided into two groups: (i) aggressive actions, which treat the discontinuity as an opportunity and (ii) defensive actions, which regard it as a threat. Some of the specific action alternatives are shown in the upper part of EXHIBIT 7.

DIRECT RESPONSE

EXTERNAL ACTION

- Optimize Timing of Response
- Seize Opportunity-Enter New Product Market
- Convert a Threat to Opportunity
- Change Competitive Strategy
- Share Risks with other Firms
- Secure Supply of Scarce Resource
- Diversify Threatened Resource Technology
- Reduce Commitment to Threatened Areas
- Divest from a Threatened Product Market

INTERNAL READINESS

- Preplan
- Adapt Structure and Systems
- Acquire Technology
- Preposition Resources
- Acquire Skills
- Build Facilities
- Develop New Product Services
- Develop Operational Capability

EXHIBIT 7: MEASURES FOR DIRECT RESPONSE STRATEGY

The lower part of Exhibit 7 shows measures available in the **internal readiness** strategy, which match the skills, the structure, and the resources of the firm to the demands of specific threats/opportunities. The result of internal readiness is a state of preparedness for external action. In strategic planning internal readiness is commonly referred to as "strategy implementation" signifying that preparedness must await selection of the course of action that it will support. In **Chandler**'s words,[8] "structure follows strategy". Thus, the prescribed sequence is: Strategic planning → internal readiness → action.

For triggering off this sequence, it is essential that the strategic information be in the highly developed state five. But many of the measures can be successfully carried out in state three, as soon as the shape of the impending threat/opportunity becomes concrete and before state five is reached. Thus, it is possible to accelerate response by reversing the sequence to: internal preparedness → strategic planning → action. The earliest possible response to an opportunity/threat is offered by the pair of **awareness strategies**, shown at the right hand of EXHIBIT 6 and elaborated into action alternatives in EXHIBITS 8 and 9 (on the next page).

[8] H. Igor Ansoff: *Corporate Strategy*, 1965.

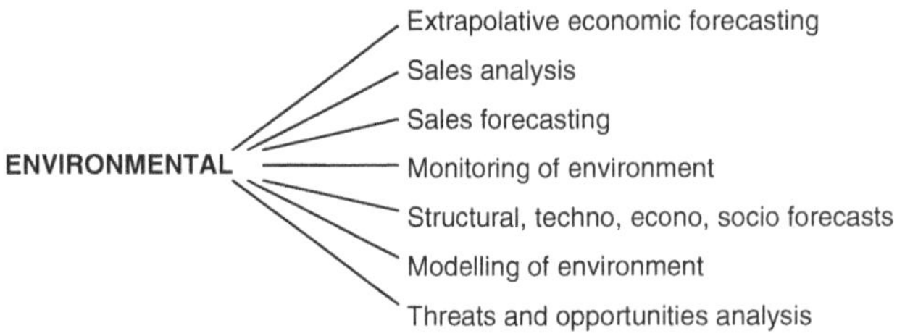

EXHIBIT 8: ALTERNATIVES FOR AWARENESS STRATEGY (1)

EXHIBIT 9: ALTERNATIVES FOR AWARENESS STRATEGY (2)

In most firms a degree of environmental awareness is provided through economic forecasting, sales forecasting and analysis of competitive behavior. All of these are essentially extrapolative, based on a smooth extension of the past into the future. To broaden the awareness to include discontinuities, the firm can add special types of environmental analysis shown in EXHIBIT 8. An examination of these shows that most of them do not require information about specific threats and opportunities. Knowledge at ignorance level one, a **sense of threat** is adequate. State one is also adequate for many of the self-awareness measures.

The **flexibility strategy** shown in the middle column of EXHIBIT 6 and elaborated in EXHIBITS 10 and 11 differs from the direct action strategies in the fact that its end product is an enhanced potential for the firm's future, rather than tangible changes in profits and in growth.

The **external flexibility** sub-strategy concerns itself with positioning the firm in the environment in a way which satisfies two criteria: (i) satisfactory expected **average** potential for profitability over the long term, (ii) adequate diversification of the firm's position to assure coping with **deviations** from the expected average: capture of attractive major opportunities and minimization of catastrophic reversals.

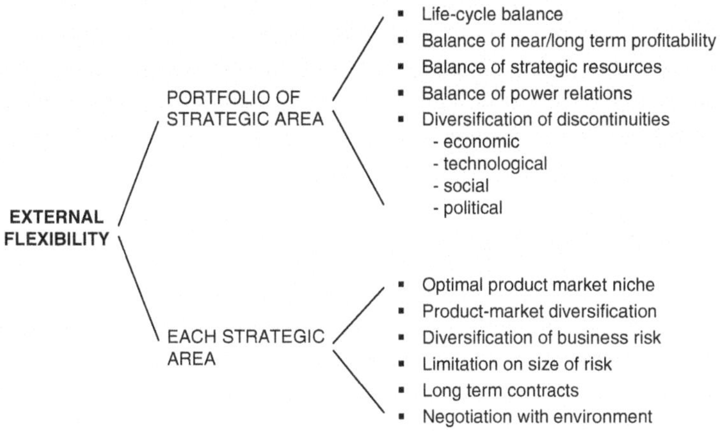

EXHIBIT 10: ALTERNATIVES FOR FLEXIBILITY STRATEGY (1)

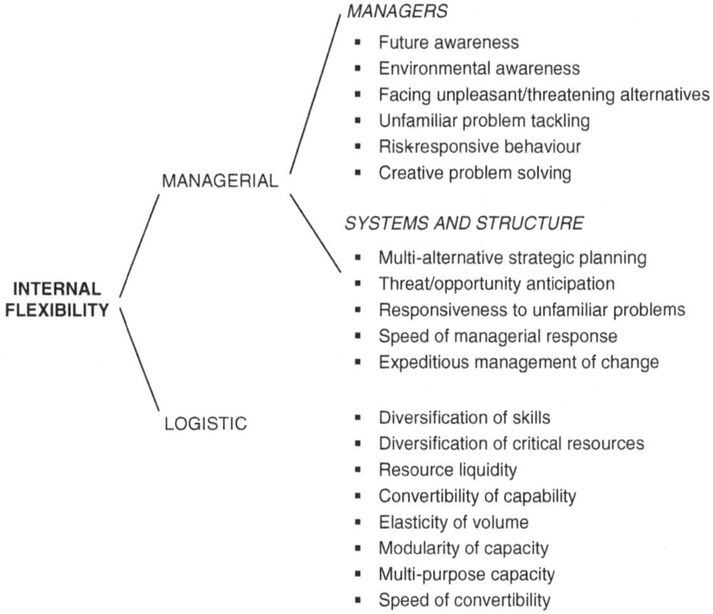

EXHIBIT 11: ALTERNATIVES FOR FLEXIBILITY STRATEGY (2)

Formulation of the external flexibility strategy (commonly known as "position strategy") and of the external action strategy ("development planning") are both a part of the strategic planning process. In most instances of strategic planning it is usually assumed that the same level five information input is needed for both planning activities. But an examination of the measures in EXHIBIT 10 shows that a flexibility strategy can be substantially planned **and implemented** if the state of ignorance is no better than level two, long before threats become concrete.

Unlike external flexibility, internal flexibility received relatively little attention from strategic planners. But recent history shows it to be a crucial ingredient in strategic preparedness. In the area of managerial flexibility the preparation of managers along the lines described in EXHIBIT 11 is now recognized as essential and vital if the firm is to anticipate and deal with the growing turbulence of the environment. Without it, efforts to introduce strategic planning typically encounter strong "resistance to planning".

In the past **logistic flexibility** (lower part of EXHIBIT 11) has received even less attention than managerial flexibility. A major reason is the fact that the concept of flexibility runs contrary to the fundamental organizational design principle of the Industrial Age, which holds that maximum profitability is to be gained through the maximum possible specialization of logistic activities. This calls for special facilities, specialized machinery, largest possible capacity, maximum capital-labor substitution, and longest possible production runs. In a world of strategic surprises the principle maximum specialization has already been repeatedly compromised, when expensive capacities had been made prematurely obsolete by unexpected technological changes, or when the length of production runs has been cut short by shrinking product lifecycles. Thus, it is safe to predict that, in the coming years, logistic flexibility will be used increasingly as a measure of strategic preparedness. We note that, as with external flexibility, the mere knowledge of the source(s) of threat/opportunities is sufficient to start a rigorous program of logistic preparedness.

7. Feasible Response Strategies

The preceding discussion shows that much can be done to prepare the firm for strategic discontinuities if the firm is ready to respond to weak signals. But it also shows that the range of feasible responses is limited by the state of ignorance. EXHIBIT 12 summarizes the relationship between states of ignorance and feasible responses. As already discussed, environmental awareness requires only a conviction that the potential threat(s) is serious enough to warrant an early presentation. An examination of the options of EXHIBIT 8 would show that not all response actions are immediately feasible. But such activities as monitoring of the environment, structural forecasting, critical capacities and its resource analyses can be instituted right away. Others can be added as the environmental information matures.

Reference to EXHIBIT 11 shows that in the internal flexibility strategy, virtually all of managerial preparedness as well as a significant number of logistic measures, can also be launched under sense of threat conditions. External flexibility work can be started, as soon as the sources of threat are identifiable.

For direct response strategy, it is necessary to have a good idea of the threats which one is proposing to attack. But even here, a sufficiently clear idea of the origin and shape of threat is sufficient to launch a substantial percentage of internal readiness measures: acquisition of necessary technological, production and marketing skills, new product development, development of sources of supply.

Even direct external action need not, an **in practice frequently does not**, await information which makes possible reliable cash and profit flow calculations. For example, adventurous firms will typically launch their entry into a new industry at stage 4, before the technology, the market, and the competition are well enough defined to permit such calculations. Other more conservative firms will prefer to wait on the sidelines until the "ball game" is better defined.

EXHIBIT 12 is a **response feasibility table**: given knowledge of the state of ignorance, the table yields the currently feasible range of strategies. But the ability to use these strategies depends also on the dynamics of the situation: the time left before the threat becomes overwhelming as compared to the time needed by the firm to respond.

8. Dynamics of Response

Each of the six response strategies makes a complementary contribution to the firm's ability to handle strategic discontinuities. But the sequence in which they are used will vary from firm to firm. As mentioned in the preceding section, conventional strategic planning suggests a sequence in which measures for flexibility and awareness are derived from the analysis of direct response. EXHIBIT 12 suggests that a reverse procedure enables the firm to start the response much earlier.

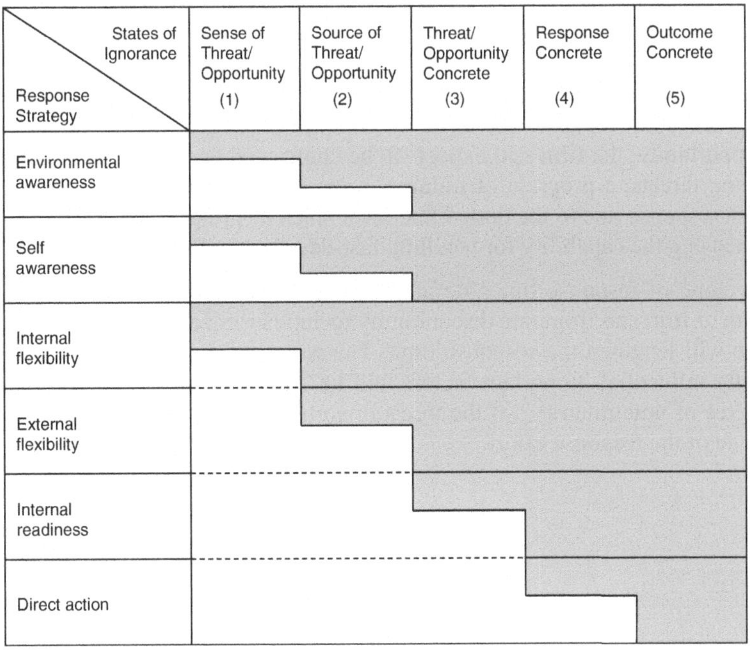

EXHIBIT 12: FEASIBLE RANGES OF RESPONSE STRATEGIES

One such reverse sequence is shown on the horizontal axis in EXHIBIT 13 (on the next page). The vertical axis measures the time needed by the firm to complete a response to an opportunity/threat. The two solid curves show that, the less prepared the firm is the more time it needs to respond, in which existing processes, structures, systems and procedures are used to handle the threat/opportunity. The lower solid curve, the "ad hoc crash response" shows the time savings that can be effected if "all the stops are pulled out", when everything that is possible is done to speed up the response: normal rules and procedures are disregarded, other priorities are pushed into the background, organizational lines are crossed, activities are duplicated, overtime is incurred.

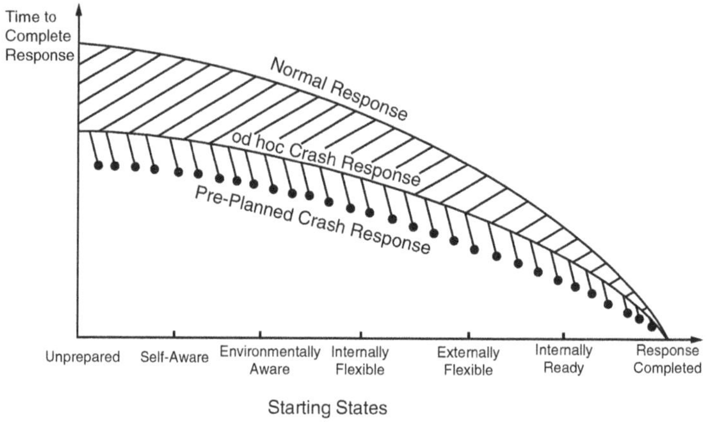

EXHIBIT 13: INTERNAL DYNAMICS OF RESPONSE

The mere "pulling out of the stops" in a crash response when a crisis looms, is not the only procedure open to the firm. If, in spite of best efforts to anticipate the threat/opportunity, the firm still expects to be confronted occasionally with very fast developing threats, a program of training in **crisis management**[9] can lower the ad hoc crash response line to the dotted line level labeled "pre-planned crash response", thus increasing the capability for handling fast-developing discontinuities.

The envelope of response times defined by the three curves will of course differ from firm to firm and from one discontinuity to another. Size, complexity, rigidity of structure will lengthen the response time. The nature of the threat/opportunity will be equally influential. Important factors will be size of the discontinuity as well as the "degree of unfamiliarity" of the threat/opportunity, both of which determine the magnitude of the response effort.

9 Op. Cit. Ref. 1.

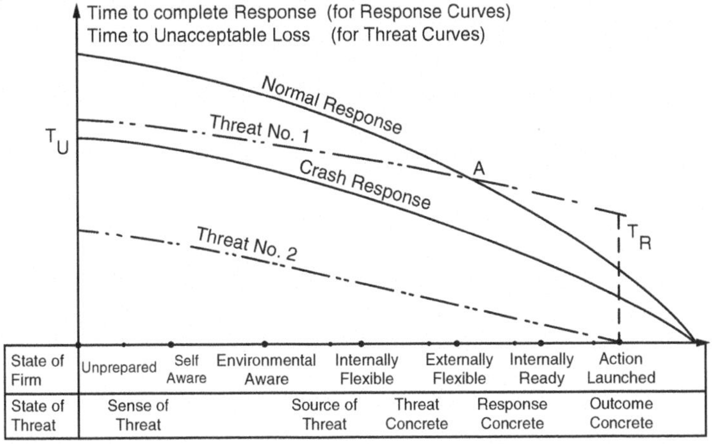

EXHIBIT 14: DYNAMICS OF RESPONSE VS. AVAILABILITY OF INFORMATION

In an earlier section, we have shown that each threat will develop over time, leaving less and less time before the critical moment, but yielding increasing information about the threat/opportunity. As the information gets better, EXHIBIT 12 can be used to determine the newly feasible state of preparedness; and EXHIBIT 13 will provide the time needed to complete the response starting from that state. This relatively simple procedure can combine the history of threats and history of responses in the manner shown in EXHIBIT 14. The vertical scale measures, using the same units of time, first the response time needed, as shown by the response curves, and second, the time available, as shown by the threat curves. Two matched horizontal scales are given. Lower shows the state of ignorance at the various stages of the threat. On the upper scale are the states of preparedness feasible in the respective states of threat.

Thus when Threat No. 1 reaches state one, the firm does not have enough time for a normal response. An urgent response is needed in its early stages. But at point A on EXHIBIT 14, response finally catches up with the threat. Management can "take a breath" and treat the remainder of the response through normal business procedures. On the other hand, Threat No. 2, as anticipated earlier, turns out to be "bad news". It arises, like a low flying attack airplane, too near the target, and moving too fast to be intercepted. Inevitably, it will lead to higher losses than the firm considers acceptable.

EXHIBIT 14 illustrates the superiority of the gradual preparedness approach over conventional strategic planning in handling fast developing T/O. The former can handle threat No. 1 with time to spare. But in conventional strategic planning the firm have remained unprepared (requiring T_U time to respond) until the threat is fully developed, leaving only T_R, which is less than T_U for reaction. To put this differently, the stringent information requirement of strategic planning limit the useable forecasting horizon and make the firm lose the benefits of early warnings about T/O. On the other hand, gradual preparedness expands the useable forecasting horizon and permits response to start early. One necessary consequence of this expansion is to broaden the firm's attention to the weak signals and trends in the environment. By responding to these the firm prepares itself to live in a climate of uncertainty and turbulence and thus respond early to threats/opportunities which arise suddenly and develop very rapidly.

II. PRACTICAL APPLICATION

Discussion of EXHIBIT 14 completes the development of a conceptual framework which matches weak responses against weak signals. Clearly, the curves sketched in the preceding exhibits are illustrative. Most actual situations will lack the information needed for constructing them. We need, therefore, to reduce concepts to practice, to construct a step-by-step approach which will enable the firm to respond to weak signals. A necessary basis for such approach is an environment surveillance system which can prepare and interpret vague trends and threats. We next turn to a brief discussion of such a system.

1. Opportunity-Vulnerability Analysis

No matter how high the state of ignorance, three types of information can be identified in each impending strategic discontinuity: (1) Timing of its probable occurrence, (2) The magnitude of the positive or negative impact on the firm, (3) Information about its specific characteristics: where it will occur, what changes it will introduce in products, in technology, in marketing, etc.

A reference back to EXHIBIT 4 shows that in the earlier states of ignorance, the third type of information is scarce, while estimates of timing and impact are based more on judgement than analysis, imprecise, and span a broad range of possibilities. Nevertheless, even such estimates can provide a basis for substantial strategic action.

To obtain such estimates, it is necessary to make a systematic analysis of the firm's environment. In the early days of strategic planning this was done by examining the strategic future of each self-contained organizational unit, usually a division. After a time it became apparent that unit-by-unit analysis gives a confusing picture of the future particularly when a division has a number of product lines and operates in many markets. Recently, an alternative approach has emerged which subdivides the firm's environment into relatively independent **strategic business areas**, each of which is a segment of the firm's environment with distinctive trends, threats, and opportunities.[10] Such a breakdown is illustrated in EXHIBIT 15, in which a firm has identified four SBAs and their respective contributions to the firm's profit.[11]

The entries in the table of EXHIBIT 15 identify the impending strategic discontinuities along four dimensions. The first is the state of ignorance within the firm. This is indicated by the choice of an appropriate column for each SBA. Next, each entry is identified as a potential threat or an opportunity, or both. This is entered into the appropriate columns, together with the ranges of the possible impacts on the profit of the firm. This is shown as a range of variation in the SBA's contribution to profit (0.2–0.5 for SBA_1). (A number of techniques is available for estimating profit impact of strategic discontinuity. The best known among them is **impact analysis**.)

		STATE OF IGNORANCE				
Strategic business areas	Profit contribution	Sense of T/O	Source of T/O	T/O concrete	Response concrete	Outcome concrete
SBA_1	50%			T 3-5 yrs. 0,2-0,5		
SBA_2	30%	T-O 10-15 yrs. 0,0-0,2				
SBA_3	15%					OF 1-2 yrs. 2,5-3,0
SBA_4	5%		O 4-8 yrs. 2,0-5,0			
Status: environmental awareness		Very poor	Poor	Excellent		Adequ.

EXHIBIT 15: OPPORTUNITY-VULNERABILITY ANALYSIS

As illustrated, the variability of the respective estimates becomes higher as ignorance increases. Thus for SBA_2 it is not even clear whether it is a threat or an opportunity, but it is clear that the impact is likely to be the very serious. Clearly this discontinuity needs close watching. On the other hand, the profit estimates for the opportunity SBA_3 can be estimated within a narrow range of both occurrence and im-

[10] Current strategic changes in the availability of key resource, as well as increasing strategic interaction between the firm and society, focus attention on **strategic resource areas** and **strategic influence areas**. Thus, a complete analysis of strategic vulnerability would include these two in addition to strategic business areas. However, we can illustrate the method of analysis by confirming our attention to the latter.

[11] In actual analysis, both present profit contributions and their trends would be analyzed. Again to keep the argument simple, we shall use only current profits.

pact. Thus the various entries in the table will be the result of different computational approaches. In the higher states of ignorance techniques based on expert judgement such as Delphi, must be used. For the lower states of ignorance it is possible to introduce quantitative modeling and cash stream projections.

2. Preparedness Diagnosis

The next step is to assess the firm's readiness to respond. This can be accomplished through a **preparedness diagnosis** which is illustrated for SBA_1 in EXHIBIT 16.

First, feasible response strategies are identified by using the state of ignorance information for each SBA from EXHIBIT 15 in conjunction with EXHIBIT 12. As shown in EXHIBIT 15, SBA_1 is an advanced state of information. Therefore all strategies short of direct action are feasible.

Next, using EXHIBITS 7, 8, 9, 10, 11 as check-off lists, the current state of readiness, (as well as of ongoing readiness – enhancing projects) is examined to determine the adequacy of the respective readiness states. The results are indicated as fat dots in EXHIBIT 16, on a scale from 0% to 100%.

SBA * 1	Feasib.	Status	Relat. Imp.	Crash		Normal	
				Time	Cost	Time	Cost
Selfawareness		0% – 100%	VH	3		6	
Environmental awareness			H	1		2	
Internal flexibility			M	2		4	
External flexibility			L	4		8	
Action readiness			M	2		4	
Action			VH	2		4	
Completed response	X	0% – 100%	X	4 yrs.	4,0	8 yrs.	8,0

EXHIBIT 16: PREPAREDNESS DIAGNOSIS

The "relative importance" in the following column assesses the importance of the respective sub-strategies to the ultimate success of the firm's response. Thus, in the illustration, external flexibility does not appear to be a problem, and internal readiness is medium important. This might signify that much of the firm's current capability appears applicable to the discontinuity.

The last four columns estimate the times and costs for normal and crash responses, respectively. A useful measure is the estimated cost of the total response program (shown in the last line of the table) as a fraction of the percentage of current profits contributed by the SBA. For example, if crash program will cost 4.0, as indicated in EXHIBIT 16 and if the response will prevent a loss of 0.2 to 0.5 loss of this profit, the "investment" will amortize itself in 8 to 20 years. This suggests that the threat will be "written-off" and allowed to run its course to avoid "sending good money after the bad". On the other hand, the normal response (if it will be timely enough), costing 1.0, will be worth the effort, because the amortization period will be only 2 to 5 years.

Preparedness diagnosis needs to be repeated for each discontinuity identified in EXHIBIT 15, because the characteristic of the discontinuity will differ, and because the firm's capabilities will be better suited to some discontinuities than to others. Therefore, a serious effort to assess the firm's preparedness would involve two steps. The first would be a general **capability profile**[12] analysis which determines the firm's capabilities and capacities. In the second step the general profiles would be compared to specific discontinuities to determine the respective compatibilities, or synergies.

3. Vulnerability Profile

The results of discontinuity and of preparedness analyses are next combined in a summary **vulnerability profile** to produce a bird's-eye view of the firm's future prospects. This is shown in EXHIBIT 17 (on the next page).

The respective rectangles enclose the "regions of probable impact" on the respective SBAs. Rectangles below the horizontal axis spell potential losses in profitability, those above indicate gains. The height of the rectangle spans the probable range of loss/gain, the base spans the probable times when the discontinuity will impact on the firm.

[12] H. Igor Ansoff: *Corporate Strategy*, 1965.

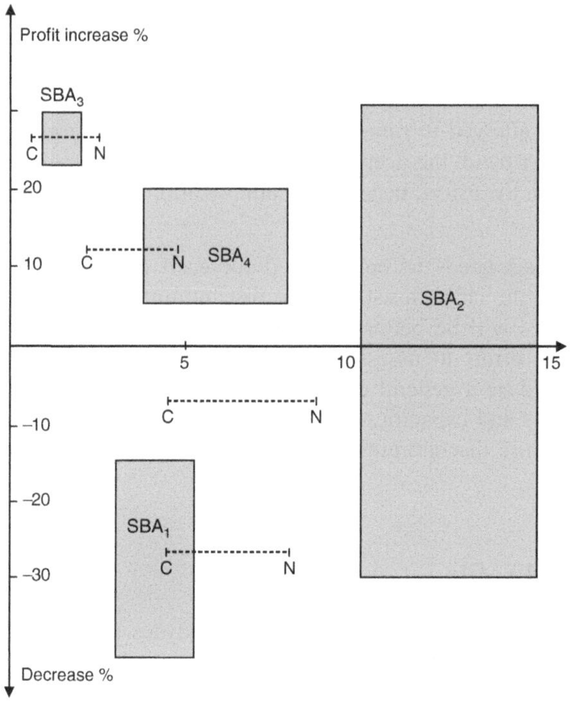

EXHIBIT 17: VULNERABILITY PROFILE

The horizontal dotted lines span the times of probable completion of successful response. Thus, the normal response for SBA$_3$ would be late, but the firm can assure itself of capturing the opportunity through a crash program. SBA$_2$ is "safe", normal response will capture it, provided the firm "moves right along". SBA$_1$ is in trouble because even a crash response may be late; it looks like a "surprise" in the making.

On a first examination, the overall picture for the firm is not bad. It has a very promising near term prospect in SBA$_2$. SBA$_1$ and SBA$_3$ may balance out. The potential problem is in SBA$_2$ which, because of the current state of ignorance may turn out either way. SBA$_2$ certainly should be carefully monitored, but a more vigorous response program can be delayed at least a year.

A closer look at the firm's position may suggest that the four SBAs of the firm, with their respective vulnerabilities, do not provide the firm enough long term security, because so much rides on the future of SBA$_2$. If this conclusion is reached, an increase in external flexibility through diversification will be indicated. But this indication is one of several suggested by the vulnerability profile. Each of the SBAs requires attention and each competes with others for management attention and for resources. Therefore, we need a systematic procedure by which the alternatives can be compared and the resources allocated.

4. Strategic Issue Analysis

For this purpose, we shall use an extended version of **Strategic Issue Analysis**, a decision technique already used in practice. This technique emerged within the last few years as a means for overcoming a number of shortcomings which were found in applications of strategic planning. One shortcoming was the inability of strategic planning systems to respond to significant discontinuities which occurred outside the planning cycle. Another was a typical long planning delay of 6–9 months between initiation and completion of planning, which made the response too slow for fast moving threats. A third shortcoming was an organizational inflexibility of strategic planning: when an issue fails to fit neatly into one of the planning units, it tends to fall "between the chairs". The fourth, and perhaps most important, shortcoming is a typical difficulty of converting plans into action.

Somewhat different varieties of Strategic Issue Analysis (SIA) have been used, sometimes as an addition, and sometimes as a replacement for annual organization-wide strategic planning. The technique has been applied both to business firms and to complex governmental organizations. When the various positive aspects are synthesized, four major features emerge:

(1) SIA embraces all potential discontinuities no matter what their source: markets of the firm, changes in characteristics of demand, changes in technology, in availability of resources, changes in political constraints, in conditions of work. For example, today the following would be scanned as potential **strategic issues**: "limits of growth", petroleum politics, environmental pollution, industrial democracy, thermonuclear fusion, technology of non-petroleum power generation, scarcity of strategic resources, etc. Thus SIA broadens the much narrower preoccupation with products-markets-technology found in typical strategic planning systems.

(2) SIA assigns responsibility for response, not according to the firm's organizational jurisdiction, but to the logic of the issue. If resolution of an issue cuts across organizational lines, a special task force is set to handle it.

(3) SIA, unlike most other management systems, does not operate on a fixed calendar basis. Rather a perpetual list of strategic issues is continually modified and updated, temporary task forces are organized and launched whenever urgency and priority dictates.

(4) Finally, SIA is treated not as a planning but as a **result oriented** activity. The charge to the task force is not limited to planning studies. Instead, the charge is to produce a change in the strategic posture of the firm.

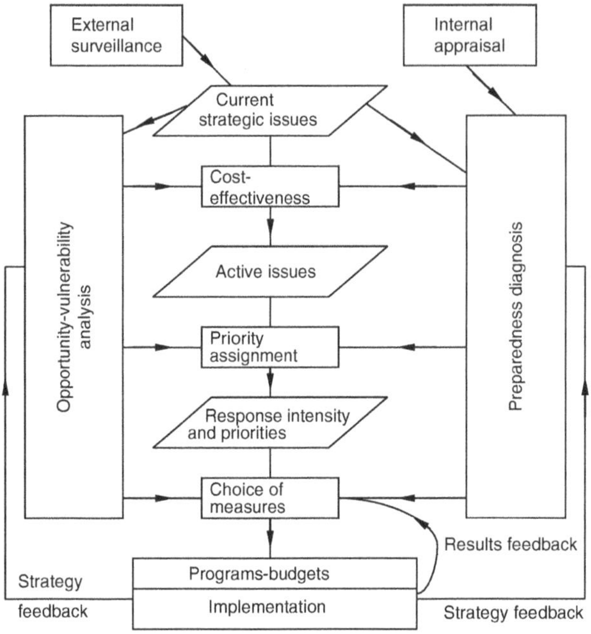

EXHIBIT 18: DYNAMIC RESPONSE TO STRATEGIC DISCONTINUITIES

In EXHIBIT 18, we present the decision flow in a SIA system which is specifically oriented towards "amplification of weak signals". As shown in EXHIBIT 18, SIA is a "real time" system triggered by two inputs: (1) a continual surveillance of the environment for "upstream" information of discontinuities; and (2) a continual appraisal and reappraisal of the changing capacities and capabilities of the firm.

The external surveillance activity yields a continually updated list of current strategic issues which have important relevance to the firm. The central stem of EXHIBIT 18 shows how this list is translated into ultimate strategic action. The issues are subjected to three successive stages of analysis. First, cost-effectiveness comparison, in the manner discussed in conjunction with EXHIBIT 17, is used to eliminate from the current list those issues whose cost would not be justified, either by the amount of potential losses saved, or the profit opportunity foregone.

The vulnerability profile (EXHIBIT 17) shows that, what matters is neither the timing of the threat, taken by itself, nor the timing of the response, but the combination of the two. An imminent near-term threat may present no problems if the firm is able to respond to it quickly. On the other hand, a long-term threat may result in a strategic crisis, if the changes needed in the firm are major and time-consuming.

This interaction between timing of threat and timing of response can be combined into a single concept of **urgency**, which is measured by the overlap of the two time periods. Thus, if we use C and N (see EXHIBIT 17) to denote the respective times required for normal and crash responses, and E and L for the earliest and latest times of probable impact of discontinuity, a scale urgency can be defined as follows (see EXHIBIT 17 for examples):

1. Low urgency if $N < E$
2. Moderate, if $E < N < L$, and increasing as N approaches L
3. High urgency if $N > L$ but $C < E$
4. Very high urgency if $E < C < L$
5. Crisis urgency if $L < C$.

But urgency alone would not, by itself determine the priority of an issue. Equally important will be the **cost-effectiveness** measured by the difference between the impact and the cost of response. Relative cost-effectiveness will differ from firm to firm, depending on the number of strategic issues and on the total budget available. A working table shown in EXHIBIT 19, can be used for assigning the relative priorities among the active strategic issues, as well as for selecting the appropriate speed of response.

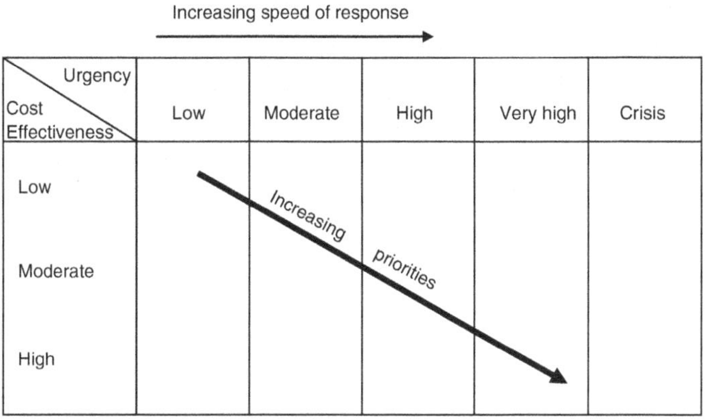

EXHIBIT 19: DETERMINING PRIORITIES FOR ACTIVE STRATEGIC ISSUES

With priorities established, the next step is to select the specific measures to be implemented. One input to this decision comes from **vulnerability-opportunity-analysis**, shown on the left-hand side of EXHIBIT 18. The key instrument is the feasibility table of EXHIBIT 12 (which would be supplemented by Strategic Resource and Societal Influence Areas) into the feasible strategies. These in turn are translated into feasible measures with the aid of EXHIBIT 7, 8, 9, 10, 11.

The second input comes from **preparedness diagnosis** shown on the right of EXHIBIT 18. The current state of readiness for dealing with each of the active issues, as well as the status of the current projects to enhance readiness can be determined using, again, EXHIBIT 7, 8, 9, 10, 11 – this time as diagnostic check-lists. The difference between feasible and existing measures are the measures to be planned and carried out, according to previously established priorities.

5. Dynamics of Strategic Issue Analysis

A two-dimensional graph, such as EXHIBIT 18, cannot properly reflect the dynamic properties of the process it represents. In the case of SIA, dynamics are very important because they distinguish SIA from the much more static strategic planning process.

The latter is typically periodic, conducted according to a set calendar. Since strategic planning embraces the totality of the firm's future and since it is typically a decentralized process, it is time consuming and usually cumbersome. The complexities of the systems, of interlocking relationships and decision responsibilities focus attention on the dynamics of the planning system. As a result, preoccupation with system dynamics leads to a neglect of the threat dynamics.

Strategic Issue Analysis is a complement to strategic planning designed to remedy this deficiency. It might be called a "real-time" component of the total system. The external surveillance and internal surveillance activities are conducted continuously providing new signals and issues for management as soon as they surface on the horizon. The current list is continuously updated and projects are activated and terminated throughout the year. The dynamic nature of such process, involving early response to vague, imperfectly understood discontinuities naturally create the possibility of over-responsiveness, of the firm going off "half-cocked" at the slightest provocation. An examination of the process described in the preceding pages shows that, while valid and necessary, this concern is anticipated in the nature of SIA. In the first place, there is a built-in protection to over-reaction in the fact that limited information permits only limited measures.

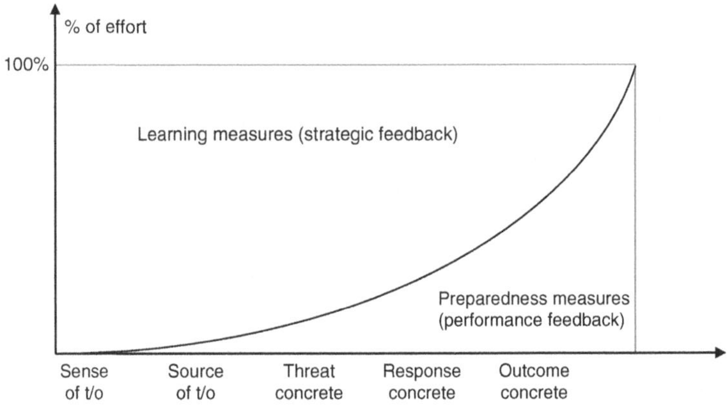

EXHIBIT 20: SELECTION OF MEASURES TO ENHANCE EARLY LEARNING

Secondarily, the strategies and measures are designed for **graduated response** which increase the firm's commitment in step with developing information. Nevertheless, maximum possible protection against over-reaction must be built in. This can be helped by proper use of feedback. Two types are shown at the bottom of EXHIBIT 18: **performance and strategic feedback**. The former informs management about progress in implementing selected measures. The latter provides information on whether the strategy pursued and measures selected are the proper ones.

When a threat is in the higher states of ignorance, the knowledge of whether one is on the right track, is more important than results. But as time passes, the firm progressively has to commit itself to a specific approach, even at the risk of error. Thus another necessary dynamic feature of SIA, not immediately evident from EXHIBIT 18, is to select measures which give the type of feedback desired. Through the life of the threat, in the manner illustrated in EXHIBIT 20, the emphasis shifts from measures which produce strategic feedback to measures which yield preparedness results.

6. SIA and Strategic Planning

As already discussed, strategic issue analysis is not a replacement but a complement to strategic planning. Strategic planning deals with the totality of the firm's future, but it lacks responsiveness to weak strategic signals. While SIA provides the latter, it is essentially an **opportunistic** approach, seeking to protect and prepare the firm for singular events. A comparison of the salient features of the two approaches is given in EXHIBIT 21. The choice of the appropriate technique is almost obvious from the preceding.

Strategic Planning	**Strategic Issue Analysis**
■ Deals with firm's total strategic future	■ Deals with probable discontinuities
■ Focussed on Products-Markets-Technology	■ Embraces discontinuities from all sources
■ Applicable when major Strategic reorientation is desired	■ Applicable when insurance against surprises is desired
■ Responds to strong signals	■ Responds to weak signals
■ Strategic information needs derived from decision	■ Feasible decisions determined by available information
■ Prepared periodically	■ A continuous process
■ Organization-focussed	■ Problem-focussed

EXHIBIT 21: COMPARISON OF STRATEGIC PLANNING AND STRATEGIC ANALYSIS

(1) A firm which confronts a strategically unsatisfactory future, or a firm which seeks to make a discontinuous improvement on the past, but which finds itself in a relatively surprise-free environment, would choose strategic planning.

(2) A firm whose overall strategic trends appear satisfactory, but whose environment is highly turbulent, can limit itself to SIA.

(3) A firm with both problems would be wise to use both.

The existence of SIA alongside strategic planning will not only provide insurance against surprises, but also, when properly used, speed up the strategic planning process. This is illustrated in EXHIBIT 22 (on the next page).

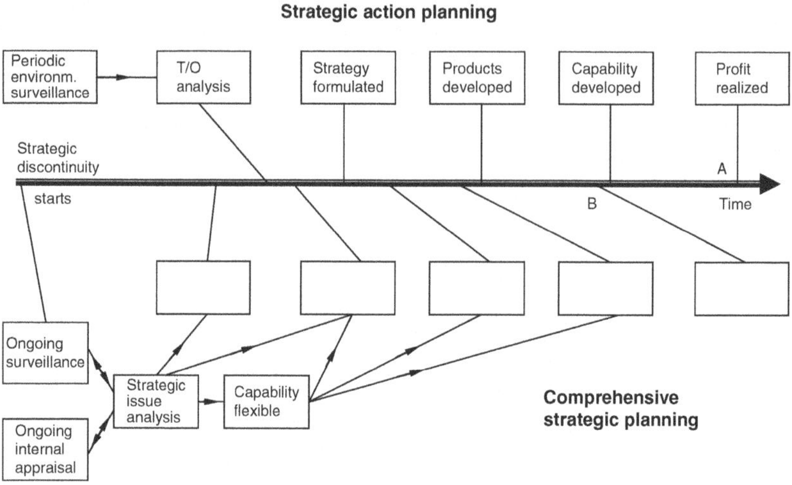

EXHIBIT 22: TIME COMPRESSION THROUGH STRATEGIC PREPAREDNESS

Shown above the horizontal time axis is the typical sequence of events involving strategic planning. Annual environmental surveillance identifies major trends, as well as threats and opportunities. These would be based on strong signals; weak signals would be filtered out. This information is translated into strategy which serves as a basis for plans and programs. The first step in implementation is to develop new product/markets, the second is to develop the needed capability, and the third is to launch the new product/markets. The strategic action sequence is completed when the new product attains a profitable position at point A.

As shown in the lower part of EXHIBIT 22, outputs of SIA can be fed to respective steps in strategic planning, thus expediting the entire chain of events and advancing the profitable position from time A to time B.

7. Application to Practice

In the preceding pages we have traveled a long road, first developing a conceptual framework and secondly, a practical procedure by which a firm in a turbulent environment can cope with "weak signals" and minimize the chances of surprise.

The result is a new planning approach, which to become effective, must be accepted and used by practicing managers. However persuasive and logical the approach, neither the acceptance, nor the use can be taken for granted. To do so, would be to disregard the numerous instances during the past twenty years, when similarly logical approaches encountered "resistance to planning", and were either rejected or emasculated by the using organizations.

To gain acceptance for this particular approach, it is necessary to assure within the firm a climate of openness to strategic risk and preparedness to face unfamiliar and threatening prospects. This problem of creating a climate of strategic decisiveness is at least equally as complex and as important as the one we have just created. Therefore, we explored decisiveness in a separate paper.[13]

In that paper we argue that in the absence of strategic decisiveness, we would find little acceptance of Strategic Issue Analysis. However, diagnosis of strategic decisiveness is not easy. Firms who lack it do not necessarily appear sleepy, environmentally inactive. Many of them will be quite active, but deeper analysis would show that their environmental engagement is **competitive** and not **strategic**, confined to dealing with familiar fluctuations in the environment[14] and unaccustomed to discontinuities. Such firms will be active in marketing, new product development, long range forecasting, even long range planning. But all these activities will cling to a historical perception of the environment and its extrapolations. As one researcher put it, such firms use environmental information to confirm their prior models of reality, and not to change and enlarge their understanding. A vivid example of such behavior is furnished by the U.S. automotive industry which, during the past fifteen years, has persisted to clinging, in spite of a flood of contrary data, to the belief that its customer is still in love with the luxurious "gas-guzzling monster".

Therefore, in a large percentage of cases, an effort to introduce strategic preparedness based on weak signals is likely either to encounter resistance, or to be rejected altogether, unless strategic decisiveness is substantially enhanced.

But this does not mean that preparedness analysis must be delayed until after the influential managers change their perceptions. The two processes can proceed in parallel, interacting and reinforcing each other. Thus, for example, enhancement of managerial flexibility will at the same time contribute to enhanced decisiveness (see EXHIBIT 11).

What is needed, therefore, is a two-way feedback organizational change process in which both the **willingness to act** and the **ability to act** are developed in a coordinated and mutually supportive manner. The technology for designing such change processes now exists and has been tested in practice,[15] and has been described in literature.

---

13 Op. Cit. Ref. 1.

14 H. Igor Ansoff, R. Hayes, and R. Declerck: From Strategic Planning to Strategic Management. In: *From Strategic Planning to Strategic Management*, 1976 (forthcoming).

15 Op. Cit. Ref. 14.

Ronald H. Coase (1937)

THE NATURE OF THE FIRM

Logistics is about the coordination of sequences of actions and functions, of dynamic flows of objects through chains and networks in order to meet customer demands and create value. This is why the "science of logistics" is in need of a set of terms, concepts, and theories related to coordination.

One of the most fundamental contributions that has been made to provide a systematic set of instruments for the analysis of coordination issues is the work of Ronald H. Coase. Born in 1910, he is a British economist and Professor Emeritus of Economics at the University of Chicago Law School. In 1991 he was awarded the Nobel Prize in Economics.

In his classical paper on the *Nature of the Firm* he notes that rational coordination in many parts of an economy is not achieved by the price mechanism alone, as had been tacitly assumed in most of the earlier economists' thinking. Coordination within firms most often comes through the orders of the entrepreneur and through contractual "master-servant" relationships. On the basis of the important distinction between coordination via markets and prices on one hand, and coordination by order, contract, and subordination on the other hand Coase finds that there are "costs of market usage", which often prevent the market mechanisms to be used. On this basis he develops a new theory of the nature of the firm. Subsequently, the fundamental alternatives for coordinating complex activities were referred to as the "market vs. hierarchy" dichotomy (cf. Williamson (1975)). Further work by Coase and the school of "Institutional Economics", which was inspired by him, led to many new terminological and conceptual contributions such as Transaction Cost, Property Rights and Principal-Agent theories (e.g. Alchian/Demsetz (1972)).

The perspectives and analytical approaches provided by Institutional Economics have become indispensable, frequently used instruments in research about the optimal structuring, sizing, coordination and control of logistical systems and supply networks.

Reference:

Ronald H. Coase: The Nature of the Firm. In: *Economica*, Vol. 4, No. 16, 1937, pp. 386–405.

Further reading:

Alchian, Armen A. and Harold Demsetz: Production, Information Costs, and Economic Organization. In: *The American Economic Review*, Vol. 62, No. 5, 1972, pp. 777–795.

Williamson, Oliver E.: *Markets and Hierarchies: Analysis and Antitrust Implications*. New York: The Free Press, 1975.

P. Klaus and S. Müller (eds.), *The Roots of Logistics*, DOI 10.1007/978-3-642-27922-5_22,

Ronald H. Coase: The Nature of the Firm

Economic theory has suffered in the past from a failure to state clearly its assumptions. Economists in building up a theory have often omitted to examine the foundations on which it was erected. This examination is, however, essential not only to prevent the misunderstanding and needles controversy which arise from a lack of knowledge of the assumptions on which a theory is based, but also because of the extreme importance for economics of good judgment in choosing between rival sets of assumptions. For instance, it is suggested that the use of the word "firm" in economics may be different from the use of the term by the "plain man".[1] Since there is apparently a trend in economic theory towards starting analysis with the individual firm and not with the industry,[2] it is all the more necessary not only that a clear definition of the word "firm" should be given but that its difference from a firm in the "real world," if it exists, should be made clear. Mrs. Robinson has said that "the two questions to be asked of a set of assumptions in economics are: Are they tractable? and: Do they correspond with the real world?"[3] Though, as Mrs. Robinson points out, "More often one set will be manageable and the other realistic," yet there may well be branches of theory where assumptions may be both manageable and realistic. It is hoped to show in the following paper that a definition of a firm may be obtained which is not only realistic in that it corresponds to what is meant by a firm in the real world, but is tractable by two of the most powerful instruments of economic analysis developed by Marshall, the idea of the margin and that of substitution, together giving the idea of substitution at the margin.[4] Our definition must, of course, "relate to formal relations which are capable of being **conceived** exactly."[5]

I

It is convenient if, in searching for a definition of a firm, we first consider the economic system as it is normally treated by the economist. Let us consider the description of the economic system given by Sir Arthur Salter.[6] "The normal economic system works itself. For its current operation it is under no central control, it needs no central survey. Over the whole range of human activity and human need, supply is adjusted to demand, and production to consumption, by a process that is auto-

1 Joan Robinson, *Economics Is a Serious Subject*, 1932, p. 12.

2 See N. Kaldor, The Equilibrium of the Firm, *Economic Journal*, March, 1934, pp. 60–76.

3 Op. cit., p. 6.

4 J.M. Keynes, *Essays in Biography*, 1933, pp. 223–224.

5 L. Robbins, *Nature and Significance of Economic Science*, 1935, p. 63.

6 This description is quoted with approval by D.H. Robertson, *Control of Industry*, 1923, p. 85, and by Professor Arnold Plant, Trends in Business Administration, *Economica*, February 1932, pp. 45–62. It appears in *Allied Shipping Control*, pp. 16–17.

matic, elastic and responsive." An economist thinks of the economic system as being co-ordinated by the price mechanism and society becomes not an organization but an organism.[7] The economic system "works itself." This does not mean that there is no planning by individuals. These exercise foresight and choose between alternatives. This is necessarily so if there is to be order in the system. But this theory assumes that the direction of resources is dependent directly on the price mechanism. Indeed, it is often considered to be an objection to economic planning that it merely tries to do what is already done by the price mechanism.[8] Sir Arthur Salter's description, however, gives a very incomplete picture of our economic system. Within a firm, the description does not fit at all. For instance, in economic theory we find that the allocation of factors of production between different uses is determined by the price mechanism. The price of factor *A* becomes higher in *X* than in *Y*. As a result, *A* moves from *Y* to *X* until the difference between the prices in *X* and *Y*, except in so far as it compensates for other differential advantages, disappears. Yet in the real world, we find that there are many areas where this does not apply. If a workman moves from department *Y* to department *X*, he does not go because of a change in relative prices, but because he is ordered to do so. Those who object to economic planning on the grounds that the problem is solved by price movements can be answered by pointing out that there is planning within our economic system which is quite different from the individual planning mentioned above and which is akin to what is normally called economic planning. The example given above is typical of a large sphere in our modem economic system. Of course, this fact has not been ignored by economists. Marshall introduces organization as a fourth factor of production; J.B. Clark gives the co-ordinating function to the entrepreneur; Professor Knight introduces managers who co-ordinate. As D.H. Robertson points out, we find "islands of conscious power in this ocean of unconscious co-operation like lumps of butter coagulating in a pail of buttermilk."[9] But in view of the fact that it is usually argued that co-ordination will be done by the price mechanism, why is such organization necessary? Why are there these "islands of conscious power"? Outside the firm, price movements direct production, which is coordinated through a series of exchange transactions on the market. Within a firm, these markets transactions are eliminated and in place of the complicated market structure with exchange transactions is substituted the entrepreneur-co-ordinator, who directs production.[10] It is clear that these are alternative methods of co-ordinating production. Yet, having regard to the fact that if production is regulated by price movements, production could be carried on without any organization at all, well might we ask, why is there any organization?

---

[7] See F.A. Hayek, The Trend of Economic Thinking, *Economica*, May 1933, pp. 121–37.

[8] See F.A. Hayek, op. cit.

[9] Op. cit., p. 85.

[10] In the rest of this paper I shall use the term entrepreneur to refer to the person or persons who, in a competitive system, take the place of the price mechanism in the direction of resources.

Of course, the degree to which the price mechanism is superseded varies greatly. In a department store, the allocation of the different sections to the various locations in the building may be done by the controlling authority or it may be the result of competitive price bidding for space. In the Lancashire cotton industry, a weaver can rent power and shop-room and can obtain looms and yarn on credit.[11] This co-ordination of the various factors of production is, however, normally carried out without the intervention of the price mechanism. As is evident, the amount of "vertical" integration, involving as it does the supersession of the price mechanism, varies greatly from industry to industry and from firm to firm.

It can, I think, be assumed that the distinguishing mark of the firm is the supersession of the price mechanism. It is, of course, as Professor Robbins points out, "related to an outside network of relative prices and costs,"[12] but it is important to discover the exact nature of this relationship. This distinction between the allocation of resources in a firm and the allocation in the economic system has been very vividly described by Mr. Maurice Dobb when discussing Adam Smith's conception of the capitalist: "It began to be seen that there was something more important than the relations inside each factory or unit captained by an undertaker; there were the relations of the undertaker with the rest of the economic world outside his immediate sphere... the undertaker busies himself with the division of labour inside each firm and he plans and organises consciously," but "he is related to the much larger economic specialisation, of which he himself is merely one specialised unit. Here, he plays his part as a single cell in a larger organism, mainly unconscious of the wider rôle he fills."[13]

In view of the fact that while economists treat the price mechanism as a coordinating instrument, they also admit the co-ordinating function of the "entrepreneur," it is surely important to enquire why co-ordination is the work of the price mechanism in one case and of the entrepreneur in another. The purpose of this paper is to bridge what appears to be a gap in economic theory between the assumption (made for some purposes) that resources are allocated by means of the price mechanism and the assumption (made for other purposes) that this allocation is dependent on the entrepreneur-co-ordinator. We have to explain the basis on which, in practice, this choice between alternatives is effected.[14]

---

[11] Survey of Textile Industries, p. 26.

[12] Op. cit., p. 71.

[13] Capitalist Enterprise and Social Progress (1925), p. 20. Cf., also, Henderson, *Supply and Demand* (1932), pp. 3–5.

[14] It is easy to see when the State takes over the direction of an industry that, in planning it, it is doing something which was previously done by the price mechanism. What is usually not realized is that any business man in organizing the relations between his departments is also doing something which could be organized through the price mechanism. There is therefore point in Mr. Durbin's answer to those who emphasize the problems involved in economic planning that the same problems have to be solved by business men in the competitive system. (See Economic Calculus in a Planned Economy, *Economic Journal*,

II

Our task is to attempt to discover why a firm emerges at all in a specialized exchange economy. The price mechanism (considered purely from the side of the direction of resources) might be superseded if the relationship which replaced it was desired for its own sake. This would be the case, for example, if some people preferred to work under the direction of some other person. Such individuals would accept less in order to work under someone, and firms would arise naturally from this. But it would appear that this cannot be a very important reason, for it would rather seem that the opposite tendency is operating if one judges from the stress normally laid on the advantage of "being one's own master."[15] Of course, if the desire was not to be controlled but to control, to exercise power over others, then people might be willing to give up something in order to direct others; that is, they would be willing to pay others more than they could get under the price mechanism in order to be able to direct them. But this implies that those who direct pay in order to be able to do this and are not paid to direct, which is clearly not true in the majority of cases.[16] Firms might also exist if purchasers preferred commodities which are produced by firms to those not so produced; but even in spheres where one would expect such preferences (if they exist) to be of negligible importance, firms are to be found in the real world.[17] Therefore there must be other elements involved.

The main reason why it is profitable to establish a firm would seem to be that there is a cost of using the price mechanism. The most obvious cost of "organizing" production through the price mechanism is that of discovering what the relevant prices are.[18] This cost may be reduced but it will not be eliminated by the emergence of specialists who will sell this information. The costs of negotiating and concluding a separate contract for each exchange transaction which takes place on a market must

---

December, 1936, pp. 676–690.) The important difference between these two cases is that economic planning is imposed on industry while firms arise voluntarily because they represent a more efficient method of organizing production. In a competitive system, there is an "optimum" amount of planning!

15 Cf. Harry Dawes, Labour Mobility in the Steel Industry, *Economic Journal*, March 1934, pp. 84–94, who instances "the trek to retail shopkeeping and insurance work by the better paid of skilled men due to the desire (often the main aim in life of a worker) to be independent" (p. 86).

16 None the less, this is not altogether fanciful. Some small shopkeepers are said to earn less than their assistants.

17 G.F. Shove, The Imperfection of the Market: A Further Note, *Economic Journal*, March 1933, pp. 116, note 1, points out that such preferences may exist, although the example he gives is almost the reverse of the instance given in the text.

18 According to N. Kaldor, A Classificatory Note of the Determinanteness of Equilibrium, *Review of Economic Studies*, February 1934, pp. 122–136, it is one of the assumptions of static theory that "All the relevant prices are known to all individuals." But this is clearly not true of the real world.

also be taken into account.[19] Again, in certain markets, e.g., produce exchanges, a technique is devised for minimizing these contract costs; but they are not eliminated. It is true that contracts are not eliminated when there is a firm but they are greatly reduced. A factor of production (or the owner thereof) does not have to make a series of contracts with the factors with whom he is co-operating within the firm, as would be necessary, of course, if this co-operation were as a direct result of the working of the price mechanism. For this series of contracts is substituted one. At this stage, it is important to note the character of the contract into which a factor enters that is employed within a firm. The contract is one whereby the factor, for a certain remuneration (which may be fixed or fluctuating), agrees to obey the directions of an entrepreneur **within certain limits.**[20] The essence of the contract is that it should only state the limits to the powers of the entrepreneur. Within these limits, he can therefore direct the other factors of production.

There are, however, other disadvantages – or costs – of using the price mechanism. It may be desired to make a long-term contract for the supply of some article or service. This may be due to the fact that if one contract is made for a longer period, instead of several shorter ones, then certain costs of making each contract will be avoided. Or, owing to the risk attitude of the people concerned, they may prefer to make a long rather than a short-term contract. Now, owing to the difficulty of forecasting, the longer the period of the contract is for the supply of the commodity or service, the less possible, and indeed, the less desirable it is for the person purchasing to specify what the other contracting party is expected to do. It may well be a matter of indifference to the person supplying the service or commodity which of several courses of action is taken, but not to the purchaser of that service or commodity. But the purchaser will not know which of these several courses he will want the supplier to take. Therefore, the service which is being provided is expressed in general terms, the exact details being left until a later date. All that is stated in the contract is the limits to what the persons supplying the commodity or service is expected to do. The details of what the supplier is expected to do is not stated in the contract but is decided later by the purchaser. When the direction of resources (within the limits of the contract) becomes dependent on the buyer in this way, that relationship which I term a "firm" may be obtained.[21] A firm is likely therefore to

---

[19] This influence was noted by Professor Usher when discussing the development of capitalism. He says: "The successive buying and selling of partly finished products were sheer waste of energy." (*Introduction to the Industrial History of England*, 1920, p. 13.) But he does not develop the idea nor consider why it is that buying and selling operations still exist.

[20] It would be possible for no limits to the powers of the entrepreneur to be fixed. This would be voluntary slavery. According to Professor Batt, *The Law of Master and Servant*, 1933, p. 18, such a contract would be void and unenforceable.

[21] Of course, it is not possible to draw a hard and fast line which determines whether there is a firm or not. There may be more or less direction. It is similar to the legal question of whether there is the relationship of master and servant or principal and agent. See the discussion of this problem below.

emerge in those cases where a very short-term contract would be unsatisfactory. It is obviously of more importance in the case of services – labor – than it is in the case of the buying of commodities. In the case of commodities, the main items can be stated in advance and the details which will be decided later will be of minor significance.

We may sum up this section of the argument by saying that the operation of a market costs something and by forming an organization and allowing some authority (an "entrepreneur") to direct the resources, certain marketing costs are saved. The entrepreneur has to carry out his function at less cost, taking into account the fact that he may get factors of production at a lower price than the market transactions which he supersedes, because it is always possible to revert to the open market if he fails to do this.

The question of uncertainty is one which is often considered to be very relevant to the study of the equilibrium of the firm. It seems improbable that a firm would emerge without the existence of uncertainty. But those, for instance, Professor Knight, who make the **mode of payment** the distinguishing mark of the firm – fixed incomes being guaranteed to some of those engaged in production by a person who takes the residual, and fluctuating, income – would appear to be introducing a point which is irrelevant to the problem we are considering. One entrepreneur may sell his services to another for a certain sum of money, while the payment to his employees may be mainly or wholly a share in profits.[22] The significant question would appear to be why the allocation of resources is not done directly by the price mechanism.

Another factor that should be noted is that exchange transactions on a market and the same transactions organized within a firm are often treated differently by Governments or other bodies with regulatory powers. If we consider the operation of a sales tax, it is clear that it is a tax on market transactions and not on the same transactions organized within the firm. Now since these are alternative methods of "organization" – by the price mechanism or by the entrepreneur – such a regulation would bring into existence firms which otherwise would have no **raison d'être**. It would furnish a reason for the emergence of a firm in a specialized exchange economy. Of course, to the extent that firms already exist, such a measure as a sales tax would merely tend to make them larger than they would otherwise be. Similarly, quota schemes, and methods of price control which imply that there is rationing, and which do not apply to firms producing such products for themselves, by allowing advantages to those who organize within the firm and not through the market, necessarily encourage the growth of firms. But it is difficult to believe that it is measures such as have been mentioned in this paragraph which have brought firms into existence. Such measures would, however, tend to have this result if they did not exist for other reasons.

These, then, are the reasons why organizations such as firms exist in a specialized exchange economy in which it is generally assumed that the distribution of resources is "organized" by the price mechanism. A firm, therefore, consists of the system of

---

22 The views of Professor Knight are examined below in more detail.

relationships which comes into existence when the direction of resources is dependent on an entrepreneur.

The approach which has just been sketched would appear to offer an advantage in that it is possible to give a scientific meaning to what is meant by saying that a firm gets larger or smaller A firm becomes larger as additional transactions (which could be exchange transactions co-ordinated through the price mechanism) are organized by the entrepreneur and becomes smaller as he abandons the organization of such transactions. The question which arises is whether it is possible to study the forces which determine the size of the firm. Why does the entrepreneur not organize one less transaction or one more? It is interesting to note that Professor Knight considers that:

> "the relation between efficiency and size is one of the most serious problems of theory, being, in contrast with the relation for a plant, largely a matter of personality and historical accident rather than of intelligible general principles. But the question is peculiarly vital because the possibility of monopoly gain offers a powerful incentive to **continuous and unlimited** expansion of the firm, which force must be offset by some decreased efficiency (in the production of money income) with growth in size, if even boundary competition is to exist."[23]

Professor Knight would appear to consider that it is impossible to treat scientifically the determinants of the size of the firm. On the basis of the concept of the firm developed above, this task will now be attempted.

It was suggested that the introduction of the firm was due primarily to the existence of marketing costs. A pertinent question to ask would appear to be (quite apart from the monopoly considerations raised by Professor Knight), why, if by organizing one can eliminate certain costs and in fact reduce the cost of production, are there any market transactions at all?[24] Why is not all production carried on by one big firm? There would appear to be certain possible explanations.

First, as a firm gets larger, there may be decreasing returns to the entrepreneur function, that is, the costs of organizing additional transactions within the firm may rise.[25] Naturally, a point must be reached where the costs of organizing an extra transaction within the firm are equal to the costs involved in carrying out the transaction in the open market, or; to the costs of organizing by another entrepreneur. Secondly, it may be that as the transactions which are organized increase, the entrepreneur fails to place the factors of production in the uses where their value is great-

[23] *Risk, Uncertainty and Profit*, Preface to the Re-issue, London School of Economics Series of Reprints, No. 16, 1933.

[24] There are certain marketing costs which could only be eliminated by the abolition of "consumers' choice" and these are the costs of retailing. It is conceivable that these costs might be so high that people would be willing to accept rations because the extra product obtained was worth the loss of their choice.

[25] This argument assumes that exchange transactions on a market can be considered as homogeneous; which is clearly untrue in fact. This complication is taken into account below.

est, that is, fails to make the best use of the factors of production. Again, a point must be reached where the loss through the waste of resources is equal to the marketing costs of the exchange transaction in the open market or to the loss if the transaction was organized by another entrepreneur. Finally, the supply price of one or more of the factors of production may rise, because the "other advantages" of a small firm are greater than those of a large firm.[26] Of course, the actual point where the expansion of the firm ceases might be determined by a combination of the factors mentioned above. The first two reasons given most probably correspond to the economists' phrase of "diminishing returns to management."[27]

The point has been made in the previous paragraph that a firm will tend to expand until the costs of organizing an extra transaction within the firm become equal to the costs of carrying out the same transaction by means of an exchange on the open market or the costs of organizing in another firm. But if the firm stops its expansion at a point below the costs of marketing in the open market and at a point equal to the costs of organizing in another firm, in most cases (excluding the case of "combination"[28]), this will imply that there is a market transaction between these two procedures, each of whom could organize it at less than the actual marketing costs. How is the paradox to be resolved? If we consider an example the reason for this will become clear. Suppose *A* is buying a product from *B* and that both *A* and *B* could organize this marketing transaction at less than its present cost. *B*, we can assume, is not organizing one process or stage of production, but several. If *A* therefore wishes to avoid a market transaction, he will have to take over all the processes of production controlled by *B*. Unless *A* takes over all the processes of production, a market transaction will still remain, although it is a different product that is bought. But we have previously assumed that as each producer expands he becomes less efficient; the additional costs of organizing extra transactions increase. It is probable that *A*'s cost of organizing the transactions previously organized by *B* will be greater than *B*'s costs of doing the same thing. *A* therefore will take over the whole of *B*'s organization only if his cost of organizing *B*'s work is not greater than *B*'s cost by an amount equal to the costs of carrying out an exchange transaction on the open mar-

---

26 For a discussion of the variation of the supply price of factors and production to firms of varying size, sec E.A.G. Robinson, *The Structure of Competitive Industry*, 1932. It is sometimes said that the supply price of organizing ability increases as the size of the firm increases because men prefer to be the heads of small independent businesses rather than the heads of departments in a large business. See Jones, *The Trust Problem*, 1921, p. 531, and Macgregor, *Industrial Combination*, 1935, p. 63. This is a common argument of those who advocate Rationalization. It is said that larger units would be more efficient, but owing to the individualistic spirit of the smaller entrepreneurs, they prefer to remain independent, apparently in spite of the higher income which their increased efficiency under Rationalization makes possible.

27 This discussion is, of course, brief and incomplete. For a more thorough discussion of this particular problem, see N. Kaldor, The Equilibrium of the Firm, *Economic Journal*, 1934, pp. 60–76, and E.A.G. Robinson, The Problem of Management and the Size of the Firm, *Economic Journal*, 1934, pp. 242–57.

28 A definition of this term is given below.

ket. But once it becomes economical to have a market transaction, it also pays to divide production in such a way that the cost of organizing an extra transaction in each firm is the same.

Up to now it has been assumed that the exchange transactions which take place through the price mechanism are homogeneous. In fact, nothing could be more diverse than the actual transactions which take place in our modem world. This would seem to imply that the costs of carrying out exchange transactions through the price mechanism will vary considerably as will also the costs of organizing these transactions within the firm. It seems therefore possible that quite apart from the question of diminishing returns the costs of organizing certain transactions within the firm may be greater than the costs of carrying out the exchange transactions in the open market. This would necessarily imply that there were exchange transactions carried out through the price mechanism, but would it mean that there would have to be more than one firm? Clearly not, for all those areas in the economic system where the direction of resources was not dependent directly on the price mechanism could be organized within one firm. The factors which were discussed earlier would seem to be the important ones, though it is difficult to say whether "diminishing returns to management" or the rising supply price of factors is likely to be the more important.

Other things being equal, therefore, a firm will tend to be larger:

a) the less the costs of organizing and the slower these costs rise with an increase in the transactions organized.

b) the less likely the entrepreneur is to make mistakes and the smaller the increase in mistakes with an increase in the transactions organized.

c) the greater the lowering (or the less the rise) in the supply price of factors of production to firms of larger size.

Apart from variations in the supply price of factors of production to firms of different sizes, it would appear that the costs of organizing and the losses through mistakes will increase with an increase in the spatial distribution of the transactions organized, in the dissimilarity of the transactions, and in the probability of changes in the relevant prices.[29] As more transactions are organized by an entrepreneur, it would appear that the transactions would tend to be either different in kind or in different places. This furnishes an additional reason why efficiency will tend to decrease as the firm gets larger. Inventions which tend to bring factors of production nearer together, by lessening spatial distribution, tend to increase the size of the firm.[30] Changes like the telephone and the telegraph which tend to reduce the cost of

[29] This aspect of the problem is emphasized by N. Kaldor, op. cit. Its importance in this connection had been previously noted by E.A.G. Robinson, *The Structure of Competitive Industry*, 1932, pp. 83–106. This assumes that an increase in the probability of price movements increases the costs of organizing within a firm more than it increases the cost of carrying out an exchange transaction on the market – which is probable.

[30] This would appear to be the importance of the treatment of the technical unit by E.A.G. Robinson, op. cit., 27–33. The larger the technical unit, the greater the concentration of factors and therefore the firm is likely to be larger.

organizing spatially will tend to increase the size of the firm. All changes which improve managerial technique will tend to increase the size of the firm.[31,32]

It should be noted that the definition of a firm which was given above can be used to give more precise meanings to the terms "combination" and "integration."[33] There is a combination when transactions which were previously organized by two or more entrepreneurs become organized by one. This becomes integration when it involves the organization of transactions which were previously carried out between the entrepreneurs on a market. A firm can expand in either or both of these two ways. The whole of the "structure of competitive industry" becomes tractable by the ordinary technique of economic analysis.

III

The problem which has been investigated in the previous section has not been entirely neglected by economists and it is now necessary to consider why the reasons given above for the emergence of a firm in a specialized exchange economy are to be preferred to the other explanations which have been offered.

It is sometimes said that the reason for the existence of a firm is to be found in the division of labor. This is the view of Professor Usher, a view which has been adopted and expanded by Mr. Maurice Dobb. The firm becomes "the result of an increasing complexity of the division of labour… The growth of this economic differentiation creates the need for some integrating force without which differentiation would collapse into chaos; and it is as the integrating force in a differentiated economy that

---

31 It should be noted that most inventions will change both the costs of organizing and the costs of using the price mechanism. In such cases, whether the invention tends to make firms larger or smaller will depend on the relative effect on these two sets of costs. For instance, if the telephone reduces the costs of using the price mechanism more than it reduces the costs of organizing, then it will have the effect of reducing the size of the firm.

32 An illustration of these dynamic forces is furnished by Maurice Dobb, *Russian Economic Development*, 1928, p. 68. "With the passing of bonded labour the factory, as an establishment where work was organised under the whip of the overseer, lost its **raison d'être** until this was restored to it with the introduction of power machinery after 1846." It seems important to realize that the passage from the domestic system to the factory system is not a mere historical accident, but is conditioned by economic forces. This is shown by the fact that it is possible to move from the factory system to the domestic system, as in the Russian example, as well as vice versa. It is the essence of serfdom that the price mechanism is not allowed to operate. Therefore, there has to be direction from some organizer. When, however, serfdom passed, the price mechanism was allowed to operate. It was not until machinery drew workers into one locality that it paid to supersede the price mechanism and the firm again emerged.

33 This is often called "vertical integration", combination being termed "lateral integration".

industrial forms are chiefly significant."[34] The answer to this argument is an obvious one. The "integrating force in a differentiated economy" already exists in the form of the price mechanism. It is perhaps the main achievement of economic science that it has shown that there is no reason to suppose that specialization must lead to chaos.[35] The reason given by Mr. Maurice Dobb is therefore inadmissible. What has to be explained is why one integrating force (the entrepreneur) should be substituted for another integrating force (the price mechanism).

The most interesting reasons (and probably the most widely accepted) which have been given to explain this fact are those to be found in Professor Knight's *Risk, Uncertainty and Profit.* His views will be examined in some detail.

Professor Knight starts with a system in which there is no uncertainty:

> "acting as individuals under absolute freedom but without collusion men are supposed to have organised economic life with the primary and secondary division of labour, the use of capital, etc., developed to the point familiar in present-day America. The principal fact which calls for the exercise of the imagination is the internal organisation of the productive groups or establishments. With uncertainty entirely absent, every individual being in possession of perfect knowledge of the situation, there would be no occasion for anything of the nature of responsible management or control of productive activity. Even marketing transactions in any realistic sense would not be found. The flow of raw materials and productive services to the consumer would be entirely automatic."[36]

Professor Knight says that we can imagine this adjustment as being "the result of a long process of experimentation worked out by trial-and-error methods alone," while it is not necessary "to imagine every worker doing exactly the right thing at the right time in a sort of 'pre-established harmony' with the work of others. There might be managers, superintendents, etc., for the purpose of co-ordinating the activities of individuals," though these managers would be performing a purely routine function, "without responsibility of any sort."[37]

Professor Knight then continues:

> "With the introduction of uncertainty – the fact of ignorance and the necessity of acting upon opinion rather than knowledge – into this Eden-like situation, its character is entirely changed... With uncertainty present doing things, the actual execution of activity, becomes in a real sense a secondary part of life; the primary problem or function is deciding what to do and how to do it."[38]

---

[34] Op. cit., p. 10. Professor Usher's views are to be found in his *Introduction to the Industrial History of England*, 1920, pp. 1–18.

[35] Cf. J.B. Clark, *Distribution of Wealth*, 1899, p. 19, who speaks of the theory of exchange as being the "theory of the organisation of industrial society."

[36] Risk, Uncertainty and Profit, p. 267.

[37] Op. cit., pp. 267–68.

[38] Op. cit., p. 268.

This fact of uncertainty brings about the two most important characteristics of social organization.

> "In the first place, goods are produced for a market, on the basis of entirely impersonal prediction of wants, not for the satisfaction of the wants of the producers themselves. The producer takes the responsibility of forecasting the consumers' wants. In the second place, the work of forecasting and at the same time a large part of the technological direction and control of production are still further concentrated upon a very narrow class of the producers, and we meet with a new economic functionary, the entrepreneur... When uncertainty is present and the task of deciding what to do and how to do it takes the ascendancy over that of execution the internal organisation of the productive groups is no longer a matter of indifference or a mechanical detail. Centralisation of this deciding and controlling function is imperative, a process of 'cephalisation' is inevitable".[39]

The most fundamental change is:

> "the system under which the confident and venturesome assume the risk or insure the doubtful and timid by guaranteeing to the latter a specified income in return for an assignment of the actual results... With human nature as we know it it would be impracticable or very unusual for one man to guarantee to another a definite result of the latter's actions without being given power to direct his work. And on the other hand the second party would not place himself under the direction of the first without such a guarantee... The result of this manifold specialisation of function is the enterprise and wage system of industry. Its existence in the world is the direct result of the fact of uncertainty."[40]

These quotations give the essence of Professor Knight's theory. The fact of uncertainty means that people have to forecast future wants. Therefore, you get a special class springing up who direct the activities of others to whom they give guaranteed wages. It acts because good judgment is generally associated with confidence in one's judgment.[41]

Professor Knight would appear to leave himself open to criticism on several grounds. First of all, as he himself points out, the fact that certain people have better judgment or better knowledge does not mean that they can only get an income from it by themselves actively taking part in production. They can sell advice or knowledge. Every business buys the services of a host of advisers. We can imagine a system where all advice or knowledge was bought as required. Again, it is possible to get a reward from better knowledge or judgment not by actively taking part in production but by making contracts with people who are producing. A merchant buying for future delivery represents an example of this. But this merely illustrates the point that it is quite possible to give a guaranteed reward providing that certain acts are performed without directing the performance of those acts. Professor Knight says

---

[39] Op. cit., pp. 268–295.

[40] Op. cit., pp. 269–270.

[41] Op. cit., p. 270.

that "with human nature as we know it, it would be impracticable or very unusual for one man to guarantee to another a definite result of the latter's actions without being given power to direct his work." This is surely incorrect. A large proportion of jobs are done to contract, that is, the contractor is guaranteed a certain sum providing he performs certain acts. But this does not involve any direction. It does mean, however, that the system of relative prices has been changed and that there will be a new arrangement of the factors of production.[42] The fact that Professor Knight mentions that the "second party would not place himself under the direction of the first without such a guarantee" is irrelevant to the problem we are considering. Finally, it seems important to notice that even in the case of an economic system where there is no uncertainty Professor Knight considers that there would be co-ordinators, though they would perform only a routine function. He immediately adds that they would be "without responsibility of any sort," which raises the question by whom are they paid and why? It seems that nowhere does Professor Knight give a reason why the price mechanism should be superseded.

IV

It would seem important to examine one further point and that is to consider the relevance of this discussion to the general question of the "cost-curve of the firm."

It has sometimes been assumed that a firm is limited in size under perfect competition if its cost curve slopes upward,[43] while under imperfect competition, it is limited in size because it will not pay to produce more than the output at which marginal cost is equal to marginal revenue.[44] But it is clear that a firm may produce more than one product and, therefore, there appears to be no **prima facie** reason why this upward slope of the cost curve in the case of perfect competition or the fact that marginal cost will not always be below marginal revenue in the case of imperfect competition should limit the size of the firm.[45] Mrs. Robinson[46] makes the simplifying

---

[42] This shows that it is possible to have a private enterprise system without the existence of firms. Though, in practice, the two functions of enterprise, which actually influences the system of relative prices by forecasting wants and acting in accordance with such forecasts, and management, which accepts the system of relative prices as being given, are normally carried out by the same persons, yet it seems important to keep them separate in theory. This point is further discussed below.

[43] See Kaldor, op. cit., and Robinson, *The Problem of Management* and the *Size of the Firm*.

[44] Mr. Robinson calls this the Imperfect Competition solution for the survival of the small firm.

[45] Mr. Robinson's conclusion, op. cit., 249, note 1, would appear to be definitely wrong. He is followed by Horace J. White, Jr, Monopolistic and Perfect Competition, *American Economic Review*, December 1936, p. 645, note 27. Mr. White states "It is obvious that the size of the firm is limited in conditions of monopolistic competition."

[46] *Economics Imperfect Competition*, 1934.

assumption that only one product is being produced. But it is clearly important to investigate how the number of products produced by a firm is determined, while no theory which assumes that only one product is in fact produced can have very great practical significance.

It might be replied that under perfect competition, since everything that is produced can be sold at the prevailing price, then there is no need for any other product to be produced. But this argument ignores the fact that there may be a point where it is less costly to organize the exchange transactions of a new product than to organize further exchange transactions of the old product. This point can be illustrated in the following way. Imagine, following von Thünen, that there is a town, the consuming center, and that industries are located around this central point in rings. These conditions are illustrated in the following diagram in which *A*, *B*, and *C* represent different industries.

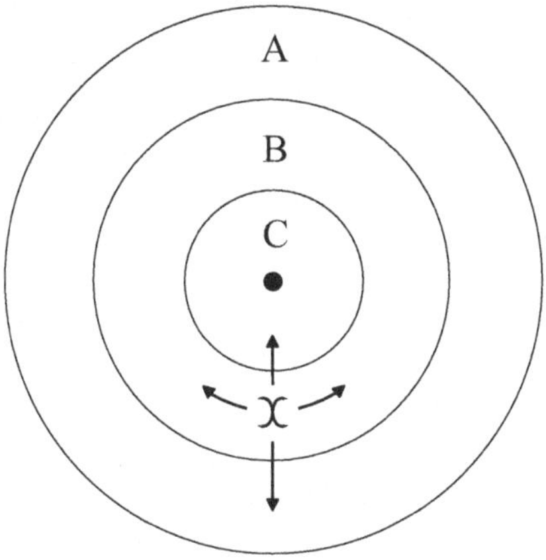

Imagine an entrepreneur who starts controlling exchange transactions from *x*. Now as he extends his activities in the same product (*B*), the cost of organizing increases until at some point it becomes equal to that of a dissimilar product which is nearer. As the firm expands, it will therefore from this point include more than one product (*A* and *C*). This treatment of the problem is obviously incomplete,[47] but it is necessary to show that merely proving that the cost curve turns upwards does not give a limitation to the size of the firm. So far we have only considered the case of perfect competition; the case of imperfect competition would appear to be obvious.

To determine the size of the firm, we have to consider the marketing costs (that is, the costs of using the price mechanism), and the costs of organizing the different entrepreneurs and then we can determine how many products will be produced by each firm and how much of each it will produce. It would, therefore, appear that

47 As has been shown above, location is only one of the factors influencing the cost of organizing.

Mr. Shove[48] in his article on "Imperfect Competition" was asking questions which Mrs. Robinson's cost curve apparatus cannot answer. The factors mentioned above would seem to be the relevant ones.

V

Only one task now remains; and that is, to see whether the concept of a firm which has been developed fits in with that existing in the real world. We can best approach the question of what constitutes a firm in practice by considering the legal relationship normally called that of "master and servant" or "employer and employee."[49] The essentials of this relationship have been given as follows:

(1) The servant must be under the duty of rendering personal services to the master or to others on behalf of the master, otherwise the contract is a contract for sale of goods or the like.

(2) The master must have the right to control the servant's work, either personally or by another servant or agent. It is this right of control or interference, of being entitled to tell the servant when to work (within the hours of service) and when not to work, and what work to do and how to do it (within the terms of such service) which is the dominant characteristic in this relation and marks off the servant from an independent contractor, or from one employed merely to give to his employer the fruits of his labour. In the latter case, the contractor or performer is not under the employer's control in doing the work or effecting the service; he has to shape and manage his work so as to give the result he has contracted to effect.[50]

We thus see that it is the fact of direction which is the essence of the legal concept of "employer and employee," just as it was in the economic concept which was developed above. It is interesting to note that Professor Batt says further:

> "That which distinguishes an agent from a servant is not the absence or presence of a fixed wage or the payment only of commission on business done, but rather the freedom with which an agent may carry out his employment."[51]

---

[48] G.F. Shove, The Imperfection of the Market, *The Economic Journal*, March 1933, p. 115. In connection with an increase in demand in the suburbs and the effect on the price charged by suppliers, Mr. Shove asks "...why do not the old firms open branches in the suburbs?" If the argument in the text is correct, this is a question which Mrs. Robinson's apparatus cannot answer.

[49] The legal concept of "employer and employee" and the economic concept of a firm are not identical, in that the firm may imply control over another person's property as well as over their labor. But the identity of these two concepts is sufficiently close for an examination of the legal concept to be of value in appraising the worth of the economic concept.

[50] Batt, *The Law of Master and Servant*, p. 6.

[51] Op. cit., p. 7.

We can therefore conclude that the definition we have given is one which approximates closely to the firm as it is considered in the real world.

Our definition is, therefore, realistic. Is it manageable? This ought to be clear. When we are considering how large a firm will be the principle of marginalism works smoothly. The question always is, will it pay to bring an extra exchange transaction under the organizing authority? At the margin, the costs of organizing within the firm will be equal either to the costs of organizing in another firm or to the costs involved in leaving the transaction to be "organized" by the price mechanism. Business men will be constantly experimenting, controlling more or less, and in this way, equilibrium will be maintained. This gives the position of equilibrium for static analysis. But it is clear that the dynamic factors are also of considerable importance, and an investigation of the effect changes have on the cost of organizing within the firm and on marketing costs generally will enable one to explain why firms get larger and smaller. We thus have a theory of moving equilibrium. The above analysis would also appear to have clarified the relationship between initiative or enterprise and management. Initiative means forecasting and operates through the price mechanism by the making of new contracts. Management proper merely reacts to price changes, rearranging the factors of production under its control. That the business man normally combines both functions is an obvious result of the marketing costs which were discussed above. Finally, this analysis enables us to state more exactly what is meant by the "marginal product" of the entrepreneur. But an elaboration of this point would take us far from our comparatively simple task of definition and clarification.

Herbert A. Simon (1962)

THE ARCHITECTURE OF COMPLEXITY

Many tasks and problems of everyday logistics – and also many academic works of the field – are connected with the question: How can logistic systems be designed in order to make them meet the complex requirements of dynamically and unpredictably changing markets and company environments? How to keep them manageable? How to deal with exploding risks inherent in overly complex systems, and how to control the transaction cost for planning, supervision, quality assurance and deficiencies in those kinds of systems? These questions raise the issue of optimal "architectures of complexity".

As logistics organizations grow in the context of industrial concentration and globalization the Supply Chain Management perspective forces logisticians to no longer restrict themselves to intra-organizational systems of purchasing, production, and distribution, but more and more raises the claim to integrate and holistically manage multi-tiered supply chains connecting many companies. The issue of the manageability of complexity has become critical to the field. Choosing an adequate architecture of logistic system complexity hence is an enormously relevant and at the same time exceedingly challenging question for the logistics discipline.

The essay about *The Architecture of Complexity* was written in 1962 by Herbert A. Simon (1916–2001), one of the most prolific writers of the 20th century on organizational behaviour and decision making, also a Nobel laureate. Simon chose the question of appropriate architectures of complexity a long time before its relevance was recognized for the young logistics discipline, anticipating the ideas of "loose coupling" under conditions of environmental turbulence (cf. the contributions by Weick (1976) and Ansoff (1975) in this volume). The parable of the two watchmakers, Hora and Tempus, which Simon used to illustrate the suitability of alternative approaches to handle complexity, is an outstanding example of how difficult issues can be presented both simply and impressively.

Reference:

Herbert A. Simon: The Architecture of Complexity. In: *Proceedings of the American Philosophical Society*, Vol. 106, 1962, pp. 62–76.

P. Klaus and S. Müller (eds.), *The Roots of Logistics*, DOI 10.1007/978-3-642-27922-5_23,

Herbert A. Simon: The Architecture of Complexity[1]

A number of proposals have been advanced in recent years for the development of "general systems theory" which, abstracting from properties peculiar to physical, biological, or social systems, would be applicable to all of them.[2] We might well feel that, while the goal is laudable, systems of such diverse kinds could hardly be expected to have any nontrivial properties in common. Metaphor and analogy can be helpful, or they can be misleading. All depends on whether the similarities the metaphor captures are significant or superficial.

It may not be entirely vain, however, to search for common properties among diverse kinds of complex systems. The ideas that go by the name of cybernetics constitute, if not a theory, at least a point of view that has been proving fruitful over a wide range of applications.[3] It has been useful to look at the behavior of adaptive systems in terms of the concepts of feedback and homeostasis, and to analyze adaptiveness in terms of the theory of selective information.[4] The ideas of feedback and information provide a frame of reference for viewing a wide range of situations, just as do the ideas of evolution, of relativism, of axiomatic method, and of operationalism.

In this paper I should like to report on some things we have been learning about particular kinds of complex systems encountered in the behavioral sciences. The developments I shall discuss arose in the context of specific phenomena, but the theoretical formulations themselves make little reference to details of structure. Instead they refer primarily to the complexity of the systems under view without specifying the exact content of that complexity. Because of their abstractness, the

---

1 The ideas in this paper have been the topic of many conversations with my colleague, Allen Newell. George W. Corner suggested important improvements in biological content as well as editorial form. I am also indebted, for valuable comments on the manuscript, to Richard H. Meier, John R. Platt, and Warren Weaver. Some of the conjectures about the nearly decomposable structure of the nucleus-atom-molecule hierarchy were checked against the available quantitative data by Andrew Schoene and William Wise. My work in this area has been supported by a Ford Foundation grant for research in organizations and a Carnegie Corporation grant for research on cognitive processes. To all of the above, my warm thanks, and the usual absolution.

2 See especially the yearbooks of the Society for General Systems Research. Prominent among the exponents of general systems theory are L. von Bertalanffy, K. Boulding, R.W. Gerard, and J.G. Miller. For a more skeptical view – perhaps too skeptical in the light of the present discussion – see H.A. Simon and A. Newell: Models: their Uses and Limitations, in L.D. White, ed: *The State of the Social Sciences*, pp. 66–83, Chicago: University of Chicago Press, 1956.

3 N. Wiener: *Cybernetics*, New York: John Wiley & Sons, 1948. For an imaginative forerunner, see A.J. Lotka: *Elements of Mathematical Biology*, New York: Dover Publications, 1951, first published in 1924 as *Elements of physical biology*.

4 C. Shannon and W. Weaver: *The Mathematical Theory of Communication*, Urbana, University of Illinois Press, 1949; W.R. Ashby: *Design for a Brain*, New York: John Wiley & Sons, 1952.

theories may have relevance – application would be too strong a term – to the other kinds of complex systems that are observed in the social, biological, and physical sciences.

In recounting these developments, I shall avoid technical detail, which can generally be found elsewhere. I shall describe each theory in the particular context in which it arose. Then, I shall cite some examples of complex systems, from areas of science other than the initial application, to which the theoretical framework appears relevant. In doing so, I shall make reference to areas of knowledge where I am not expert – perhaps not even literate. I feel quite comfortable in doing so before the members of this society, representing as it does the whole span of the scientific and scholarly endeavor. Collectively you will have little difficulty, I am sure, in distinguishing instances based on idle fancy or sheer ignorance from instances that cast some light on the ways in which complexity exhibits itself wherever it is found in nature. I shall leave to you the final judgment of relevance in your respective fields.

I shall not undertake a formal definition of "complex systems".[5] Roughly, by a complex system I mean one made up of a large number of parts that interact in a non-simple way. In such systems, the whole is more than the sum of the parts, not in an ultimate, metaphysical sense, but in the important pragmatic sense that, given the properties of the parts and the laws of their interaction, it is not a trivial matter to infer the properties of the whole. In the face of complexity, an in-principle reductionist may be at the same time a pragmatic holist.[6]

The four sections that follow discuss four aspects of complexity. The first offers some comments on the frequency with which complexity takes the form of hierarchy – the complex system being composed of subsystems that, in turn, have their own subsystems, and so on. The second section theorizes about the relation between the structure of a complex system and the time required for it to emerge through evolutionary processes: specifically, it argues that hierarchic systems will evolve far more quickly than non-hierarchic systems of comparable size. The third section explores the dynamic properties of hierarchically organized systems, and shows how they can be decomposed into subsystems in order to analyze their behavior. The fourth section examines the relation between complex systems and their descriptions.

---

5 W. Weaver, in: Science and complexity, *American Scientist*, Vol. 36, 1948, p. 536, has distinguished two kinds of complexity, disorganized and organized. We shall be primarily concerned with organized complexity.

6 See also John R. Platt: Properties of large molecules that go beyond the properties of their chemical sub-groups, *Journal of Theoretical Biology*, Vol. 1, 1961, pp. 342–358. Since the reductionism-holism issue is a major **cause de guerre** between scientists and humanists, perhaps we might even hope that peace could be negotiated between the two cultures along the lines of the compromise just suggested. As I go along, I shall have a little to say about complexity in the arts as well as in the natural sciences. I must emphasize the pragmatism of my holism to distinguish it sharply from the position taken by W.M. Elsasser in *The physical foundation of biology*, New York: Pergamon Press, 1958.

Thus, the central theme that runs through my remarks is that complexity frequently takes the form of hierarchy, and that hierarchic systems have some common properties that are independent of their specific content. Hierarchy, I shall argue, is one of the central structural schemes that the architect of complexity uses.

HIERARCHIC SYSTEMS

By a **hierarchic system**, or hierarchy, I mean a system that is composed of interrelated subsystems, each of the latter being, in turn, hierarchic in structure until we reach some lowest level of elementary subsystem. In most systems in nature, it is somewhat arbitrary as to where we leave off the partitioning, and what subsystems we take as elementary. Physics makes much use of the concept of "elementary particle" although particles have a disconcerting tendency not to remain elementary very long. Only a couple of generations ago, the atoms themselves were elementary particles; today, to the nuclear physicist they are complex systems. For certain purposes of astronomy, whole stars, or even galaxies, can be regarded as elementary subsystems. In one kind of biological research, a cell may be treated as an elementary subsystem; in another, a protein molecule; in still another, an amino acid residue.

Just why a scientist has a right to treat as elementary a subsystem that is in fact exceedingly complex is one of the questions we shall take up. For the moment, we shall accept the fact that scientists do this all the time, and that if they are careful scientists they usually get away with it.

Etymologically, the word "hierarchy" has had a narrower meaning than I am giving it here. The term has generally been used to refer to a complex system in which each of the subsystems is subordinated by an authority relation to the system it belongs to. More exactly, in a hierarchic formal organization, each system consists of a "boss" and a set of subordinate subsystems. Each of the subsystems has a "boss" who is the immediate subordinate of the boss of the system. We shall want to consider systems in which the relations among subsystems are more complex than in the formal organizational hierarchy just described. We shall want to include systems in which there is no relation of subordination among subsystems. (In fact, even in human organizations, the formal hierarchy exists only on paper; the real flesh-and-blood organization has many inter-part relations other than the lines of formal authority.) For lack of a better term, I shall use hierarchy in the broader sense introduced in the previous paragraphs, to refer to all complex systems analyzable into successive sets of subsystems, and speak of "formal hierarchy" when I want to refer to the more specialized concept.[7]

7 The mathematical term "partitioning" will not do for what I call here a hierarchy; for the set of subsystems, and the successive subsets in each of these defines the partitioning, independently of any systems of relations among the subsets. By hierarchy I mean the partitioning in conjunction with the relations that hold among its parts.

Social Systems

I have already given an example of one kind of hierarchy that is frequently encountered in the social sciences: a formal organization. Business firms, governments, universities all have a clearly visible parts-within-parts structure. But formal organizations are not the only, or even the most common, kind of social hierarchy. Almost all societies have elementary units called families, which may be grouped into villages or tribes, and these into larger groupings, and so on. If we make a chart of social interactions, of who talks to whom, the clusters of dense interaction in the chart will identify a rather well-defined hierarchic structure. The groupings in this structure may be defined operationally by some measure of frequency of interaction in this sociometric matrix.

Biological and Physical Systems

The hierarchical structure of biological systems is a familiar fact. Taking the cell as the building block, we find cells organized into tissues, tissues into organs, organs into systems. Moving downward from the cell, well-defined subsystems – for example, nucleus, cell membrane, microsomes, mitochondria, and so on – have been identified in animal cells.

The hierarchic structure of many physical systems is equally clear-cut. I have already mentioned the two main series. At the microscopic level we have elementary particles, atoms, molecules, macromolecules. At the macroscopic level we have satellite systems, planetary systems, galaxies. Matter is distributed throughout space in a strikingly non-uniform fashion. The most nearly random distributions we find, gases, are not random distributions of elementary particles but random distributions of complex systems, i.e. molecules.

A considerable range of structural types is subsumed under the term hierarchy as I have defined it. By this definition, a diamond is hierarchic, for it is a crystal structure of carbon atoms that can be further decomposed into protons, neutrons, and electrons. However, it is a very "flat" hierarchy, in which the number of first-order subsystems belonging to the crystal can be indefinitely large. A volume of molecular gas is a flat hierarchy in the same sense. In ordinary usage, we tend to reserve the word hierarchy for a system that is divided into a **small or moderate number** of subsystems, each of which may be further subdivided. Hence, we do not ordinarily think of or refer to a diamond or a gas as a hierarchic structure. Similarly, a linear polymer is simply a chain, which may be very long, of identical subparts, the monomers. At the molecular level it is a very flat hierarchy.

In discussing formal organizations, the number of subordinates who report directly to a single boss is called his **span of control**. I will speak analogously of the **span** of a system, by which I shall mean the number of subsystems into which it is partitioned. Thus, a hierarchic system is flat at a given level if it has a wide span at that level. A diamond has a wide span at the crystal level, but not at the next level down, the molecular level.

In most of our theory construction in the following sections we shall focus our attention on hierarchies of moderate span, but from time to time I shall comment on the extent to which the theories might or might not be expected to apply to very flat hierarchies.

There is one important difference between the physical and biological hierarchies, on the one hand, and social hierarchies, on the other. Most physical and biological hierarchies are described in spatial terms. We detect the organelles in a cell in the way we detect the raisins in a cake – they are "visibly" differentiated substructures localized spatially in the larger structure. On the other hand, we propose to identify social hierarchies not by observing who lives close to whom but by observing who interacts with whom. These two points of view can be reconciled by defining hierarchy in terms of intensity of interaction, but observing that in most biological and physical systems relatively intense interaction implies relative spatial propinquity. One of the interesting characteristics of nerve cells and telephone wires is that they permit very specific strong interactions at great distances. To the extent that interactions are channeled through specialized communications and transportation systems, spatial propinquity becomes less determinative of structure.

Symbolic Systems

One very important class of systems has been omitted from my examples thus far: systems of human symbolic production. A book is a hierarchy in the sense in which I am using that term. It is generally divided into chapters, the chapters into sections, the sections into paragraphs, the paragraphs into sentences, the sentences into clauses and phrases, the clauses and phrases into words. We may take the words as our elementary units, or further subdivide them, as the linguist often does, into smaller units. If the book is narrative in character, it may divide into "episodes" instead of sections, but divisions there will be.

The hierarchic structure of music, based on such units as movements, parts, themes, phrases, is well known. The hierarchic structure of products of the pictorial arts is more difficult to characterize, but I shall have something to say about it later.

THE EVOLUTION OF COMPLEX SYSTEMS

Let me introduce the topic of evolution with a parable. There once were two watchmakers, named Hora and Tempus, who manufactured very fine watches. Both of them were highly regarded, and the phones in their workshops rang frequently – new customers were constantly calling them. However, Hora prospered, while Tempus became poorer and poorer and finally lost his shop. What was the reason?

The watches the men made consisted of about 1,000 parts each. Tempus had so constructed his that if he had one partly assembled and had to put it down – to answer the phone say – it immediately fell to pieces and had to be reassembled from the elements. The better the customers liked his watches, the more they phoned him, the more difficult it became for him to find enough uninterrupted time to finish a watch.

The watches that Hora made were no less complex than those of Tempus. But he had designed them so that he could put together subassemblies of about ten elements each. Ten of these subassemblies, again, could be put together into a larger subassembly; and a system of ten of the latter subassemblies constituted the whole watch. Hence, when Hora had to put down a partly assembled watch in order to answer the phone, he lost only a small part of his work, and he assembled his watches in only a fraction of the man-hours it took Tempus.

It is rather easy to make a quantitative analysis of the relative difficulty of the tasks of Tempus and Hora: Suppose the probability that an interruption will occur while a part is being added to an incomplete assembly is p. Then the probability that Tempus can complete a watch he has started without interruption is $(1-p)^{1000}$ – a very small number unless p is .001 or less. Each interruption will cost, on the average, the time to assemble $1/p$ parts (the expected number assembled before interruption). On the other hand, Hora has to complete one hundred eleven sub-assemblies of ten parts each. The probability that he will not be interrupted while completing any one of these is $(1-p)^{10}$, and each interruption will cost only about the time required to assemble five parts.[8]

Now if p is about .01 – that is, there is one chance in a hundred that either watchmaker will be interrupted while adding any one part to an assembly – then a straightforward calculation shows that it will take Tempus, on the average, about four thousand times as long to assemble a watch as Hora.

---

[8] The speculations on speed of evolution were first suggested by H. Jacobson's application of information theory to estimating the time required for biological evolution. See his paper, Information, reproduction, and the origin of life, in *American Scientist*, Vol. 43, January 1055, pp. 119–127. From thermodynamic considerations it is possible to estimate the amount of increase in entropy that occurs when a complex system decomposes into its elements. (See, for example, R.B. Setlow and E.C. Pollard: *Molecular biophysics*, pp 63–65, Reading, Mass., Addison-Wesley Puhlishing Co., 1962, and references cited there.) But entropy is the logarithm of a probability, hence information, the negative of entropy, can be interpreted as the logarithm of the reciprocal of the probability – the "improbability", so to speak. The essential idea in Jacobson's model is that the expected time required for the system to reach a particular state is inversely proportional to the probability of the state – hence increases exponentially with the amount of information (negentropy) of the state.

Following this line of argument, but not introducing the notion of levels and stable subassemblies, Jacobson arrived at estimates of the time required for evolution so large as to make the event rather improbable. Our analysis, carried through in the same way, but with attention to the stable intermediate forms, produces very much smaller estimates.

We arrive at the estimate as follows:

1) Hora must make 111 times as many complete assemblies per watch as Tempus; but,

2) Tempus will lose on the average 20 times as much work for each interrupted assembly as Hora [100 parts, on the average, as against 5]; and,

3) Tempus will complete an assembly only 44 times per million attempts ($.99^{1000} = 44 \times 10^{-6}$) while Hora will complete nine out of ten ($.99^{10} = 9 \times 10^{-1}$). Hence Tempus will have to make 20,000 as many attempts per completed assembly as Hora. $(9 \times 10^{-1}) / (44 \times 10^{-6}) = 2 \times 10^{4}$. Multiplying these three ratios, we get:

$$1/111 \times 100/5 \times .99^{10} / .99^{1000}$$
$$= 1/111 \times 20 \times 20{,}000 \sim 4{,}000.$$

Biological Evolution

What lessons can we draw from our parable for biological evolution? Let us interpret a partially completed subassembly of k elementary parts as the coexistence of k parts in a small volume – ignoring their relative orientations. The model assumes that parts are entering the volume at a constant rate, but that there is a constant probability, p, that the part will be dispersed before another is added, unless the assembly reaches a stable state. These assumptions are not particularly realistic. They undoubtedly underestimate the decrease in probability of achieving the assembly with increase in the size of the assembly. Hence the assumptions understate – probably by a large factor – the relative advantage of a hierarchic structure.

Although we cannot, therefore, take the numerical estimate seriously the lesson for biological evolution is quite clear and direct. The time required for the evolution of a complex form from simple elements depends critically on the numbers and distribution of potential intermediate stable forms. In particular, if there exists a hierarchy of potential stable "subassemblies", with about the same span, s, at each level of the hierarchy, then the time required for a subassembly can be expected to be about the same at each level – that is proportional to $1/(1-p)^{8}$. The time required for the assembly of a system of n elements will be proportional to $\log_8 n$, that is, to the number of levels in the system. One would say – with more illustrative than literal intent – that the time required for the evolution of multi-celled organisms from single-celled organisms might be of the same order of magnitude as the time required for the evolution of single-celled organisms from macromolecules. The same argument could be applied to the evolution of proteins from amino acids, of molecules from atoms, of atoms from elementary particles.

A whole host of objections to this oversimplified scheme will occur, I am sure, to every working biologist, chemist, and physicist. Before turning to matters I know more about, I shall mention three of these problems, leaving the rest to the attention of the specialists.

First, in spite of the overtones of the watchmaker parable, the theory assumes no teleological mechanism. The complex forms can arise from the simple ones by purely random processes. (I shall propose another model in a moment that shows this clearly.) Direction is provided to the scheme by the stability of the complex forms, once these come into existence. But this is nothing more than survival of the fittest – i.e., of the stable.

Second, not all large systems appear hierarchical. For example, most polymers – e.g., nylon – are simply linear chains of large numbers of identical components, the monomers. However, for present purposes we can simply regard such a structure as a hierarchy with a span of one – the limiting case. For a chain of any length represents a state of relative equilibrium.[9]

Third, the evolution of complex systems from simple elements implies nothing, one way or the other, about the change in entropy of the entire system. If the process absorbs free energy, the complex system will have a smaller entropy than the elements; if it releases free energy, the opposite will be true. The former alternative is the one that holds for most biological systems, and the net inflow of free energy has to be supplied from the sun or some other source if the second law of thermodynamics is not to be violated. For the evolutionary process we are describing, the equilibria of the intermediate states need have only local and not global stability, and they may be stable only in the steady state – that is, as long as there is an external source of free energy that may be drawn upon.[10]

Because organisms are not energetically closed systems, there is no way to deduce the direction, much less the rate, of evolution from classical thermodynamic considerations. All estimates indicate that the amount of entropy, measured in physical units, involved in the formation of a one-celled biological organism is trivially small – about – 10^{-11} cal/degree.[11] The "improbability" of evolution has nothing to do with this quantity of entropy, which is produced by every bacterial cell every generation. The irrelevance of quantity of information, in this sense, to speed of evolution can also be seen from the fact that exactly as much information is required to "copy" a cell through the reproductive process as to produce the first cell through evolution.

---

9 There is a well-developed theory of polymer size, based on models of random assembly. See for example P.J. Flory: *Principles of polymer chemistry*, ch. 8, Ithaca: Cornell University Press, 1953. Since **all** subassemblies in the polymerization theory are stable, limitation of molecular growth depends on "poisoning" of terminal groups by impurities or formation of cycles rather than upon disruption of partially-formed chains.

10 This point has been made many times before, but it cannot be emphasized too strongly. For further discussion, see Setlow and Pollard, op. cit., pp. 49–64; E. Schrodinger: *What is life?* Cambridge University Press, 1945; and H. Linschitz: The Information Content of a Bacterial Cell, in H. Questler, (ed.: *Information theory in biology*, pp. 251–262, Urbana: University of Illinois Press, 1953.

11 See Linschitz, op. cit. This quantity, 10^{-11} cal/degree, corresponds to about 10^{13} bits of information.

The effect of the existence of stable intermediate forms exercises a powerful effect on the evolution of complex forms that may be likened to the dramatic effect of catalysts upon reaction rates and steady state distribution of reaction products in open systems.[12] In neither case does the entropy change provide us with a guide to system behavior.

Problem Solving as Natural Selection

Let us turn now to some phenomena that have no obvious connection with biological evolution: human problem-solving processes. Consider, for example, the task of discovering the proof for a difficult theorem. The process can be – and often has been – described as a search through a maze. Starting with the axioms and previously proved theorems, various transformations allowed by the rules of the mathematical systems are attempted, to obtain new expressions. These are modified in turn until, with persistence and good fortune, a sequence or path of transformations is discovered that leads to the goal.

The process usually involves a great deal of trial and error. Various paths are tried; some are abandoned, others are pushed further. Before a solution is found, a great many paths of the maze may be explored. The more difficult and novel the problem, the greater is likely to be the amount of trial and error required to find a solution. At the same time, the trial and error is not completely random or blind; it is, in fact, rather highly selective. The new expressions that are obtained by transforming given ones are examined to see whether they represent progress toward the goal. Indications of progress spur further search in the same direction; lack of progress signals the abandonment of a line of search. Problem solving requires **selective** trial and error.[13]

A little reflection reveals that cues signaling progress play the same role in the problem-solving process that stable intermediate forms play in the biological evolutionary process. In fact, we can take over the watchmaker parable and apply it also to problem solving. In problem solving, a partial result that represents recognizable progress toward the goal plays the role of a stable subassembly.

Suppose that the task is to open a safe whose lock has ten dials, each with one hundred possible settings, numbered from 0 to 99. How long will it take to open the safe

---

[12] See H. Kacser: Some Physico-chemical Aspects of Biological Organization, Appendix, pp. 191–249, in C.H. Waddington: *The Strategy of the Genes*, London: George Allen & Unwin, 1957.

[13] See A. Newell, J.C. Shaw, and H.A. Simon: Empirical Explorations of the Logic Theory Machine, *Proceedings of the 1957 Western Joint Computer Conference*, February, 1957, New York: Institute of Radio Engineers; Chess-playing programs and the problem of complexity, *IBM Journal of Research and Development 2*: pp. 320–335, October, 1958; and for a similar view of problem solving, W.R. Ashby: Design for an intelligence amplifier, pp. 215–233 in C.E. Shannon and J. McCarthy: *Automata studies*, Princeton: Princeton University Press, 1956.

by a blind trial-and-error search for the correct setting? Since there are 100 possible settings, we may expect to examine about one-half of these, on the average, before finding the correct one – that is, fifty billion billion settings. Suppose, however, that the safe is defective, so that a click can be heard when any one dial is turned to the correct setting. Now each dial can be adjusted independently, and does not need to be touched again while the others are being set. The total number of settings that has to he tried is only 10 × 50, or five hundred. The task of opening the safe has been altered, by the cues the clicks provide, from a practically impossible one to a trivial one.[14]

A considerable amount has been learned in the past five years about the nature of the mazes that represent common human problem-solving tasks – proving theorems, solving puzzles, playing chess, making investments, balancing assembly lines, to mention a few. All that we have learned about these mazes points to the same conclusion: that human problem solving, from the most blundering to the most insightful, involves nothing more than varying mixtures of trial and error and selectivity. The selectivity derives from various rules of thumb, or heuristics, that suggest which paths should be tried first and which leads are promising. We do not need to postulate processes more sophisticated than those involved in organic evolution to explain how enormous problem mazes are cut down to quite reasonable size.[15]

The sources of selectivity

When we examine the sources from which the problem-solving system, or the evolving system, as the case may be, derives its selectivity, we discover that selectivity can always be equated with some kind of feedback of information from the environment.

Let us consider the case of problem solving first. There are two basic kinds of selectivity. One we have already noted: various paths are tried out, the consequences of following them are noted, and this information is used to guide further search. In the same way, in organic evolution, various complexes come into being, at least evanescently, and those that are stable provide new building blocks for further construc-

---

[14] The clicking safe example was supplied by D.P. Simon. Ashby, op. cit., p. 230, has called the selectivity involved in situations of this kind "selection by components". The even greater reduction in time produced by hierarchization in the clicking safe example, as compared with the watchmaker's metaphor, is due to the fact that a random **search** for the correct combination is involved in the former case, while in the latter the parts come together in the right order. It is not clear which of these metaphors provides the better model for biological evolution, but we may be sure that the watchmaker's metaphor gives an exceedingly conservative estimate of the savings due to hierarchization. The safe may give an excessively high estimate because it assumes all possible arrangements of the elements to be equally probable.

[15] A. Newell and H.A. Simon: Computer Simulation of Human Thinking, in: *Science,* Vol. 134, December 22, 1961, pp. 2011–2017.

tion. It is this information about stable configurations, and not free energy or negentropy from the sun, that guides the process of evolution and provides the selectivity that is essential to account for its rapidity.

The second source of selectivity in problem solving is previous experience. We see this particularly clearly when the problem to be solved is similar to one that has been solved before. Then, by simply trying again the paths that led to the earlier solution, or their analogues, trial-and-error search is greatly reduced or altogether eliminated. What corresponds to this latter kind of information in organic evolution? The closest analogue is reproduction. Once we reach the level of selfreproducing systems, a complex system, when it has once been achieved, can be multiplied indefinitely. Reproduction in fact allows the inheritance of acquired characteristics, but at the level of genetic material, of course; i.e., only characteristics acquired by the genes can be inherited. We shall return to the topic of reproduction in the final section of this paper.

On Empires and Empire-Building

We have not exhausted the categories of complex systems to which the watchmaker argument can reasonably be applied. Philip assembled his Macedonian empire and gave it to his son, to be later combined with the Persian subassembly and others into Alexander's greater system. On Alexander's death, his empire did not crumble to dust, but fragmented into some of the major subsystems that had composed it.

The watchmaker argument implies that if one would be Alexander, one should be born into a world where large stable political systems already exist. Where this condition was not fulfilled, as on the Scythian and Indian frontiers, Alexander found empire building a slippery business. So too, T.E. Lawrence's organizing of the Arabian revolt against the Turks was limited by the character of his largest stable building blocks, the separate, suspicious desert tribes.

The profession of history places a greater value upon the validated particular fact than upon tendentious generalization. I shall not elaborate upon my fancy, therefore, but will leave it to historians to decide whether anything can be learned for the interpretation of history from an abstract theory of hierarchic complex systems.

Conclusion: The Evolutionary Explanation of Hierarchy

We have shown thus far that complex systems will evolve from simple systems much more rapidly if there are stable intermediate forms than if there are not. The resulting complex forms in the former case will be hierarchic. We have only to turn the argument around to explain the observed predominance of hierarchies among the complex systems nature presents to us. Among possible complex forms, hierarchies are the ones that have the time to evolve. The hypothesis that complexity will be hierarchic makes no distinction among very flat hierarchies, like crystals, and tissues, and polymers, and the intermediate forms. Indeed, in the complex systems we encounter in nature, examples of both forms are prominent. A more complete theory than the one we have developed here would presumably have something to say about the determinants of width of span in these systems.

NEARLY DECOMPOSABLE SYSTEMS

In hierarchic systems, we can distinguish between the interactions **among** subsystems, on the one hand, and the interactions **within** subsystems – i.e., among the parts of those subsystems – on the other. The interactions at the different levels may be, and often will be, of different orders of magnitude. In a formal organization there will generally be more interaction, on the average, between two employees who are members of the same department than between two employees from different departments. In organic substances, intermolecular forces will generally be weaker than molecular forces, and molecular forces than nuclear forces.

In a rare gas, the intermolecular forces will be negligible compared to those binding the molecules – we can treat the individual particles, for many purposes, as if they were independent of each other. We can describe such a system as **decomposable** into the subsystems comprised of the individual particles. As the gas becomes denser, molecular interactions become more significant. But over some range, we can treat the decomposable case as a limit, and as a first approximation. We can use a theory of perfect gases, for example, to describe approximately the behavior of actual gases if they are not too dense. As a second approximation, we may move to a theory of **nearly decomposable** systems, in which the interactions among the subsystems are weak, but not negligible.

At least some kinds of hierarchic systems can be approximated successfully as nearly decomposable systems. The main theoretical findings from the approach can be summed up in two propositions: *(a)* in a nearly decomposable system, the short-run behavior of each of the component subsystems is approximately independent of the short-run behavior of the other components; *(b)* in the long run, the behavior of any one of the components depends in only an aggregate way on the behavior of the other components.

Let me provide a very concrete simple example of a nearly decomposable system.[16] Consider a building whose outside walls provide perfect thermal insulation from the environment. We shall take these walls as the boundary of our system. The building

[16] This discussion of near-decomposability is based upon H.A. Simon and A. Ando: Aggregation of Variables in Dynamic Systems, in: *Econometrica*, Vol. 29, April 1961, pp. 111–138. The example is drawn from the same source, pp. 117–118. The theory has been further developed and applied to a variety of economic and political phenomena by Ando and F.M. Fisher. See F. M. Fisher: On the Cost of Approximate Specification in Simultaneous Equation Estimation, in: *Econometrica*, Vol. 29, April 1961, pp. 139–170, and F.M. Fisher and A. Ando: Two Theorems on *Ceteris Paribus* in the Analysis of Dynamic Systems, in: *American Political Science Review*, Vol. 61, March 1961, pp. 103–113.

is divided into a large number of rooms, the walls between them being good, but not perfect, insulators. The walls between rooms are the boundaries of our major subsystems. Each room is divided by partitions into a number of cubicles, but the partitions are poor insulators. A thermometer hangs in each cubicle. Suppose that at the time of our first observation of the system there is a wide variation in temperature from cubicle to cubicle and from room to room – the various cubicles within the building are in a state of thermal disequilibrium. When we take new temperature readings several hours later, what shall we find? There will be very little variation in temperature among the cubicles within each single room, but there may still be large temperature variations **among** rooms. When we take readings again several days later, we find an almost uniform temperature throughout the building; the temperature differences among rooms have virtually disappeared.

We can describe the process of equilibration formally by setting up the usual equations of heat flow. The equations can be represented by the matrix of their coefficients, r_{ij}, where r_{ij} is the rate at which heat flows from the ith cubicle to the jth cubicle per degree difference in their temperatures. If cubicles i and j do not have a common wall, r_{ij} will be zero. If cubicles i and j have a common wall, and are in the same room, r_{ij}, will be large. If cubicles i and j are separated by the wall of a room, r_{ij} will be nonzero but small. Hence, by grouping all the cubicles together that are in the same room, we can arrange the matrix of coefficients so that all its large elements lie inside a string of square submatrices along the main diagonal. All the elements outside these diagonal squares will be either zero or small (see EXHIBIT 1). We may take some small number, ε, as the upper bound of the extradiagonal elements. We shall call a matrix having these properties a **nearly decomposable matrix**.

	A1	A2	A3	B1	B2	C1	C2	C3
A1	—	100	—	2	—	—	—	—
A2	100	—	100	1	1	—	—	—
A3	—	100	—	—	2	—	—	—
B1	2	1	—	—	100	2	1	—
B2	—	1	2	100	—	—	1	2
C1	—	—	—	2	—	—	100	—
C2	—	—	—	1	—	100	—	100
C3	—	—	—	—	2	—	100	—

EXHIBIT 1 A hypothetical nearly-decomposable system. In terms of the heat-exchange example on the text, A1, A2, and A3 may be interpreted as cubicles in one room, B1 and B2 as cubicles in a second room, and C1, C2, and C3 as cubicles in a third. The matrix entries then are the heat diffusion coefficients between cubicles.

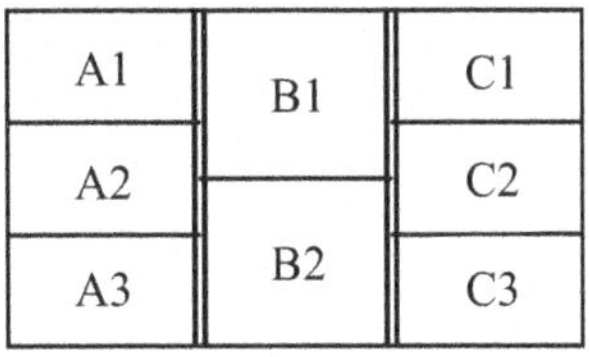

Now it has been proved that a dynamic system that can be described by a nearly decomposable matrix has the properties, stated above, of nearly decomposable system. In our simple example of heat flow this means that in the short run each room will reach an equilibrium temperature (an average of the initial temperatures of its offices) nearly independently of the others; and that each room will remain approximately in a state of equilibrium over the longer period during which an over-all temperature equilibrium is being established throughout the building. After the intraroom short-run equilibria have been reached, a single thermometer in each room will be adequate to describe the dynamic behavior of the entire system – separate thermometers in each cubicle will be superfluous.

Near Decomposability of Social Systems

As a glance at EXHIBIT 1 shows, near decomposability is a rather strong property for a matrix to possess, and the matrices that have this property will describe very special dynamic systems – vanishingly few systems out of all those that are thinkable. How few they will be depends, of course, on how good an approximation we insist upon. If we demand that epsilon be very small, correspondingly few dynamic systems will fit the definition. But we have already seen that in the natural world nearly decomposable systems are far from rare. On the contrary, systems in which each variable is linked with almost equal strength with almost all other parts of the system are far rarer and less typical.

In economic dynamics, the main variables are the prices and quantities of commodities. It is empirically true that the price of any given commodity and the rate at which it is exchanged depend to a significant extent only on the prices and quantities of a few other commodities, together with a few other aggregate magnitudes, like the average price level or some over-all measure of economic activity. The large linkage coefficients are associated, in general, with the main flows of raw materials and semi-finished products within and between industries. An input-output matrix of the economy, giving the magnitudes of these flows, reveals the nearly decomposable structure of the system – with one qualification. There is a consumption subsystem of the economy that is linked strongly to variables in most of the other subsystems. Hence, we have to modify our notions of decomposability slightly to accommodate the special role of the consumption subsystem in our analysis of the dynamic behavior of the economy.

In the dynamics of social systems, where members of a system communicate with and influence other members, near decomposability is generally very prominent. This is most obvious in formal organizations, where the formal authority relation connects each member of the organization with one immediate superior and with a small number of subordinates. Of course many communications in organizations follow other channels than the lines of formal authority. But most of these channels lead from any particular individual to a very limited number of his superiors, subordinates, and associates. Hence, departmental boundaries play very much the same role as the walls in our heat example.

Physico-Chemical Systems

In the complex systems familiar in biological chemistry, a similar structure is clearly visible. Take the atomic nuclei in such a system as the elementary parts of the system, and construct a matrix of bond strengths between elements. There will be matrix elements of quite different orders of magnitude. The largest will generally correspond to the covalent bonds, the next to the ionic bonds, the third group to hydrogen bonds, still smaller linkages to van der Waals forces.[17] If we select an epsilon just a little smaller than the magnitude of a covalent bond, the system will decompose into subsystems – the constituent molecules. The smaller linkages will correspond to the intermolecular bonds.

[17] For a survey of the several classes of molecular and inter-molecular forces, and their dissociation energies see Setlow and Pollard, op. cit., chapter 6. The energies of typical covalent bonds are of the order of 80–100 k cal/mole, of the hydrogen bonds, 10 k cal/mole. Ionic bonds generally lie between these two levels, the bonds due to van der Waals forces are lower in energy.

It is well known that high-energy, high-frequency vibrations are associated with the smaller physical subsystems, low-frequency vibrations with the larger systems into which the subsystems are assembled. For example, the radiation frequencies associated with molecular vibrations are much lower than those associated with the vibrations of the planetary electrons of the atoms; the latter, in turn, are lower than those associated with nuclear processes.[18] Molecular systems are nearly decomposable systems, the short-run dynamics relating to the internal structures of the subsystems; the long-run dynamics to the interactions of these subsystems.

A number of the important approximations employed in physics depend for their validity on the near-decomposability of the systems studied. The theory of the thermodynamics of irreversible processes, for example, requires the assumption of macroscopic disequilibrium but microscopic equilibrium,[19] exactly the situation described in our heat-exchange example. Similarly computations in quantum mechanics are often handled by treating weak interactions as producing perturbations on a system of strong interactions.

Some Observations on Hierarchic Span

To understand why the span of hierarchies is sometimes very broad – as in crystals – sometimes narrow, we need to examine more detail of the interactions. In general, the critical consideration is the extent to which interaction between two (or a few) subsystems excludes interaction of these subsystems with the others. Let us examine first some physical examples.

Consider a gas of identical molecules, each of which can form covalent bonds, in certain ways, with others. Let us suppose that we can associate with each atom a specific number of bonds that it is capable of maintaining simultaneously. (This number is obviously related to the number we usually call its valence.) Now suppose that two atoms join, and that we can also associate with the combination a specific number of external bonds it is capable of maintaining. If this number is the same as the number associated with the individual atoms, the bonding process can go on indefinitely – the atoms can form crystals or polymers of indefinite extent. If the

---

18 Typical wave numbers for vibrations associated with various systems (the wave number is the reciprocal of wave length hence proportional to frequency):

steel wire under tension – 10^{-10} to 10^{-9} cm^{-1}

molecular rotations – 10^{0} to 10^{2} cm^{-1}

molecular vibrations – 10^{2} to 10^{3} cm^{-1}

planetary electrons – 10^{4} to 10^{5} cm^{-1}

nuclear rotations – 10^{9} to 10^{10} cm^{-1}

nuclear surface vibrations – 10^{11} to 10^{12} cm^{-1}.

19 S.R. de Groot: *Thermodynamics of Irreversible Processes*, 11–12, New York, Interscience Publishers, 1951.

number of bonds of which the composite is capable is less than the number associated with each of the parts, then the process of agglomeration must come to a halt.

We need only mention some elementary examples. Ordinary gases show no tendency to agglomerate because the multiple bonding of atoms "uses up" their capacity to interact. While each oxygen atom has a valence of two, the O_2 molecules have a zero valence. Contrariwise, indefinite chains of single-bonded carbon atoms can be built up because a chain of any number of such atoms, each with two side groups, has a valence of exactly two.

Now what happens if we have a system of elements that possess both strong and weak interaction capacities, and whose strong bonds are exhaustible through combination? Subsystems will form, until all the capacity for strong interaction is utilized in their construction. Then these subsystems will be linked by the weaker second-order bonds into larger systems. For example, a water molecule has essentially a valence of zero – all the potential covalent bonds are fully occupied by the interaction of hydrogen and oxygen molecules. But the geometry of the molecule creates an electric dipole that permits weak interaction between the water and salts dissolved in it – whence such phenomena as its electrolytic conductivity.[20]

Similarly, it has been observed that, although electrical forces are much stronger than gravitational forces, the latter are far more important than the former for systems on an astronomical scale. The explanation, of course, is that the electrical forces, being bipolar, are all "used up" in the linkages of the smaller subsystems, and that significant net balances of positive or negative charges are not generally found in regions of macroscopic size.

In social as in physical systems there are generally limits on the simultaneous interaction of large numbers of subsystems. In the social case, these limits are related to the fact that a human being is more nearly a serial than a parallel information-processing system. He can carry on only one conversation at a time, and although this does not limit the size of the audience to which a mass communication can be addressed, it does limit the number of people simultaneously involved in most other forms of social interaction. Apart from requirements of direct interaction, most roles impose tasks and responsibilities that are time consuming. One cannot, for example, enact the role of "friend" with large numbers of other people.

It is probably true that in social as in physical systems, the higher frequency dynamics are associated with the subsystems, the lower frequency dynamics with the larger systems. It is generally believed, for example, that the relevant planning horizon of executives is longer the higher their location in the organizational hierarchy. It is probably also true that both the average duration of an interaction between executives and the average interval between interactions is greater at higher than at lower levels.

---

[20] See, for example, L. Pauling: *General chemistry*, ch. 15.

Summary: Near Decomposability

We have seen that hierarchies have the property of near-decomposability. Intra-component linkages are generally stronger than intercomponent linkages. This fact has the effect of separating the high-frequency dynamics of a hierarchy – involving the internal structure of the components – from the low frequency dynamics – involving interaction among components. We shall turn next to some important consequences of this separation for the description and comprehension of complex systems.

THE DESCRIPTION OF COMPLEXITY

If you ask a person to draw a complex object – e.g., a human face – he will almost always proceed in a hierarchic fashion.[21] First he will outline the face. Then he will add or insert features: eyes, nose, mouth, ears, hair. If asked to elaborate, he will begin to develop details for each of the features – pupils, eyelids, lashes for the eyes, and so on – until he reaches the limits of his anatomical knowledge. His information about the object is arranged hierarchically in memory, like a topical outline.

When information is put in outline form, it is easy to include information about the relations among the major parts and information about the internal relations of parts in each of the suboutlines. Detailed information about the relations of subparts belonging to different parts has no place in the outline and is likely to be lost. The loss of such information and the preservation mainly of information about hierarchic order is a salient characteristic that distinguishes the drawings of a child or someone untrained in representation from the drawing of a trained artist. (I am speaking of an artist who is striving for representation.)

Near Decomposability and Comprehensibility

From our discussion of the dynamic properties of nearly decomposable systems, we have seen that comparatively little information is lost by representing them as hierarchies. Subparts belonging to different parts only interact in an aggregative fashion – the detail of their interaction can be ignored. In studying the interaction of two large molecules, generally we do not need to consider in detail the interactions of nuclei of the atoms belonging to the one molecule with the nuclei of the atoms belonging to the other. In studying the interaction of two nations, we do not need to

---

[21] George A. Miller has collected protocols from subjects who were given the task of drawing faces, and finds that they behave in the manner described here (private communication). See also E.H. Gombrich: *Art and Illusion*, New York, Pantheon Books, 1960, pp. 291–296.

study in detail the interactions of each citizen of the first with each citizen of the second.

The fact, then, that many complex systems have a nearly decomposable, hierarchic structure is a major facilitating factor enabling us to understand, to describe, and even to "see" such systems and their parts. Or perhaps the proposition should be put the other way round. If there are important systems in the world that are complex without being hierarchic, they may to a considerable extent escape our observation and our understanding. Analysis of their behavior would involve such detailed knowledge and calculation of the interactions of their elementary parts that it would be beyond our capacities of memory or computation.[22]

I shall not try to settle which is chicken and which is egg: whether we are able to understand the world because it is hierarchic, or whether it appears hierarchic because those aspects of it which are not elude our understanding and observation. I have already given some reasons for supposing that the former is at least half the truth – that evolving complexity would tend to be hierarchic – but it may not be the whole truth.

Simple Descriptions of Complex Systems

One might suppose that the description of a complex system would itself be a complex structure of symbols – and indeed, it may be just that. But there is no conservation law that requires that the description be as cumbersome as the object described. A trivial example will show how a system can be described economically. Suppose the system is a two-dimensional array like this:

[22] I believe the fallacy in the central thesis of W.M. Elsasser's *The Physical Foundation of Biology*, mentioned earlier, lies in his ignoring the simplification in description of complex systems that derives from their hierarchic structure. Thus (p. 155): "If we now apply similar arguments to the coupling of enzymatic reactions with the substratum of protein molecules, we see that over a sufficient period of time, the information corresponding to the structural details of these molecules will be communicated to the dynamics of the cell, to higher levels of organization as it were, and may influence such dynamics. While this reasoning is only qualitative, it lends credence to the assumption that in the living organism, unlike the inorganic crystal, the effects of microscopic structure cannot be simply averaged out; as time goes on this influence will pervade the behavior of the cell 'at all levels'."

But from our discussion of near-decomposability it would appear that those aspects of microstructure that control the slow developmental aspects of organismic dynamics can be separated out from the aspects that control the more rapid cellular metabolic processes. For this reason we should not despair of unravelling the web of causes. See also J.R. Platt's review of Elsasser's book in *Perspectives in Biology and Medicine* 2, 1959, pp. 243–245.

$$\begin{matrix} A & B & M & N & R & S & H & I \\ C & D & O & P & T & U & J & K \\ M & N & A & B & H & I & R & S \\ O & P & C & D & J & K & T & U \\ R & S & H & I & A & B & M & N \\ T & U & J & K & C & D & O & P \\ H & I & R & S & M & N & A & B \\ J & K & T & U & O & P & C & D \end{matrix}$$

Let us call the array $\begin{vmatrix} A & B \\ C & D \end{vmatrix}$ a, the array $\begin{vmatrix} M & N \\ O & P \end{vmatrix}$ m, the array $\begin{vmatrix} R & S \\ T & U \end{vmatrix}$ r, and the array $\begin{vmatrix} H & I \\ J & K \end{vmatrix}$ h. Let us call the array $\begin{vmatrix} a & m \\ m & a \end{vmatrix}$ w, and the array $\begin{vmatrix} r & h \\ h & r \end{vmatrix}$ x. Then the entire array is simply $\begin{vmatrix} w & x \\ x & w \end{vmatrix}$. While the original structure consisted of 64 symbols, it requires only 35 to write down its description:

$$S = \begin{matrix} w & x \\ x & w \end{matrix}$$

$$w = \begin{matrix} a & m \\ m & a \end{matrix} \qquad x = \begin{matrix} r & h \\ h & r \end{matrix}$$

$$a = \begin{matrix} A & B \\ C & D \end{matrix} \qquad m = \begin{matrix} M & N \\ O & P \end{matrix} \qquad r = \begin{matrix} R & S \\ T & U \end{matrix} \qquad h = \begin{matrix} H & I \\ J & K \end{matrix}$$

We achieve the abbreviation by making use of the redundancy in the original structure. Since the pattern $\begin{matrix} A & B \\ C & D \end{matrix}$, for example, occurs four times in the total pattern, it is economical to represent it by the single symbol, a.

If a complex structure is completely unredundant – if no aspect of its structure can be inferred from any other – then it is its own simplest description. We can exhibit it, but we cannot describe it by a simpler structure. The hierarchic structures we have been discussing have a high degree of redundancy, hence can often be described in economical terms. The redundancy takes a number of forms, of which I shall mention three:

a. Hierarchic systems are usually composed of only a few different kinds of subsystems, in various combinations and arrangements. A familiar example is the proteins, their multitudinous variety arising from arrangements of only twenty different amino acids. Similarly, the ninety-odd elements provide all the kinds of building blocks needed for an infinite variety of molecules. Hence, we can construct our description from a restricted alphabet of elementary terms corresponding to the basic set of elementary subsystems from which the complex system is generated.

b. Hierarchic systems are, as we have seen, often nearly decomposable. Hence only aggregative properties of their parts enter into the description of the interactions of those parts. A generalization of the notion of near-decomposability might be called the "empty world hypothesis" – most things are only weakly connected with most other things; for a tolerable description of reality only a tiny fraction of all possible interactions needs to be taken into account. By adopting a descriptive language that allows the absence of something to go unmentioned, a nearly empty world can be described quite concisely. Mother Hubbard did not have to check off the list of possible contents to say that her cupboard was bare.

c. By appropriate "recoding", the redundancy that is present but unobvious in the structure of a complex system can often be made patent. The most common recoding of descriptions of dynamic systems consists in replacing a description of the time path with a description of a differential law that generates that path. The simplicity, that is, resides in a constant relation between the state of the system at any given time and the state of the system a short time later. Thus, the structure of the sequence, 1 3 5 7 9 11 . . ., is most simply expressed by observing that each member is obtained by adding 2 to the previous one. But this is the sequence that Galileo found to describe the velocity at the end of successive time intervals of a ball rolling down an inclined plane.

It is a familiar proposition that the task of science is to make use of the world's redundancy to describe that world simply. I shall not pursue the general methodological point here, but shall instead take a closer look at two main types of description that seem to be available to us in seeking an understanding of complex systems. I shall call these **state description** and **process description**, respectively.

State Descriptions and Process Descriptions

"A circle is the locus of all points equidistant from a given point." "To construct a circle, rotate a compass with one arm fixed until the other arm has returned to its starting point." It is implicit in Euclid that if you carry out the process specified in the second sentence, you will produce an object that satisfies the definition of the first. The first sentence is a state description of a circle, the second a process description.

These two modes of apprehending structure are the warp and weft of our experience. Pictures, blueprints, most diagrams, chemical structural formulae are state descriptions. Recipes, differential equations, equations for chemical reactions are process descriptions. The former characterize the world as sensed; they provide the criteria for identifying objects, often by modeling the objects themselves. The latter characterize the world as acted upon; they provide the means for producing or generating objects having the desired characteristics.

The distinction between the world as sensed and the world as acted upon defines the basic condition for the survival of adaptive organisms. The organism must develop correlations between goals in the sensed world and actions in the world of process. When they are made conscious and verbalized, these correlations correspond to what we usually call means-end analysis. Given a desired state of affairs and an existing state of affairs, the task of an adaptive organism is to find the difference between these two states, and then to find the correlating process that will erase the difference.[23]

Thus, problem solving requires continual translation between the state and process descriptions of the same complex reality. Plato, in the *Meno*, argued that all learning is remembering. He could not otherwise explain how we can discover or recognize the answer to a problem unless we already know the answer.[24] Our dual relation to the world is the source and solution of the paradox. We pose a problem by giving the state description of the solution. The task is to discover a sequence of processes that will produce the goal state from an initial state. Translation from the process description to the state description enables us to recognize when we have succeeded. The solution is genuinely new to us – and we do not need Plato's theory of remembering to explain how we recognize it.

There is now a growing body of evidence that the activity called human problem solving is basically a form of means-end analysis that aims at discovering a process description of the path that leads to a desired goal. The general paradigm is: given a blueprint, to find the corresponding recipe. Much of the activity of science is an application of that paradigm: given the description of some natural phenomena, to find the differential equations for processes that will produce the phenomena.

The Description of Complexity in Self-Reproducing Systems

The problem of finding relatively simple descriptions for complex systems is of interest not only for an understanding of human knowledge of the world but also for an explanation of how a complex system can reproduce itself. In my discussion of the evolution of complex systems, I touched only briefly on the role of self-reproduction.

Atoms of high atomic weight and complex inorganic molecules are witnesses to the fact that the evolution of complexity does not imply self-reproduction. If evolution of complexity from simplicity is sufficiently probable, it will occur repeatedly; the statistical equilibrium of the system will find a large fraction of the elementary particles participating in complex systems.

---

23 See H.A. Simon and A. Newell, Simulation of human thinking, in M. Greenberger (ed.), *Management and the Computer of the Future*, New York: Wiley, 1962, pp. 95–114, esp. pp. 110ff.

24 *The Works of Plato*, B. Jowett, trans., 3, New York, Dial Press, pp. 26–35.

If, however, the existence of a particular complex form increased the probability of the creation of another form just like it, the equilibrium between complexes and components could be greatly altered in favor of the former. If we have a description of an object that is sufficiently clear and complete, we can reproduce the object from the description. Whatever the exact mechanism of reproduction, the description provides us with the necessary information.

Now we have seen that the descriptions of complex systems can take many forms. In particular, we can have state descriptions or we can have process descriptions; blueprints or recipes. Reproductive processes could be built around either of these sources of information. Perhaps the simplest possibility is for the complex system to serve as a description of itself – a template on which a copy can be formed. One of the most plausible current theories, for example, of the reproduction of deoxyribonucleic acid (DNA) proposes that a DNA molecule, in the form of a double helix of matching parts (each essentially a "negative" of the other), unwinds to allow each half of the helix to serve as a template on which a new matching half can form.

On the other hand, our current knowledge of how DNA controls the metabolism of the organism suggests that reproduction by template is only one of the processes involved. According to the prevailing theory, DNA serves as a template both for itself and for the related substance ribonucleic acid (RNA). RNA, in turn, serves as a template for protein. But proteins – according to current knowledge – guide the organism's metabolism not by the template method but by serving as catalysts to govern reaction rates in the cell. While RNA is a blueprint for protein, protein is a recipe for metabolism.[25]

Ontogeny Recapitulates Phylogeny

The DNA in the chromosomes of an organism contains some, and perhaps most, of the information that is needed to determine its development and activity. We have seen that, if current theories are even approximately correct, the information is recorded not as a state description of the organism but as a series of "instructions" for the construction and maintenance of the organism from nutrient materials. I have already used the metaphor of a recipe; I could equally well compare it with a computer program, which is also a sequence of instructions, governing the construction of symbolic structures. Let me spin out some of the consequences of the latter comparison.

If genetic material is a program – viewed in its relation to the organism – it is a program with special and peculiar properties. First, it is a self-reproducing program; we have already considered its possible copying mechanism. Second, it is a program that has developed by Darwinian evolution. On the basis of our watchmaker's argu-

[25] C.B. Anfinsen: *The Molecular Basis of Evolution*, chs. 3 and 10, New York: Wiley, 1959, will qualify this sketchy, oversimplified account. For an imaginative discussion of some mechanisms of process description that could govern molecular structure, see H.H. Pattee, On the Origin of Macromolecular Sequences, *Biophysical Journal* 1, 1961, pp. 683–710.

ment, we may assert that many of its ancestors were also viable programs – programs for the subassemblies.

Are there any other conjectures we can make about the structure of this program? There is a well-known generalization in biology that is verbally so neat that we would be reluctant to give it up even if the facts did not support it: ontogeny recapitulates phylogeny. The individual organism, in its development, goes through stages that resemble some of its ancestral forms. The fact that the human embryo develops gill bars and then modifies them for other purposes is a familiar particular belonging to the generalization. Biologists today like to emphasize the qualifications of the principle – that ontogeny recapitulates only the grossest aspects of phylogeny, and these only crudely. These qualifications should not make us lose sight of the fact that the generalization does hold in rough approximation – it does summarize a very significant set of facts about the organism's development. How can we interpret these facts?

One way to solve a complex problem is to reduce it to a problem previously solved – to show what steps lead from the earlier solution to a solution of the new problem. If, around the turn of the century, we wanted to instruct a workman to make an automobile, perhaps the simplest way would have been to tell him how to modify a wagon by removing the singletree and adding a motor and transmission. Similarly, a genetic program could be altered in the course of evolution by adding new processes that would modify a simpler form into a more complex one – to construct a gastrula, take a blastula and alter it!

The genetic description of a single cell may, therefore, take a quite different form from the genetic description that assembles cells into a multi-celled organism. Multiplication by cell division would require, as a minimum, a state description (the DNA, say), and a simple "interpretive process" – to use the term from computer language – that copies this description as a part of the larger copying process of cell division. But such a mechanism clearly would not suffice for the differentiation of cells in development. It appears more natural to conceptualize that mechanism as based on a process description, and a somewhat more complex interpretive process that produces the adult organism in a sequence of stages, each new stage in development representing the effect of an operator upon the previous one.

It is harder to conceptualize the interrelation of these two descriptions. Interrelated they must be, for enough has been learned of gene-enzyme mechanisms to show that these play a major role in development as in cell metabolism. The single clue we obtain from our earlier discussion is that the description may itself be hierarchical, or nearly decomposable, in structure, the lower levels governing the fast, "high-frequency" dynamics of the individual cell, the higher level interactions governing the slow, "low-frequency" dynamics of the developing multi-cellular organism.

There are only bits of evidence, apart from the facts of recapitulation, that the genetic program is organized in this way, but such evidence as exists is compatible with this notion.[26] To the extent that we can differentiate the genetic information that governs cell metabolism from the genetic information that governs the development of differentiated cells in the multi-cellular organization, we simplify enormously – as we have already seen – our task of theoretical description. But I have perhaps pressed this speculation far enough.

The generalization that in evolving systems whose descriptions are stored in a process language, we might expect ontogeny partially to recapitulate phylogeny has applications outside the realm of biology. It can be applied as readily, for example, to the transmission of knowledge in the educational process. In most subjects, particularly in the rapidly advancing sciences, the progress from elementary to advanced courses is to a considerable extent a progress through the conceptual history of the science itself. Fortunately, the recapitulation is seldom literal – any more than it is in the biological case. We do not teach the phlogiston theory in chemistry in order later to correct it. (I am not sure I could not cite examples in other subjects where we do exactly that.) But curriculum revisions that rid us of the accumulations of the past are infrequent and painful. Nor are they always desirable – partial recapitulation may, in many instances, provide the most expeditious route to advanced knowledge.

Summary: The Description of Complexity

How complex or simple a structure is depends critically upon the way in which we describe it. Most of the complex structures found in the world are enormously redundant, and we can use this redundancy to simplify their description. But to use it, to achieve the simplification, we must find the right representation.

The notion of substituting a process description for a state description of nature has played a central role in the development of modern science. Dynamic laws, expressed in the form of systems of differential or difference equations, have in a large number of cases provided the clue for the simple description of the complex. In the preceding paragraphs I have tried to show that this characteristic of scientific inquiry

[26] There is considerable evidence that successive genes along a chromosome often determine enzymes controlling successive stages of protein syntheses. For a review of some of this evidence, see P.E. Hartman: Transduction: A Comparative Review, in: W.D. McElroy and B. Glass (eds.), *The Chemical Basis of Heredity*, Baltimore: Johns Hopkins Press, 1957, at pp. 442–454. Evidence for differential activity of genes in different tissues and at different stages of development is discussed by J.G. Gall: Chromosomal Differentiation, in: W.D. McElroy and B. Glass (eds.), *The Chemical Basis of Development*, Baltimore: Johns Hopkins Press, 1958, at pp. 103–135. Finally, a model very like that proposed here has been independently, and far more fully, outlined by J.R. Platt: A "Book Model" of Genetic Information Transfer in Cells and Tissues, in: Kasha and Pullman (eds.): *Horizons in Biochemistry*, New York: Academic Press, forthcoming. Of course, this kind of mechanism is not the only one in which development could be controlled by a process description. Induction, in the form envisaged in Spemann's organizer theory, is based on process description, in which metabolites in already formed tissue control the next stages of development.

is not accidental or superficial. The correlation between state description and process description is basic to the functioning of any adaptive organism, to its capacity for acting purposefully upon its environment. Our present-day understanding of genetic mechanisms suggests that even in describing itself the multi-cellular organism finds a process description – a genetically encoded program – to be the parsimonious and useful representation.

CONCLUSION

Our speculations have carried us over a rather alarming array of topics, but that is the price we must pay if we wish to seek properties common to many sorts of complex systems. My thesis has been that one path to the construction of a nontrivial theory of complex systems is by way of a theory of hierarchy. Empirically, a large proportion of the complex systems we observe in nature exhibit hierarchic structure. On theoretical grounds we could expect complex systems to be hierarchies in a world in which complexity had to evolve from simplicity. In their dynamics, hierarchies have a property, near-decomposability, that greatly simplifies their behavior. Near-decomposability also simplifies the description of a complex system, and makes it easier to understand how the information needed for the development or reproduction of the system can be stored in reasonable compass.

In both science and engineering, the study of "systems" is an increasingly popular activity. Its popularity is more a response to a pressing need for synthesizing and analyzing complexity than it is to any large development of a body of knowledge and technique for dealing with complexity. If this popularity is to be more than a fad, necessity will have to mother invention and provide substance to go with the name. The explorations reviewed here represent one particular direction of search for such substance.

Robert U. Ayres/Allen V. Kneese (1969)

PRODUCTION, CONSUMPTION, AND EXTERNALITIES

In the 1970s, with energy crises and the Club of Rome a discussion about the "Limits to Growth" was initiated. Increasingly, the media and politics recognized the necessity of sustainable activity and responsible handling of the non-regenerative resources of our planet. Awareness rose for the need for protection of environment, sustainable management, and a "circular" economy, which systematically considers the "fundamental law of conservation of mass" (cf. Ayres/Kneese (1969), p. 283): Sustainability and the vision of "greening" logistics have since become major issues of practical concern and scientific interest.

The paper by Ayres and Kneese, in several ways, provides the basis for a broader, truly holistic view of material flows which has become an essential element of modern logistics. They offer a model of the flows of materials throughout the economy and introduce the critical concept of externalities as an inevitable effect of any economic activity. With the help of their conceptual framework added by empirical data of the US economy the authors point out that environmentally relevant externalities originate from all transformation processes on the production and the consumption side and that these externalities are hence to be integrated into any considerations regarding welfare theory and politics.

Both aspects – a conceptualization of externalities as a relevant factor for the optimization of comprehensive logistical systems and a stringent orientation at end-to-end material flows – make this paper a classical contribution, which belongs to the conceptual toolbox of modern logisticians in ways which may become even more important in the future.

Reference:

Robert U. Ayres and Allen V. Kneese: Production, Consumption, and Externalities. In: *The American Economic Review*, Vol. 59, No. 3, 1969, pp. 282–297.

P. Klaus and S. Müller (eds.), *The Roots of Logistics*, DOI 10.1007/978-3-642-27922-5_24,

Robert U. Ayres/Allen V. Kneese: Production, Consumption, and Externalities[1]

> "For all that, welfare economics can no more reach conclusions applicable to the real world without some knowledge of the real world than can positive economics" (Mishan (1965)).

Despite tremendous public and governmental concern with problems such as environmental pollution, there has been a tendency in the economics literature to view externalities as exceptional cases. They may distort the allocation of resources but can be dealt with adequately through simple ad hoc arrangements. To quote Pigou:

> When it was urged above, that in certain industries a wrong amount of resources is being invested because the value of the marginal social net product there differs from the value of the marginal private net product, it was tacitly assumed that in the main body of industries these two values are equal (Pigou (1952)).[2]

And Scitovsky, after having described his cases two and four which deal with technological externalities affecting consumers and producers respectively, says:

> The second case seems exceptional, because most instances of it can be and usually are eliminated by zoning ordnances and industrial regulations concerned with public health and safety. The fourth case seems unimportant, simply because examples of it seem to be few and exceptional (Scitovsky (1954)).

We believe that at least one class of externalities – those associated with the disposal of residuals resulting from the consumption and production process – must be viewed quite differently.[3] They are a normal, indeed, inevitable part of these proc-

---

[1] The authors are respectively visiting scholar and director, Quality of the Environment Program, Resources for the Future, Inc. We are indebted to our colleagues Blair Bower, Orris Herfindahl, Charles Howe, John Krutilla, and Robert Steinberg for comments on an earlier draft. We have also benefited from comments by James Buchanan, Paul Davidson, Robert Dorfman, Otto Eckstein, Myrick Freeman, Mason Gaffney, Lester Lave, Herbert Mohring, and Gordon Tullock.

[2] Even Baumol who saw externalities as a rather pervasive feature of the economy tends to discuss external diseconomies like "smoke nuisance" entirely in terms of particular examples (Baumol (1967)). A perspective more like that of the present paper is found in Kapp (1950).

[3] We by no means wish to imply that this is the only important class of externalities associated with production and consumption. Also, we do not wish to imply that there has been a lack of theoretical attention to the externalities problem. In fact, the past few years have seen the publication of several excellent articles which have gone far toward systematizing definitions and illuminating certain policy issues. Of special note are Coase (1960), Davis and Whinston (1962), Buchanan/Stubblebine (1962), and Turvey (1962). However, all

esses. Their economic significance tends to increase as economic development proceeds, and the ability of the ambient environment to receive and assimilate them is an important natural resource of increasing value.[4] We will argue below that the common failure to recognize these facts may result from viewing the production and consumption processes in a manner that is somewhat at variance with the fundamental law of conservation of mass.

Modern welfare economics concludes that if (1) preference orderings of consumers and production functions of producers are independent and their shapes appropriately constrained, (2) consumers maximize utility subject to given income and price parameters, and (3) producers maximize profits subject to the price parameters; a set of prices exists such that no individual can be made better off without making some other individual worse off. For a given distribution of income this is an efficient state. Given certain further assumptions concerning the structure of markets, this "Pareto optimum" can be achieved via a pricing mechanism and voluntary decentralized exchange.

If waste assimilative capacity of the environment is scarce, the decentralized voluntary exchange process cannot be free of uncompensated technological external diseconomies unless (1) all inputs are fully converted into outputs, with no unwanted material residuals along the way,[5] and all final outputs are utterly destroyed in the process of consumption, or (2) property rights are so arranged that all relevant environmental attributes are in private ownership and these rights are exchanged in competitive markets. Neither of these conditions can be expected to hold in an actual economy and they do not.

Nature does not permit the destruction of matter except by annihilation with antimatter, and the means of disposal of unwanted residuals which maximizes the internal

---

these contributions deal with externality as a comparatively minor aberration from Pareto optimality in competitive markets and focus upon externalities between two parties. Mishan, after a careful review of the literature, has commented on this as follows: "The form in which external effects have been presented in the literature is that of partial equilibrium analysis; a situation in which a single industry produces an equilibrium output, usually under conditions of perfect competition, some form of intervention being required in order to induce the industry to produce an 'ideal' or 'optimal' output. If the point is not made explicitly, it is tacitly understood that unless the rest of the economy remains organized in conformity with optimum conditions, one runs smack into Second Best problems" (Mishan (1965)).

4 That external diseconomies are integrally related to economic development and increasing congestion has been noted in passing in the literature. Mishan (1965) has commented: "The attention given to external effects in the recent literature is, I think, fully justified by the unfortunate, albeit inescapable, fact that as societies grow in material wealth, the incidence of these effects grows rapidly..."; and Buchanan/Tullock (1965) have stated that as economic development proceeds, "congestion" tends to replace "cooperation" as the underlying motive force behind collective action, i.e., controlling external diseconomies tends to become more important than cooperation to realize external economies.

5 Or any residuals which occur must be stored on the producer's premises.

return of decentralized decision units is by discharge to the environment, principally, watercourses and the atmosphere. Water and air are traditionally examples of free goods in economics. But in reality, in developed economies they are common property resources of great and increasing value presenting society with important and difficult allocation problems which exchange in private markets cannot resolve. These problems loom larger as increased population and industrial production put more pressure on the environment's ability to dilute and chemically degrade waste products. Only the crudest estimates of present external costs associated with residuals discharge exist but it would not be surprising if these costs were in the tens of billions of dollars annually.[6] Moreover, as we shall emphasize again, technological means for processing or purifying one or another type of waste discharge do not destroy the residuals but only alter their form. Thus, given the level, patterns, and technology of production and consumption, recycle of materials into productive uses or discharge into an alternative medium are the only general options for protecting a particular environmental medium such as water. Residual problems must be seen in a broad regional or economy-wide context rather than as separate and isolated problems of disposal of gas, liquid, and solid wastes.

Frank Knight perhaps provides a key to why these elementary facts have played so small a role in economic theorizing and empirical research.

> The next heading, to be mentioned ties up with the question of dimensions from another angle, and relates to the second main error mentioned earlier as connected with taking food and eating as the type of economic activity. The basic economic magnitude (value or utility) is service, not good. It is inherently a stream or flow in time… (Knight (1921)).[7]

Almost all of standard economic theory is in reality concerned with services. Material objects are merely the vehicles which carry some of these services, and they are exchanged because of consumer preferences for the services associated with their use or because they can help to add value in the manufacturing process. Yet we persist in referring to the "final consumption" of goods as though material objects such as fuels, materials, and finished goods somehow disappeared into the void – a practice which was comparatively harmless so long as air and water were almost

---

6 It is interesting to compare this with estimates of the cost of another well known misallocation of resources that has occupied a central place in economic theory and research. In 1954, Harberger (1954) published an estimate of the welfare cost of monopoly which indicated that it amounted to about .07 percent of GNP. In a later study, Schwartzman (1960) calculated the allocative cost at only .01 percent of GNP. Leibenstein (1966) generalized studies such as these to the statement that "…in a great many instances the amount to be gained by increasing allocative efficiency is trivial…". But Leibenstein did not consider the allocative costs associatedw ith environmental pollution.

7 The point was also clearly made by Fisher: "The only true method, in our view, is to regard uniformly as income the service of a dwelling to its owner (shelter or money rental), the service of a piano (music), and the service of food (nourishment)…" (Fisher (1906), emphasis in original).

literally free goods.[8] Of course, residuals from both the production and consumption processes remain and they usually render disservices (like killing fish, increasing the difficulty of water treatment, reducing public health, soiling and deteriorating buildings, etc.) rather than services. Control efforts are aimed at eliminating or reducing those disservices which flow to consumers and producers whether they want them or not and which, except in unusual cases, they cannot control by engaging in individual exchanges.[9]

I. THE FLOW OF MATERIALS

To elaborate on these points, we find it useful initially to view environmental pollution and its control as a materials balance problem for the entire economy.[10] The inputs to the system are fuels, foods, and raw materials which are partly converted into final goods and partly become waste residuals. Except for increases in inventory, final goods also ultimately enter the waste stream. Thus goods which are "consumed" really only render certain services. Their material substance remains in existence and must either be reused or discharged to the ambient environment.

In an economy which is closed (no imports or exports) and where there is no net accumulation of stocks (plant, equipment, inventories, consumer durables, or residential buildings), the amount of residuals inserted into the natural environment must be approximately equal to the weight of basic fuels, food, and raw materials entering the processing and production system, plus oxygen taken from the atmosphere.[11] This result, while obvious upon reflection, leads to the, at first rather surprising, corollary that residuals disposal involves a greater tonnage of materials than basic materials processing, although many of the residuals, being gaseous, require no physical "handling."

8 We are tempted to suggest that the word consumption be dropped entirely from the economist's vocabulary as being basically deceptive. It is difficult to think of a suitable substitute, however. At least, the word consumption should not be used in connection with goods, but only with regard to services or flows of "utility."

9 There is a substantial literature dealing with the question of under what conditions individual exchanges can optimally control technological external diseconomies. A discussion of this literature, as it relates to waterborne residuals, is found in Kneese and Bower (1968).

10 As far as we know, the idea of applying materials balance concepts to waste disposal problems was first expressed by Smith (1967). We also benefitted from an unpublished paper by Joseph Headley in which a pollution "matrix" is suggested. We have also found references by Boulding (1966) to a "spaceship economy" suggestive. One of the authors has previously used a similar approach in ecological studies of nutrient interchange among plants and animals; see Ayres (1967).

11 To simplify our language, we will not repeat this essential qualification at each opportunity, but assume it applies throughout the following discussion. In addition, we must include residuals such as NO and NO_2 arising from reactions between components of the air itself but occurring as combustion by-products.

EXHIBIT 1 shows a materials flow of the type we have in mind in greater detail and relates it to a broad classification of economic sectors for convenience in our later discussion, and for general consistency with the Standard Industrial Classification. In an open (regional or national) economy, it would be necessary to add flows representing imports and exports. In an economy undergoing stock or capital accumulation, the production of residuals in any given year would be less by that amount than the basic inputs. In the entire U.S. economy, accumulation accounts for about 10–15 percent of basic annual inputs, mostly in the form of construction materials, and there is some net importation of raw and partially processed materials amounting to 4 or 5 percent of domestic production. TABLE 1 shows estimates of the weight of raw materials produced in the United States in several recent years, plus net imports of raw and partially processed materials.

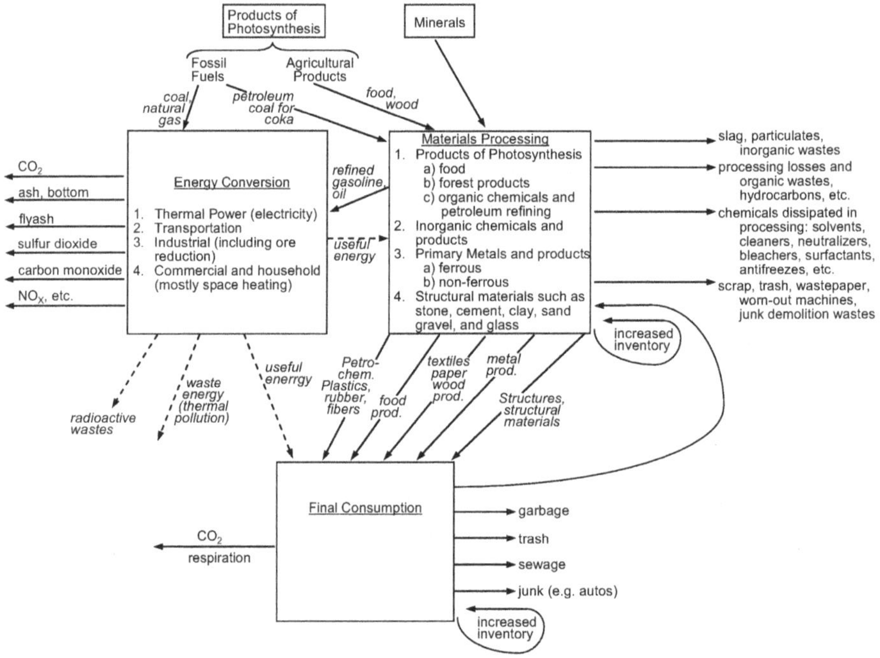

EXHIBIT 1: MATERIALS FLOW

	1963	1964	1965
Agricultural (incl. Fishery and wildlife and forest) products			
Food: Crops (excl. livestock feed)	125	128	130
Food: Livestock	100	103	102
Other products	5	6	6
Fishery	3	3	3
Forestry products (85 per cent dry wt. basis			
Sawlogs	53	55	56
Pulpwood	107	116	120
Other	42	41	42
Total	125	128	130
Mineral fuels	1,337	1,399	1,448
Other minerals			
Iron ore	204	237	245
Other metal ores	161	171	191
Other nonmetals	125	133	149
Total	490	541	585
Grand total[a]	2.261	2,392	2,492

[a] Excluding construction materials, stone, sand, gravel, and other minerals used for structural purposes, ballast, fillers, insulation, etc. Gangue and mine tailings are also excluded from this totasl. These materials account for enormous tonnages but undergo essentially no chemical change. Hence, their use is more or less tantamount to physically moving them from one location to another. If this were to be included, there is no logical reason to exclude material shifted in highway cut and fill operations, harbor dredging, land-fill, plowing, and even silt moved by rivers. Since a line must be drawn somewhere, we chose to draw it as indicated above.

Reference: Ayres/Kneese (1968), p. 630.

TABLE 1: WEIGHT OF BASIC MATERIALS PRODUCTION IN THE UNITED STATES PLUS NET IMPORTS 1963 (10^4 TONS)

Of the active inputs,[12] perhaps three-quarters of the overall weight is eventually discharged to the atmosphere as carbon (combined with atmospheric oxygen in the form of CO or CO_2) and hydrogen (combined with atmospheric oxygen as H_2O) under current conditions. This results from combustion of fossil fuels and from animal respiration. Discharge of carbon dioxide can be considered harmless in the short run. There are large "sinks" (in the form of vegetation and large water bodies, mainly the oceans) which reabsorb this gas, although there is evidence of net accumulation of CO_2 in the atmosphere. Some experts believe that the latter is likely to show a large relative increase, as much as 50 per cent by the end of the century, possibly giving rise to significant – and probably, on balance, adverse – weather changes.[13] Thus continued combustion of fossil fuels at a high rate could produce externalities affecting the entire world. The effects associated with most residuals will normally be more confined, however, usually limited to regional air and water sheds.

The remaining residuals are either gases (like carbon monoxide, nitrogen dioxide, and sulfur dioxide – all potentially harmful even in the short run), dry solids (like rubbish and scrap), or wet solids (like garbage, sewage, and industrial wastes suspended or dissolved in water). In a sense, the dry solids are an irreducible, limiting form of waste. By the application of appropriate equipment and energy, most undesirable substances can, in principle, be removed from water and air streams[14] – but what is left must be disposed of in solid form, transformed, or reused. Looking at the matter in this way clearly reveals a primary interdependence between the various waste streams which casts into doubt the traditional classification of air, water, and land pollution as individual categories for purposes of planning and control policy.

Residuals do not necessarily have to be discharged to the environment. In many instances, it is possible to recycle them back into the productive system. The materials balance view underlines the fact that the throughput of new materials necessary to maintain a given level of production and consumption decreases as the technical

---

[12] See footnote to TABLE 1.

[13] Cf. Conservation Foundation (1963). There is strong evidence that discharge of residuals has already affected the climate of individual cities; cf. Lowry (1967).

[14] Except CO_2 which may be harmful in the long run, as noted.

efficiency of energy conversion and materials utilization increases. Similarly, other things being equal, the longer that cars, buildings, machinery, and other durables remain in service, the fewer new materials are required to compensate for loss, wear, and obsolescence – although the use of old or worn machinery (e.g. automobiles) tends to increase other residuals problems. Technically efficient combustion of (desulfurized) fossil fuels would leave only water, ash, and carbon dioxide as residuals, while nuclear energy conversion need leave only negligible quantities of material residuals (although thermal pollution and radiation hazards cannot be dismissed by any means).

Given the population, industrial production, and transport service in an economy (a regional rather than a national economy would normally be the relevant unit), it is possible to visualize combinations of social policy which could lead to quite different relative burdens placed on the various residuals – receiving environmental media; or, given the possibilities for recycle and less residual – generating production processes, the overall burden to be placed upon the environment as a whole. To take one extreme, a region which went in heavily for electric space heating and wet scrubbing of stack gases (from steam plants and industries), which ground up its garbage and delivered it to the sewers and then discharged the raw sewage to watercourses, would protect its air resources to an exceptional degree. But this would come at the sacrifice of placing a heavy residuals load upon water resources. On the other hand, a region which treated municipal and industrial waste water streams to a high level and relied heavily on the incineration of sludges and solid wastes would protect its water and land resources at the expense of discharging waste residuals predominantly to the air. Finally, a region which practiced high level recovery and recycle of waste materials and fostered low residual production processes to a far reaching extent in each of the economic sectors might discharge very little residual waste to any of the environmental media.

Further complexities are added by the fact that sometimes it is possible to modify an environmental medium through investment in control facilities so as to improve its assimilative capacity. The clearest, but far from only, example is with respect to watercourses where reservoir storage can be used to augment low river flows that ordinarily are associated with critical pollution (high external cost situations).[15] Thus internalization of external costs associated with particular discharges, by means of taxes or other restrictions, even if done perfectly, cannot guarantee Pareto optimality. Investments involving public good aspects must enter into an optimal solution.[16]

[15] Careful empirical work has shown that this technique can fit efficiently into water quality management systems. See Davis (1968).

[16] A discussion of the theory of such public investments with respect to water quality management is found in Boyd (1968).

To recapitulate our main points briefly: (1) Technological external diseconomies are not freakish anomalies in the processes of production and consumption but an inherent and normal part of them. (2) These external diseconomies are quantitatively negligible in a low-population or economically undeveloped setting, but they become progressively (nonlinearly) more important as the population rises and the level of output increases (i.e., as the natural reservoirs of dilution and assimilative capacity become exhausted).[17] (3) They cannot be properly dealt with by considering environmental media such as air and water in isolation. (4) Isolated and ad hoc taxes and other restrictions are not sufficient for their optimum control, although they are essential elements in a more systematic and coherent program of environmental quality management. (5) Public investment programs, particularly including transportation systems, sewage disposal, and river flow regulation, arc intimately related to the amounts and effects of residuals and must be planned in light of them.

It is important to develop not only improved measures of the external costs resulting from differing concentrations and duration of residuals in the environment but more systematic methods for forecasting emissions of external-cost-producing residuals, technical and economic trade-offs between them, and the effects of recycle on environmental quality.

In the hope of contributing to this effort and of revealing more clearly the types of information which would be needed to implement such a program, we set forth a more formal model of the materials balance approach in the following sections and relate it to some conventional economic models of production and consumption. The main objective is to make some progress toward defining a system in which flows of services and materials are simultaneously accounted for and related to welfare.

II. BASIC MODEL

The take off point for our discussion is the Walras-Cassel general equilibrium model,[18] extended to include intermediate consumption, which involve the following quantities:

[17] Externalities associated with residuals discharge may appear only at certain threshold values which are relevant only at some stage of economic development and industrial and population concentrations. This may account for their general treatment as "exceptional" cases in the economics literature. These threshold values truly were exceptional cases for less developed economies.

[18] The original references are Walras (1954) and Cassel (1932). Our own treatment is largely based on Dorfman et al. (1958).

$$\overbrace{r_1, \dots\dots\dots\dots, r_M}^{\text{resources and services}}$$

$$\overbrace{X_1, \dots\dots\dots\dots, X_N}^{\text{products or commodities}}$$

$$\overbrace{v_1, \dots\dots\dots\dots, v_M}^{\text{resource prices}}$$

$$\overbrace{p_1, \dots\dots\dots\dots, p_N}^{\text{product or commodity prices}}$$

$$\overbrace{Y_1, \dots\dots\dots\dots, Y_N}^{\text{final demands}}$$

The M basic resources are allocated among the N sectors as follows:

$$\begin{aligned} r_1 &= a_{11}X_1 + a_{12}X_2 + \dots + a_{1N}X_N \\ r_2 &= a_{21}X_1 + a_{22}X_2 + \dots + a_{2N}X_N \\ &\vdots \\ r_M &= a_{M1}X_1 + a_{M2}X_2 + \dots + a_{MN}X_N \end{aligned}$$

(1a) or

$$r_j = \sum_{k=1}^{N} a_{jk} X_k \qquad j = 1, \dots, M$$

In (1a) we have implicitly assumed that there is no possibility of factor or process substitution and no joint production. These conditions will be discussed later. In matrix notation we can write:

(1b) $$[r_{j1}]_{M,1} = [a_{jk}]_{M,N} \times [X_{k1}]_{N,1}$$

where [a] is an $M \times N$ matrix.

A similar set of equations describes the relations between commodity production and final demand:

(2a) $$X_k = \sum_{l=1}^{N} A_{kl} Y_l \qquad k = 1, \dots, N$$

(2b) $$[X_{k1}]_{N,1} = [A_{kl}]_{N,N} \times [Y_{l1}]_{N,1}$$

and the matrix $[A]$ is given by

$$(3) \qquad [A] = [I - C]^{-1}$$

where $[I]$ is the unit diagonal matrix and the elements C_{ij}, of the matrix $[C]$ are essentially the well known Leontief input coefficients. In principle these are functions of the existing technology and, therefore, are fixed for any given situation.

By combining (1) and (2), we obtain a set of equations relating resource inputs directly to final demand, viz.,

$$(4a) \qquad \begin{aligned} r_j &= \sum_{k=1}^{N} a_{jk} \sum_{l=1}^{N} A_{kl} Y_l = \sum_{k.l=1}^{N} a_{jk} A_{kl} Y_l \\ &= \sum_{l=1}^{N} b_{jl} Y_l \qquad j = 1, \ldots, M \end{aligned}$$

or, of course, in matrix notation (4b).

$$(4b) \qquad \begin{aligned} [r_{j1}]_{N,1} &= [a_{jk}]_{M,N} \times [A_{kl}]_{N,N} \times [Y_{l1}]_{N,I} \\ &= [b_{jl}]_{M,N} \times [Y_{l1}]_{N,1} \end{aligned}$$

We can also impute the prices of N intermediate goods and commodities to the prices of the M basic resources, as follows:

$$(5a) \qquad p_k = \sum_{j=1}^{M} v_j b_{jk} \qquad k = 1, \ldots, N$$

$$(5b) \qquad [p_{1k}]_{1,N} = [v_{1j}]_{1,M} \times [b_{jk}]_{M,N}$$

To complete the system, it may be supposed that demand and supply relationships are given, a priori, by Pareto-type preference functions:

$$(6) \quad \text{Demand:} \qquad Y_k = F_k(p_1, \ldots, p_N) \qquad k = 1, \ldots, N$$

$$(7) \quad \text{Supply:} \qquad r_k = G_k(v_1, \ldots, v_M) \qquad k = 1, \ldots, M$$

where, of course, the p_j are functions of the v_i as in (5b).

In order to interpret the X's as physical production, it is necessary for the sake of consistency to arrange that outputs and inputs always balance, which implies that the

C_{ij} must comprise all materials exchanges including residuals. To complete the system so that there is no net gain or loss of physical substances, it is also convenient to introduce two additional sectors, viz., an "environmental" sector whose (physical) output is X_0 and a "final consumption" sector whose output is denoted X_f. The system is then easily balanced by explicitly including flows both to and from these sectors.

To implement this further modification of the Walras-Cassel model, it is convenient to subdivide and relabel the resource category into tangible raw materials $\{r^m\}$ and services $\{r^s\}$:

$$\begin{bmatrix} r_1 \\ r_2 \\ \cdot \\ \cdot \\ \cdot \\ \cdot \\ r_L \end{bmatrix} \text{ becomes } \left.\begin{bmatrix} r_1^m \\ r_2 \\ \cdot \\ \cdot \\ \cdot \\ \cdot \\ r_L^m \end{bmatrix}\right\} \text{raw materials (units)}$$

$$\begin{bmatrix} r_{L+1} \\ \cdot \\ \cdot \\ \cdot \\ r_M \end{bmatrix} \text{ becomes } \left.\begin{bmatrix} r_1^s \\ \cdot \\ \cdot \\ \cdot \\ r_p^s \end{bmatrix}\right\} \text{service (units)}$$

where, of course,

$$L + P = M \tag{8}$$

It is understood that services, while not counted in tons, can be measured in meaningful units, such as man-days, with well defined prices. Thus, we similarly relabel the price variables as follows:

$$\begin{bmatrix} V_1 \\ \cdot \\ \cdot \\ \cdot \\ V_L \end{bmatrix} \quad \begin{bmatrix} V_1^m \\ \cdot \\ \cdot \\ \cdot \\ V_L^m \end{bmatrix} \Bigg\} \text{raw material (prices)}$$

becomes

$$\begin{bmatrix} V_{L+1} \\ \cdot \\ \cdot \\ \cdot \\ V_M \end{bmatrix} \quad \begin{bmatrix} V_1^s \\ \cdot \\ \cdot \\ \cdot \\ V_L^s \end{bmatrix} \Bigg\} \text{labor and service (prices)}$$

The coefficients $\{a_{ij}\}$, $\{b_{ij}\}$ are similarly partitioned into two groups,

$$\text{e.g.,} \quad \begin{matrix} b_{1j} \\ \cdot \\ \cdot \\ \cdot \\ b_{Lj} \\ b_{L+1,j} \\ \cdot \\ \cdot \\ \cdot \\ b_{Mj} \end{matrix} \quad \text{becomes} \quad \begin{matrix} b_{1j}^m \\ \cdot \\ \cdot \\ \cdot \\ b_{Lj}^m \\ b_{1j}^s \\ \cdot \\ \cdot \\ \cdot \\ b_{pj}^s \end{matrix}$$

These notational changes have no effect whatever on the substance of the model, although the equations become somewhat more cumbersome. The partitioned matrix notation simplifies the restatement of the basic equations. Thus (1b) becomes (9), while (5b) becomes (10).

$$(9)\qquad M = \left\{ \begin{bmatrix} \vdots \\ r \\ \vdots \end{bmatrix} \equiv \begin{bmatrix} r^m \\ \cdots \\ r^s \end{bmatrix} \begin{matrix} \}L \\ \\ \}P \end{matrix} = M \left\{ \overbrace{\begin{bmatrix} \}L & b^m \\ & \cdots \\ \}P & b^s \end{bmatrix}}^{N} \begin{bmatrix} \\ Y \\ \\ \end{bmatrix} \right\} N \right.$$

$$(10)\qquad \underset{N}{[p_1, \ldots, p_N]} = \underbrace{[v^m \vdots v^s]}_{\substack{L \quad P \\ M}} \underbrace{\begin{bmatrix} b^m \\ \cdots \\ b^s \end{bmatrix}}_{N} \Bigg\} M$$

$$= [\cdots v^m \cdots] \begin{bmatrix} \vdots & b^m & \vdots \\ \cdot & \cdots & \cdot \end{bmatrix} + [\cdots v^s \cdots] \begin{bmatrix} \vdots & b^s & \vdots \\ \cdot & \cdots & \cdot \end{bmatrix}$$

The equivalent of (5a) is:

$$(11)\qquad p_k = \underbrace{\sum_{j=1}^{L} b_{jk}^m v_j^m}_{\substack{\text{prices imputed} \\ \text{to cost of raw} \\ \text{materials}}} + \underbrace{\sum_{j=1}^{P} b_{jk}^s v_j^s}_{\substack{\text{prices imputed} \\ \text{to cost of} \\ \text{services}}}$$

where $k = 1, \ldots, N$

We wish to focus attention explicitly on the flow of materials through the economy. By definition of the Leontief input coefficients (now related to materials flow), we have:

$C_{kj} X_j$ (physical) quantity transferred from k to j

$C_{jk} X_k$ quantity transferred from j to k

Hence, material flows from the environment to all other sectors are given by:

$$(12)\qquad \sum_{k=1}^{N} C_{0k} X_k = \sum_{j=1}^{L} r_j^m = \sum_{j=1}^{L} \sum_{k=1}^{N} a_{jk}^m X_k = \sum_{j=1}^{L} \sum_{k=1}^{N} b_{jk}^m Y_k$$

using equation (1), as modified.[19] Obviously, comparing the first and third terms,

$$(13)\quad \underbrace{C_{0k}}_{\text{total material flow (0 to } k)} = \underbrace{\sum_{j=1}^{L} a_{jk}^{m}}_{\text{all raw materials (0 to } k)}$$

Flows into and out of the environmental sector must be in balance:

$$(14)\quad \underbrace{\sum_{k=1}^{N} C_{0k} X_k}_{\text{sum of all raw material flows}} = \underbrace{\sum_{k=1}^{N} C_{k0} X_k + C_{f0} X_0}_{\text{sum of all return (waste) flows}}$$

Material flows to and from the final sector must also balance:[20]

$$(15)\quad \underbrace{\sum_{k=1}^{N} C_{kf} X_f}_{\text{sum of all final goods}} \qquad \underbrace{\sum_{k=1}^{N} C_{fk} X_k}_{\text{sum of all materials recycled}} + \underbrace{C_{f0} X_0}_{\text{waste residuals (plus accumulation)}}$$

Of course, by definition, X_f is the sum of the final demands:

$$(16)\quad X_f = \sum_{j=1} Y_j$$

Substituting (16) into the left side of (15) and (2a) into the right side of (15), we obtain an expression for the waste flow in terms of final demands:

$$(17)\quad C_{j0} X_0 = \sum_{j=1}^{N} \sum_{k=1}^{N} (C_{jf} - C_{fj} A_{jk})\, Y_k$$

[19] Ignoring, for convenience, any materials flow from the environment **directly** to the final consumption sector.

[20] For convenience, we can treat accumulation in the final sector as a return flow to the environment. In truth, structures actually **become** part of our environment, although certain disposal costs may be deferred.

The treatment could be simplified slightly if we assumed that there is no recycling per se. Thus, in the context of the model, we could suppose that all residuals return to the environmental sector,[21] where some of them (e.g., waste paper) become "raw materials." They would then be indistinguishable from new raw materials, however, and price differentials between the two would be washed out. In principle, this is an important distinction to retain.

III. INCLUSION OF EXTERNALITIES

The physical flow of materials between various intermediate (production) sectors and the final (consumption) sector tends to be accompanied by, and correlated with, a (reverse) flow of dollars.[22] However, the physical flow of materials from and back to the environment is only partly reflected by actual dollar flows, namely, land rents and payments for raw materials. There are three classes of physical exchange for which there exist no counterpart economic transactions. These are: (1) private use for production inputs of "common property" resources, notably air, streams, lakes, and the ocean; (2) private use of the assimilative capacity of the environment to "dispose of" or dilute wastes and residuals; (3) inadvertent or unwanted material inputs to productive processes-diluents and pollutants.

All these goods (or "bads") are physically transferred at zero price, not because they are not scarce relative to demand – they often are in developed economies – or because they confer no service or disservice on the user – since they demonstrably do so – but because there exist no social institutions that permit the resources in question to be "owned," and exchanged in the market.

The allocation of resources corresponding to a Pareto optimum cannot be attained without subjecting the above-mentioned nonmarket and involuntary exchanges to the moderation of a market or a surrogate thereof. In principle, the influence of a market might be simulated, to a first approximation, by introducing a set of shadow (or virtual) prices.[23] These may well be zero, where supply truly exceeds demand, or negative (i.e., costs) in some instances; they will be positive in others. The exchanges are, of course, real.

The Walras-Cassel model can be generalized to handle these effects in the following way:

---

[21] In calculating actual quantities, we would (by convention) ignore the weight of oxygen taken free from the atmospherein combustion and return as CO_2. However, such inputs will be treated explicitly later.

[22] To be precise, the dollar flow governs and is governed by a combined flow of materials and services (value added).

[23] A similar concept exists in mechanics where the forces producing "reaction" (to balance action and reaction) are commonly described as "virtual forces."

1. One can introduce a set of R common-property resources or services of raw materials $\{r_1^{cp}, \ldots, r_R^{cp}\}$ as a subset of the set $\{r_j\}$; these will have corresponding virtual prices $\{v_j^{cp}\}$, which would constitute an "income" from the environment. Such resources include the atmosphere; streams, lakes, and oceans; landscape; wildlife and biological diversity; and the indispensable assimilative capacity of the environment (its ability to accept and neutralize or recycle residuals).[24]

2. One can introduce a set of *S* environmental **disservices** imposed on consumers of material resources, by forcing them to accept unwanted inputs $\{r_1^u, \ldots, r_s^u\}$ (pollutants, contaminants, etc.); these disservices would have negative value, giving rise to **negative** virtual prices $\{u_j\}$.[25]

The matrix coefficients $\{a_{ij}\}$ and $\{b_{ij}\}$ can be further partitioned to take account of this additional refinement, and equations analogous to (9), (10), and (11) can be generalized in the obvious way. Equation (6) carries over unchanged, but (7) must be appropriately generalized to take account of the altered situation. Actually, (7) breaks up into several groups of equations:

$$r_k^m = G_k^m(p_1, \ldots, p_N) \qquad k = 1, \ldots, L \tag{18}$$

$$r_k^s = G_k^s(p_1, \ldots, p_N) \qquad k = 1, \ldots, P \tag{19}$$

However, as we have noted at the outset, the supplies of common-property resources and environmental services or disservices are not regulated directly by market prices of other goods and services. In the case of common-property resources, the supplies are simply constants fixed by nature or otherwise determined by accident or noneconomic factors.

The total value of these services performed by the environment cannot be calculated but it is suggestive to consider the situation if the natural reservoir of air, water, minerals, etc., were very much smaller, as if the earth were a submarine or "spaceship" (i.e., a vehicle with no assimilative and/or regenerative capacity). In such a

---

[24] Economists have previously suggested generalization of the Walras-Cassel model to take account of public goods. One of the earliest appears to be Schlesinger (1933). We are indebted to Otto Eckstein for calling our attention to this key reference.

[25] The notion of introducing the possibility of negative prices in general equilibrium theory has apparently been discussed before, although we are not aware of any systematic development of the idea in the published literature. In this connection, it is worth pointing out the underlying similarity of negative prices and effluent taxes – which have been, and still are being considered as an attractive alternative to subsidies and federal standard-setting as a means of controlling air and water pollution. Such taxes would, of course, be an explicit attempt to rectify an imbalance caused by a market failure.

case, all material resources would have to be recycled,[26] and the cost of all goods would necessarily reflect the cost of reprocessing residuals and wastes for reuse. In this case, incidentally, the ambient level of unrecovered wastes continuously circulating through the resource inventory of the system (i.e., the spaceship) would in general be nonzero because of the difficulty of 100 percent efficient waste-removal of air and water. However, although the quantity of waste products in constant circulation may fluctuate within limits, it cannot be allowed to increase monotonically with time, which means that as much material must be recycled, on the average, as is discarded. The value of common resources plus the assimilation services performed by the environment, then, is only indirectly a function of the ambient level of untreated residuals per se, or the disutility caused thereby, which depend on the cost efficiency of the available treatment technology. Be this as it may, of course, the bill of goods produced in a spaceship economy would certainly be radically different from that we are familiar with. For this reason, no standard economic comparison between the two situations is meaningful. The measure of worth we are seeking is actually the difference between the total welfare "produced" by a spaceship economy, where 100 percent of all residuals are promptly recycled, vis-à-vis the existing welfare output on earth, where resource inventories are substantial and complete recycling need not be contemplated for a very long time to come.

This welfare difference might well be very large, although we possess no methodological tools for quantifying it. In any case, the resource inventory and assimilative capacity of the environment probably contribute very considerably to our standard of living.

If these environmental contributions were paid for, the overall effect on prices would presumably be to push them generally upward. However, the major **differential** effect of undervaluing the environmental contribution is that goods produced by high residual-producing processes, such as papermaking, are substantially underpriced vis-à-vis goods which involve more economical uses of basic resources. This is, however, not socially disadvantageous per se: that is, it causes no misallocation of resources unless, or until, the large resource inventory and/or the assimilative capacity of the environment are used up. When this happens, however, as it now has in most highly industrialized regions, either a market must be allowed to operate or some other form of decision rule must be introduced to permit a rational choice to be made, e.g., between curtailing or controlling the production of residuals or tolerating the effects (disservice) thereof.

It appears that the natural inventory of most common resources used as inputs (e.g., air as an input to combustion and respiratory processes) is still ample,[27] but the as-

---

[26] Any consistent deviation from this 100 per cent rule implies an accumulation of waste products, on the average, which, by definition, is inconsistent with maintaining an equilibrium.

[27] Water is an exception in arid regions; in humid regions, however, water "shortages" are misnomers: they are really consequences of excessive use of watercourses as cheap means of waste disposal. But some ecologists have claimed that oxygen depletion may be a very serious long-run problem; see Cole (1967).

similative capacity of the environment has already been exceeded in many areas, with important external costs resulting. This suggests a compromise treatment. If an appropriate price could be charged to the producers of the residuals and used to compensate the inadvertent recipients – with the price determined by appropriate Pareto preference criteria – there would be no particular analytic purpose in keeping books on the exchange of the other environmental benefits mentioned, although they are quantitatively massive. We will, therefore, in the remainder of the discussion omit the common-property variables $\{r_j^{cp}\}$ and the corresponding virtual-price variables $\{v_j^{cp}\}$ defined previously, retaining only the terms $\{r_j^u\}$ and $\{u_{jk}\}$. The variable $\{r_j^u\}$ represents a physical quantity of the *j*th unwanted input. There are S such terms, by assumption, whose magnitudes are proportional to the levels of consumption of basic raw materials, subject to the existing technology. However, residuals production is not immutable: it can be increased or decreased by investment, changes in materials processing technology, raw material substitutions, and so forth.

At first glance it might seem entirely reasonable to assert that the **supplies** of unwanted residuals received will be functions of the (negative) prices (i.e., compensation) paid for them, in analogy with (7). Unfortunately, this assertion immediately introduces a theoretical difficulty, since the assumption of unique coefficients $\{a_{ij}\}$ and $\{C_{ij}\}$[28] is not consistent with the possibility of factor or process substitution or joint-production, as stated earlier. To permit such substitutions, one would have to envision a very large collection of alternative sets of coefficients: one complete set of *a*'s and *C*'s for each specific combination of factors and processes. Maximization of any objective function (such as GNP) would involve solving the entire system of equations as many times as there are combinations of factors and processes, and picking out that set of solutions which yields the largest value. Alternatively, if the *a*'s and *C*'s are assumed to be continuously variable functions (of each other), the objective function could also, presumably, be parameterized. However, as long as the *a*'s and *C*'s are uniquely given, the supply of the kth unwanted residual is only marginally under the control of the producer, since it will be produced in strict relationship to the composition of the bill of final goods $\{Y_j\}$.

Hence, for the present model it is only correct to assume

$$(20) \qquad r_k^u = G_k^u\,(Y_1, \ldots, Y_N)$$

This limitation does not affect the existence of an equilibrium solution for the system of equations; it merely means that the shadow prices $\{u_{jk}\}$ which would emerge from such a solution for given coefficients $\{a_{ij}\}$, $\{b_{ij}\}$, and $\{C_{ij}\}$ might be considerably higher than the real economic optimum, since the latter could only be achieved by introducing factor and process changes.

Of course, the physical inputs are also related to the physical outputs of goods, as in (21):

---

[28] Or $\{b_{ij}\}$ and $\{A_{ij}\}$.

$$[r] = \begin{bmatrix} r^m \\ r^s \\ r^u \end{bmatrix} = M \left\{ \underbrace{\begin{bmatrix} a^m \\ \cdots \\ a^s \\ \cdots \\ a^u \end{bmatrix}}_{N} \begin{bmatrix} \\ X \\ \\ \end{bmatrix} \right\} N = \begin{bmatrix} b^m \\ \cdots \\ b^s \\ \cdots \\ b^u \end{bmatrix} \begin{bmatrix} \\ Y \\ \\ \end{bmatrix} \tag{21}$$

Written out in full detail (21) is equivalent to:

(22) raw materials
$$r_k^m = \sum_{j=1}^{N} a_{kj}^m X_j = \sum_{j=1}^{N} b_{kj}^m Y_j$$
$$k = 1, \ldots, L$$

(23) labor and technical services
$$r_k^s = \sum_{j=1}^{N} a_{kj}^s X_j \quad \sum_{j=1}^{N} b_{kj}^s Y_j$$
$$k = 1, \ldots, P$$

(24) unwanted inputs
$$r_k^u = \sum_{j=1}^{N} a_{kj}^u X_j \quad \sum_{j=1}^{N} b_{kj}^u Y_j$$
$$k = 1, \ldots, S$$

where, of course,

$$L + P + S = M \tag{25}$$

The corresponding matrix equation for the prices of goods, in terms of production costs, is

$$[p_1, \ldots, p_N] = [v^m \vdots v^s \vdots u] \begin{bmatrix} b^m \\ \cdots \\ b^s \\ \cdots \\ b^u \end{bmatrix} \tag{26}$$

Written out in the standard form, we obtain

$$p_k = \underbrace{\sum_{j=1}^{L} b_{jk}^m v_j^m}_{\text{cost of raw materials}} + \underbrace{\sum_{j=1}^{P} b_{jk}^s v_j^s}_{\text{cost of labor and technical services}}$$

$$(27) \qquad + \underbrace{\sum_{j=1}^{S} b_{jk}^u v_j^u}_{\text{cost (compensation) for providing environmental disservices}} \qquad k = 1, \ldots, N$$

Evidently, the coefficients b_{jk}^u are empirically determined by the structure of the regional economy and its geography. It is assumed that a single overall (negative) price for each residual has meaning, even though each productive sector – and even each consumer – has his own individual utility function. Much the same assumption is conventionally made, and accepted, in the case of positive real prices. All of the additional variables now fit into the general framework of the original Walras-Cassel analysis. Indeed, we have $2N+2M-1$ variables (r_j, Y_i, p_i, v_i) (allowing an arbitrary normalization factor for the price level) and $2N+2M-1$ independent equations.[29] If solutions exist for the Walras-Cassel system of equations, the arguments presumably continue to hold true for the generalized model. In any case, a discussion of such mathematical questions would carry us too far from our main theme.

IV. CONCLUDING COMMENTS

The limited economics literature currently available which is devoted to environmental pollution problems has generally taken a partial equilibrium view of the matter, as well as treated the pollution of particular environmental media, such as air and water, as separate problems.[30] This no doubt reflects the propensity of the theoretical literature to view externalities as exceptional and minor. Clearly, the partial equilibrium approach in particular is very convenient theoretically and empirically for it permits external damage and control cost functions to be defined for each particular case without reference to broader interrelationships and adjustments in the economy.

We have argued in this paper that the production of residuals is an inherent and general part of the production and consumption process and, moreover, that there are

[29] There is one redundant equation in the system, which expresses the identity between gross product and gross income for the system as a whole (sometimes called "Walras law").

[30] See, for example, the essays in Wolozin (1966).

important trade-offs between the gaseous, liquid, or solid forms that these residuals may take. Further, we have argued that under conditions of intensive economic and population development the environmental media which can receive and assimilate residual wastes are not free goods but natural resources of great value with respect to which voluntary exchange cannot operate because of their common property characteristics. We have also noted, in passing, that the assimilative capacity of environmental media can sometimes be altered and that therefore the problem of achieving Pareto optimality reaches beyond devising appropriate shadow prices and involves the planning and execution of investments with public goods aspects.

We have exhibited a formal mathematical framework for tracing residuals flows in the economy and related it to the general equilibrium model of resources allocation, altered to accommodate recycle and containing unpriced sectors to represent the environment. This formulation, in contrast to the usual partial equilibrium treatments, implies knowledge of all preference and production functions including relations between residuals discharge and external cost and all possible factor and process substitutions. While we feel that it represents reality with greater fidelity than the usual view, it also implies a central planning problem of impossible difficulty, both from the standpoint of data collection and computation.

What, if any, help can the general interdependency approach we have outlined offer in dealing with pollution problems effectively and reasonably efficiently? A minimal contribution is its warning that partial equilibrium approaches, while more tractable, may lead to serious errors. Second, in projecting waste residuals for an economy – a regional economy would usually be the most relevant for planning and control – the inter-industry materials flow model can provide a much more conceptually satisfying and accurate tool for projecting future residuals production than the normal aggregative extrapolations.[31] The latter not only treat gaseous, liquid, and solid wastes separately, but do not take account of input-output relations and the fact that the materials account for the region must balance.

We think that in the next few years it will be possible to make improved regional projections of residuals along the lines sketched above. Undoubtedly, there will also be further progress in empirically estimating external costs associated with residuals discharge and in estimating control costs via various alternative measures. On the basis of this kind of information, a control policy can be devised. However, this approach will still be partial. Interrelations between the regional and national economy must be treated simplistically and to be manageable, the analysis must confine itself to a specific projected bill of goods.

---

[31] Some efforts to implement these concepts are already underway. Walter Isard and his associates have prepared an input-output table for Philadelphia which includes coefficients representing waterborne wastes (unpublished). The recent study of waste management in the New York Metropolitan region by the Regional Plan Association (1968) took a relatively broad view of the waste residuals problem. Relevant data on several industries are being gathered. Richard Frankel's not yet published study of thermal power in which the range of technical options for controlling residuals, and their costs, is being explored is notable in this regard. His and other salient studies are described in Ayres/ Kneese (1968).

The basic practical question which remains to be answered is whether an iterated series of partial equilibrium treatments – e.g., focusing on one industry or region at a time, ceteris paribus – would converge toward the general equilibrium solution, or not. We know of no theoretical test of convergence which would be applicable in this case but, in the absence of such a criterion, would be willing to admit the possible relevance of an empirical sensitivity test more or less along the following lines: take a major residuals-producing industry (such as electric power) and parametrize its cost structure in terms of emission control levels, allowing all technically feasible permutations of factor (fuel) inputs and processes. It would be a straight-forward, but complicated, operations research problem to determine the minimum cost solution as a function of the assumed (negative) price of the residuals produced. If possible industry patterns – factor and process combinations – exist which would permit a high level of emission control at only a small increase in power production cost, then it might be possible to conclude that for a significant range of (negative) residuals prices the effect on power prices – and therefore on the rest of the economy – would not be great. Such a conclusion would support the convergence hypothesis. If, on the other hand, electric power prices are very sensitive to residuals prices, then one would at least have to undertake a deeper study of consumer preference functions to try to determine what residuals prices would actually be if a market mechanism existed. If people prove to have a strong antipathy to soot and sulfur dioxide, for instance, resulting in a high (negative) price for these unwanted inputs, then one would be forced to suspect that the partial equilibrium approach is probably not convergent to the general equilibrium solution and that much more elaborate forms of analysis will be required.

REFERENCES

Ayres (1967): Ayres, R.U.: *Stability of Biosystems in Sea Water*, Tech. Rept. No. 142, Hudson Laboratories, Columbia University, New York, 1967.

Ayres/Kneese (1968): Ayres, R.U. and A.V. Kneese: Environmental Pollution, in U.S. Congress, Joint Economic Committee, *Federal Programs for the Development of Human Resources*, Vol. 2, Washington, 1968.

Baumol (1967): Baumol, W.J.: Welfare Economics and the Theory of the State. Cambridge, 1967.

Boulding (1966): Boulding, K.E.: The Economics of the Coming Spaceship Earth, in: H. Jarrett (ed.): *Environmental Quality in a Growing Economy*, Baltimore, 1966, pp. 3–14.

Boyd (1968): Boyd, J.H.: Collective Facilities in Water Quality Management, Appendix to: Kneese, A.V. and B.T. Bower: *Managing Water Quality: Economics, Technology*, Institutions. Baltimore, 1968.

Buchanan/Stubblebine (1962): Buchanan, J.W. and W.C. Stubblebine: Externality. In: *Economica*, Vol. 29, 1962, pp. 371–384.

Buchanan/Tullock (1965): Buchanan, J.W. and G. Tullock: Public and Private Interaction under Reciprocal Externality. In: J. Margolis (ed.): *The Public Economy of Urban Communities*, Baltimore, 1965, pp. 52–73.

Cassel (1932): Cassel, G.: *The Theory of Social Economy*. New York, 1932.

Coase (1960): Coase, R.H.: The Problem of Social Cost, *Journal of Law and Economics*, October 1960, No. 3, pp. 1–44.

Cole (1967): Cole, L.: *Can the World be Saved?* Paper presented at the 134th Meeting of the American Association for the Advancement of Science, December 27th, 1967.

Conservation Foundation (1963): Conservation Foundation: *Implications of Rising Carbont Dioxide Content of the Atmosphere*. New York, 1963.

Davis (1968): Davis, R.K.: *The Range of Choice in Water Management*. Baltimore, 1968.

Davis/Whinston (1962): Davis, O.A. and A. Whinston: Externalities, Welfare, and the Theory of Games. Journal of Political Economy, Vol. 70, June 1962, pp. 241–62.

Dorfman et al. (1958): Dorfman, R., P. Samuelson and R.M. Solov: *Linear Programming and Economic Analysis*. New York, 1958.

Fisher (1906): Fisher, I.: *Nature of Capital and Income*. New York, 1906.

Harberger (1954): Harberger, A.C.: Monopoly and Resources Allocation, *American Economic Review*, Vol. 44, May 1954, pp. 77–87.

Kapp (1950): Kapp, K.W.: *The Social Costs of Private Enterprise*. Cambridge, 1950.

Kneese/Bower (1968): Kneese, A.V. and B.T. Bower: *Managing Water Quality: Economics, Technology, Institutions*. Baltimore, 1968.

Knight (1921): Knight, F.H.: *Risk, Uncertainty, and Profit*. Boston/New York, 1921.

Leibenstein (1966): Leibenstein, H.: Allocative Efficiency vs. 'X-Efficiency,' *American Economic Review*, Vol. 56, June 1966, pp. 392–415.

Lowry (1967): Lowry, W.P.: The Climate of Cities, *Scientific American*, Vol. 217, Aug. 1967, pp. 15–23.

Mishan (1965): Mishan, E.J.: Reflections on Recent Developments in the Concept of External Effects, *Canadian Journal of Economics and Political Science*, Vol. 31, Feb. 1965, pp. 1–34.

Pigou (1952): Pigou, A.C.: *Economics of Welfare*. London, 1952.

Regional Plan Association (1968): Regional Plan Association: *Waste Management, a Report of the Second Regional Plan*. New York, 1968.

Schlesinger (1933): Schlesinger, K.: *Über die Produktionsgleichungen der ökonomischen Wertlehre*, Ergebnisse eines mathematischen Kolloquiums, No. 6., F. Denticke, Vienna, 1933.

Schwarzman (1960): Schwarzman, D.: The Burden of Monopoly, *Journal of Political Economy*, Vol. 68, December 1960, pp. 627–30.

Scitovsky (1954): Scitovsky, T.: Two Concepts of External Economies, *Journal of Political Economy*, Vol. 62, Apr. 1954, pp. 143–51.

Smith (1967): Smith, F.: *The Economic Theory of Industrial Waste Production and Disposal*, draft of a doctoral dissertation, Northwestern Univ., 1967.

Turvey (1962): Turvey, R.: On Divergencies between Social Cost and Private Cost, Economica, Vol. 30, November 1962, pp. 309–13.

Walras (1954) :Walras, L.: *Elements d'economie politique pure*, Jaffé translation, London, 1954.

Wolozin (1966): Wolozin, H. (ed.): *The Economics of Air Pollution*. New York, 1966.

Karl E. Weick (1976)

EDUCATIONAL ORGANIZATIONS AS LOOSELY COUPLED SYSTEMS

> "Imagine that you're either the referee, coach, player or spectator at an unconventional soccer match: the field for the game is round; there are several goals scattered haphazardly around the circular field; people can enter and leave the game whenever they want to; they can throw balls in whenever they want; they can say 'that's my goal' whenever they want to, as many times as they want to, and for as many goals as they want to; the entire game takes place on a sloped field; and the game is played as if it makes sense."

The image Karl Weick (born 1936, meanwhile retired professor of psychology and organizational behavior at Cornell University, Ithaca, NY) uses as an introduction for his "classic" essay on loosely coupled systems is at the same time bizarre and fascinating. Had the economic discussion on supply chain management and value creation systems already taken place in the 1970s, maybe he would have transferred his unconventional football metaphor to the "logistics playing field" of modern supply networks.

The analogies are impressive: A hardly determinable number of actors enters and leaves the arena at random moments; every player uses his own playing equipment, standards, and routines. The objectives and success criteria of the different players – which, in reality, could be customers, retailers, manufacturers, suppliers, or service providers – are heterogeneous. Inclined playing fields favour certain players and handicap others. As far as strategists, coordinators, and impulse generators try to influence the interaction, their commands and instructions are often ignored, misunderstood or complied in an imperfect way.

With this picture Weick gives the reader a new perspective pointing out that the nature of complex systems is different from conceptions that are to be found in organization and management theory, where organizations are seen as tightly coupled information and order chains, featuring firm dependencies of causes and effects and being characterized by clearly determined, durable structures. Instead, he spotlights the concept of loosely coupled organizations in which relations of various consistency between means and ends, between actions and targets, and between processes and their outputs are to be found. He shows that within complex systems (like modern supply chains) loose couplings need not be negative, but provides a convincing list of potentially positive functions of loose coupling between organizational units.

P. Klaus and S. Müller (eds.), *The Roots of Logistics*, DOI 10.1007/978-3-642-27922-5_25,

Jointly with related earlier and later contributions to organization theory (such as those included in this volume, i.e. Coase (1937), Simon (1962), and Ouchi (1982), also Granovetter (1973)), the unconventional perspective, the terminology and the suggestions for further research enrich the toolset of the "Science of Logistics" in ways which still have not been realized in full.

Reference:

Karl E. Weick: Educational Organizations as Loosely Coupled Systems. In: *Administrative Science Quarterly*, Vol. 21, March 1976, pp. 1–19.

Further reading:

Granovetter, Mark S.: The Strength of Weak Ties. In: *American Journal of Sociology*, Vol. 78, No. 6, pp. 1360–1380.

Karl E. Weick: Educational Organizations as Loosely Coupled Systems

In contrast to the prevailing image that elements in organizations are coupled through dense, tight linkages, it is proposed that elements are often tied together frequently and loosely. Using educational organizations as a case in point, it is argued that the concept of loose coupling incorporates a surprising number of disparate observations about organizations, suggests novel functions, creates stubborn problems for methodologists, and generates intriguing questions for scholars. Sample studies of loose coupling are suggested and research priorities are posed to foster cumulative work with this concept.[1]

Imagine that you're either the referee, coach, player or spectator at an unconventional soccer match: the field for the game is round; there are several goals scattered haphazardly around the circular field; people can enter and leave the game whenever they want to; they can throw balls in whenever they want; they can say "that's my goal" whenever they want to, as many times as they want to, and for as many goals as they want to; the entire game takes place on a sloped field; and the game is played as if it makes sense (March, Personal Communication).

If you now substitute in that example principals for referees, teachers for coaches, students for players, parents for spectators and schooling for soccer, you have an equally unconventional depiction of school organizations. The beauty of this depiction is that it captures a different set of realities within educational organizations than are caught when these same organizations are viewed through the tenets of bureaucratic theory.

Consider the contrast in images. For some time people who manage organizations and people who study this managing have asked, "How does an organization go about doing what it does and with what consequences for its people, processes, products, and persistence?" And for some time they've heard the same answers. In paraphrase the answers say essentially that an organization does what it does be-

---

1 This paper is the result of a conference held at La Jolla, California, February 2-4, 1975 with support from the National Institute of Education (NIE). Participants in the conference were, in addition to the author, W.W. Charters, Center for Educational Policy and Management, University of Oregon; Craig Lundberg, School of Business, Oregon State University; John Meyer, Dept. of Sociology, Stanford University; Miles Meyers, Dept. of English, Oakland (Calif.) High School; Karlene Roberts, School of Business, University of California, Berkeley; Gerald Salancik, Dept. of Business Administration, University of Illinois; and Robert Wentz, Superintendent of Schools, Pomona (Calif.) Unified School District. James G. March, School of Education, Stanford University, a member of the National Council on Educational Research, and members of the NIE staff were present as observers. This conference was one of several on organizational processes in education which will lead to a report that will be available from the National Institute of Education, Washington, D.C. 20208. The opinions expressed in this paper do not necessarily reflect the position or policy of the National Institute of Education or the Department of Health, Education, and Welfare.

cause of plans, intentional selection of means that get the organization to agree upon goals, and all of this is accomplished by such rationalized procedures as cost-benefit analyses, division of labor, specified areas of discretion, authority invested in the office, job descriptions, and a consistent evaluation and reward system. The only problem with that portrait is that it is rare in nature. People in organizations, including educational organizations, find themselves hard pressed either to find actual instances of those rational practices or to find rationalized practices whose outcomes have been as beneficent as predicted, or to feel that those rational occasions explain much of what goes on within the organization. Parts of some organizations are heavily rationalized but many parts also prove intractable to analysis through rational assumptions.

It is this substantial unexplained remainder that is the focus of this paper. Several people in education have expressed dissatisfaction with the prevailing ideas about organizations supplied by organizational theorists. Fortunately, they have also made some provocative suggestions about newer, more unconventional ideas about organizations that should be given serious thought. A good example of this is the following observation Stephens:

> "[There is a] remarkable constancy of educational results in the face of widely differing deliberate approaches. Every so often we adopt new approaches or new methodologies and place our reliance on new panaceas. At the very least we seem to chorus new slogans. Yet the academic growth within the classroom continues at about the same rate, stubbornly refusing to cooperate with the bright new dicta emanating from the conference room… [These observations suggest that] we would be making a great mistake in regarding the management of schools as similar to the process of constructing a building or operating a factory. In these latter processes deliberate decisions play a crucial part, and the enterprise advances or stands still in proportion to the amount of deliberate effort exerted. If we must use a metaphor or model in seeking to understand the process of schooling, we should look to agriculture rather than to the factory. In agriculture we do not start from scratch, and we do not direct our efforts to inert and passive materials. We start, on the contrary, with a complex and ancient process, and we organize our efforts around what seeds, plants, and insects are likely to do anyway... The crop, once planted, may undergo some development even while the farmer sleeps or loafs. No matter what he does, **some** aspects of the outcome will remain constant. When teachers and pupils foregather, some education may proceed even while the Superintendent disports himself in Atlantic City."[2]

It is crucial to highlight what is important in the examples of soccer and schooling viewed as agriculture. To view these examples negatively and dismiss them by observing that "the referee should tighten up those rules," "superintendents don't do that," "schools are more sensible than that," or "these are terribly sloppy organizations" is to miss the point. The point is although researchers don't know what these

[2] Cf. Stephens (1967), pp. 9–11.

kinds of structures are like but researchers do know they exist and that each of the negative judgments expressed above makes sense only if the observer assumes that organizations are constructed and managed according to rational assumptions and therefore are scrutable only when rational analyses are applied to them. This paper attempts to expand and enrich the set of ideas available to people when they try to make sense out of their organizational life. From this standpoint, it is unproductive to observe that fluid participation in schools and soccer is absurd. But it can be more interesting and productive to ask, how can it be that even though the activities in both situations are only modestly connected, the situations are still recognizable and nameable? The goals, player movements, and trajectory of the ball are still recognizable and can be labeled "soccer." And despite variations in class size, format, locations, and architecture, the results are still recognized and can be labeled "schools." How can such loose assemblages retain sufficient similarity and permanence across time that they can be recognized, labeled, and dealt with? The prevailing ideas in organization theory do not shed much light on how such "soft" structures develop, persist, and impose crude orderliness among their elements.

The basic premise here is that concepts such as loose coupling serve as sensitizing devices. They sensitize the observer to notice and question things that had previously been taken for granted. It is the intent of the program described here to develop a language for use in analyzing complex organizations, a language that may highlight features that have previously gone unnoticed. The guiding principle is a reversal of the common assertion, "I'll believe it when I see it" and presumes an epistemology that asserts, "I'll see it when I believe it." Organizations as loosely coupled systems may not have been seen before because nobody believed in them or could afford to believe in them. It is conceivable that preoccupation with rationalized, tidy, efficient, coordinated structures has blinded many practitioners as well as researchers to some of the attractive and unexpected properties of less rationalized and less tightly related clusters of events. This paper intends to eliminate such blindspots.

THE CONCEPT OF COUPLING

The phrase "loose coupling" has appeared in the literature[3] and it is important to highlight the connotation that is captured by this phrase and by no other. It might seem that the word coupling is synonymous with words like connection, link, or interdependence, yet each of these latter terms misses a crucial nuance.

By loose coupling, the author intends to convey the image that coupled events are responsive, **but** that each event also preserves its own identity and some evidence of its physical or logical separateness. Thus, in the case of an educational organization,

---

3 Cf. Glassman (1973) and March/Olsen (1975).

it may be the case that the counselor's office is loosely coupled to the principal's office. The image is that the principal and the counselor are somehow attached, but that each retains some identity and separateness and that their attachment may be circumscribed, infrequent, weak in its mutual affects, unimportant, and/or slow to respond. Each of those connotations would be conveyed if the qualifier loosely were attached to the word coupled. Loose coupling also carries connotations of impermanence, dissolvability, and tacitness all of which are potentially crucial properties of the "glue" that holds organizations together.

Glassman (1973) categorizes the degree of coupling between two systems on the basis of the activity of the variables which the two systems share. To the extent that two systems either have few variables in common or share weak variables, they are independent of each other. Applied to the educational situation, if the principal-vice-principal-superintendent is regarded as one system and the teacher-classroom-pupil-parent-curriculum as another system, then by Glassman's argument if we did not find many variables in the teacher's world to be shared in the world of a principal and/or if the variables held in common were unimportant relative to the other variables, then the principal can be regarded as being loosely coupled with the teacher.

A final advantage of coupling imagery is that it suggests the idea of building blocks that can be grafted onto an organization or severed with relatively little disturbance to either the blocks or the organization. Simon (1969) has argued for the attractiveness of this feature in that most complex systems can be decomposed into stable subassemblies and that these are the crucial elements in any organization or system. Thus, the coupling imagery gives researchers access to one of the more powerful ways of talking about complexity now available.

But if the concept of loose coupling highlights novel images heretofore unseen in organizational theory, what is it about these images that is worth seeing?

COUPLED ELEMENTS

There is no shortage of potential coupling elements, but neither is the population infinite.

At the outset the two most commonly discussed coupling mechanisms are the technical core of the organization and the authority of office. The relevance of those two mechanisms for the issue of identifying elements is that in the case of technical couplings, each element is some kind of technology, task, subtask, role, territory and person, and the couplings are task-induced. In the case of authority as the coupling mechanism, the elements include positions, offices, responsibilities, opportunities, rewards, and sanctions and it is the couplings among these elements that presumably hold the organization together. A compelling argument can be made that **neither** of these coupling mechanisms is prominent in educational organizations found in the United States. This leaves one with the question what **does** hold an educational organization together?

A short list of potential elements in educational organizations will provide background for subsequent propositions. March/Olsen (1975) utilize the elements of intention and action. There is a developing position in psychology which argues that intentions are a poor guide for action, intentions often follow rather than precede action, and that intentions and action are loosely coupled. Unfortunately, organizations continue to think that planning is a good thing, they spend much time on planning, and actions are assessed in terms of their fit with plans. Given a potential loose coupling between the intentions and actions of organizational members, it should come as no surprise that administrators are baffled and angered when things never happen the way they were supposed to.

Additional elements may consist of events like yesterday and tomorrow (what happened yesterday may be tightly or loosely coupled with what happens tomorrow) or hierarchical positions, like, top and bottom, line and staff, or administrators and teachers. An interesting set of elements that lends itself to the loose coupling imagery is means and ends. Frequently, several different means lead to the same outcome. When this happens, it can be argued that any one means is loosely coupled to the end in the sense that there are alternative pathways to achieve that same end. Other elements that might be found in loosely coupled educational systems are teachers-materials, voters-schoolboard, administrators-classroom, process-outcome, teacher-teacher, parent-teacher, and teacher-pupil.

While all of these elements are obvious, it is not a trivial matter to specify which elements are coupled. As the concept of coupling is crucial because of its ability to highlight the identity and separateness of elements that are momentarily attached, that conceptual asset puts pressure on the investigator to specify clearly the identity, separateness, and boundaries of the elements coupled. While there is some danger of reification when that kind of pressure is exerted, there is the even greater danger of portraying organizations in inappropriate terms which suggest an excess of unity, integration, coordination, and consensus. If one is nonspecific about boundaries in defining elements then it is easy – and careless – to assemble these ill-defined elements and talk about integrated organizations. It is not a trivial issue explaining how elements persevere over time. Weick, for example, has argued that elements may appear or disappear and may merge or become separated in response to need-deprivations within the individual, group, and/or organization.[4] This means that specification of elements is not a one-shot activity. Given the context of most organizations, elements both appear and disappear over time. For this reason a theory of how elements become loosely or tightly coupled may also have to take account of the fact that the nature and intensity of the coupling may itself serve to create or dissolve elements.

The question of what is available for coupling and decoupling within an organization is an eminently practical question for anyone wishing to have some leverage on a system.

---

[4] Cf. Weick (1974), pp. 363–364.

STRENGTH OF COUPLING

Obviously there is no shortage of meanings for the phrase loose coupling. Researchers need to be clear in their own thinking about whether the phenomenon they are studying is described by two words or three. A researcher can study "loose coupling" in educational organizations or "loosely coupled systems." The shorter phrase, "loose coupling," simply connotes things, "anythings," that may be tied together either weakly or infrequently or slowly or with minimal interdependence. Whether those things that are loosely coupled exist in a system is of minor importance. Most discussions in this paper concern loosely coupled systems rather than loose coupling since it wishes to clarify the concepts involved in the perseverance of sets of elements across time.

The idea of loose coupling is evoked when people have a variety of situations in mind. For example, when people describe loosely coupled systems they are often referring to (1) slack times – times when there is an excessive amount of resources relative to demands; (2) occasions when any one of several means will produce the same end; (3) richly connected networks in which influence is slow to spread and/or is weak while spreading; (4) a relative lack of coordination, slow coordination or coordination that is dampened as it moves through a system; (5) a relative absence of regulations; (6) planned unresponsiveness; (7) actual causal independence; (8) poor observational capabilities on the part of a viewer; (9) infrequent inspection of activities within the system; (10) decentralization; (11) delegation of discretion; (12) the absence of linkages that should be present based on some theory – for example, in educational organizations the expected feedback linkage from outcome back to inputs is often nonexistent; (13) the observation that an organization's structure is not coterminous with its activity; (14) those occasions when no matter what you do things always come out the same – for instance, despite all kinds of changes in curriculum, materials, groupings, and so forth the outcomes in an educational situation remain the same; and (15) curricula or courses in educational organizations for which there are few prerequisites – the longer the string of prerequisites, the tighter the coupling.

POTENTIAL FUNCTIONS AND DYSFUNCTIONS OF LOOSE COUPLING

It is important to note that the concept of loose coupling need not be used normatively. People who are steeped in the conventional literature of organizations may regard loose coupling as a sin or something to be apologized for. This paper takes a neutral, if not mildly affectionate, stance toward the concept. Apart from whatever affect one might feel toward the idea of loose coupling, it does appear a priori that certain functions can be served by having a system in which the elements are loosely coupled. Below are listed seven potential functions that could be associated with loose coupling plus additional reasons why each advantage might also be a liability.

The dialectic generated by each of these oppositions begins to suggest dependent variables that should be sensitive to variations in the tightness of coupling.

The basic argument of Glassman (1973) is that loose coupling allows some portions of an organization to persist. Loose coupling lowers the probability that the organization will have to – or be able to – respond to each little change in the environment that occurs. The mechanism of voting, for example, allows elected officials to remain in office for a full term even though their constituency at any moment may disapprove of particular actions. Some identity and separateness of the element "elected official" is preserved relative to a second element, "constituency," by the fact of loosely coupled accountability which is measured in two, four, or six year terms. While loose coupling may foster perseverance, it is not selective in what is perpetuated. Thus archaic traditions as well as innovative improvisations may be perpetuated.

A second advantage of loose coupling is that it may provide a sensitive sensing mechanism. This possibility is suggested by Fritz Heider's perceptual theory of things and medium. Heider (1959) argues that perception is most accurate when a medium senses a thing and the medium contains many independent elements that can be externally constrained. When elements in a medium become either fewer in number and/or more internally constrained and/or more interdependent, their ability to represent some remote thing is decreased. Thus sand is a better medium to display wind currents than are rocks, the reason being that sand has more elements, more independence among the elements, and the elements are subject to a greater amount of external constraint than is the case for rocks. Using Heider's formulation metaphorically, it could be argued that loosely coupled systems preserve many independent sensing elements and therefore "know" their environments better than is true for more tightly coupled systems which have fewer externally constrained, independent elements. Balanced against this improvement in sensing is the possibility that the system would become increasingly vulnerable to producing faddish responses and interpretations. If the environment is known better, then this could induce more frequent changes in activities done in response to this "superior intelligence."

A third function is that a loosely coupled system may be a good system for localized adaptation. If all of the elements in a large system are loosely coupled to one another, then any one element can adjust to and modify a local unique contingency without affecting the whole system. These local adaptations can be swift, relatively economical, and substantial. By definition, the antithesis of localized adaptation is standardization and to the extent that standardization can be shown to be desirable, a loosely coupled system might exhibit fewer of these presumed benefits. For example, the localized adaptation characteristic of loosely coupled systems may result in a lessening of educational democracy.

Fourth, in loosely coupled systems where the identity, uniqueness, and separateness of elements is preserved, the system potentially can retain a greater number of mutations and novel solutions than would be the case with a tightly coupled system. A loosely coupled system could preserve more "cultural insurance" to be drawn upon in times of radical change than in the case for more tightly coupled systems. Loosely coupled systems may be elegant solutions to the problem that adaptation can preclude adaptability. When a specific system fits into an ecological niche and does so with great success, this adaptation can be costly. It can be costly because resources which are useless in a current environment might deteriorate or disappear even though they could be crucial in a modified environment. It is conceivable that loosely coupled systems preserve more diversity in responding than do tightly coupled systems, and therefore can adapt to a considerably wider range of changes in the environment than would be true for tightly coupled systems. To appreciate the possible problems associated with this abundance of mutations, reconsider the dynamic outlined in the preceding discussion of localized adaptation. If a local set of elements can adapt to local idiosyncracies without involving the whole system, then this same loose coupling could also forestall the spread of advantageous mutations that exist somewhere in the system. While the system may contain novel solutions for new problems of adaptation, the very structure that allows these mutations to flourish may prevent their diffusion.

Fifth, if there is a breakdown in one portion of a loosely coupled system then this breakdown is sealed off and does not affect other portions of the organization. Previously we had noted that loosely coupled systems are an exquisite mechanism to adapt swiftly to local novelties and unique problems. Now we are carrying the analysis one step further, and arguing that when any element misfires or decays or deteriorates, the spread of this deterioration is checked in a loosely coupled system. While this point is reminiscent of earlier functions, the emphasis here is on the localization of trouble rather than the localization of adaptation. But even this potential benefit may be problematic. A loosely coupled system can isolate its trouble spots and prevent the trouble from spreading, but it should be difficult for the loosely coupled system to repair the defective element. If weak influences pass from the defective portions to the functioning portions, then the influence back from these functioning portions will also be weak and probably too little, too late.

Sixth, since some of the most important elements in educational organizations are teachers, classrooms, principals, and so forth, it may be consequential that in a loosely coupled system there is more room available for self-determination by the actors. If it is argued that a sense of efficacy is crucial for human beings, then a sense of efficacy might be greater in a loosely coupled system with autonomous units than it would be in a tightly coupled system where discretion is limited. A further comment can be made about self-determination to provide an example of the kind of imagery that is invoked by the concept of loose coupling.

It is possible that much of the teacher's sense of – and actual – control comes from the fact that diverse interested parties expect the teacher to link their intentions with teaching actions. Such linking of diverse intentions with actual work probably involves considerable negotiation. A parent complains about a teacher's action and the teacher merely points out to the parent how the actions are really correspondent with the parent's desires for the education of his or her children. Since most actions have ambiguous consequences, it should always be possible to justify the action as fitting the intentions of those who complain. Salancik (1975) goes even farther and suggests the intriguing possibility that when the consequences of an action are ambiguous, the stated **intentions** of the action serve as surrogates for the consequences. Since it is not known whether reading a certain book is good or bad for a child, the fact that it is intended to be good for the child itself becomes justification for having the child read it. The potential trade-off implicit in this function of loose coupling is fascinating. There is an increase in autonomy in the sense that resistance is heightened, but this heightened resistance occurs at the price of shortening the chain of consequences that will flow from each autonomous actor's efforts. Each teacher will have to negotiate separately with the same complaining parent.

Seventh, a loosely coupled system should be relatively inexpensive to run because it takes time and money to coordinate people. As much of what happens and should happen inside educational organizations seems to be defined and validated outside the organization, schools are in the business of building and maintaining categories, a business that requires coordination only on a few specific issues – for instance, assignment of teachers. This reduction in the necessity for coordination results in fewer conflicts, fewer inconsistencies among activities, fewer discrepancies between categories and activity. Thus, loosely coupled systems seem to hold the costs of coordination to a minimum. Despite this being an inexpensive system, loose coupling is also a nonrational system of fund allocation and therefore, unspecifiable, unmodifiable, and incapable of being used as means of change.

When these several sets of functions and dysfunctions are examined, they begin to throw several research issues into relief. For example, oppositions proposed in each of the preceding seven points suggest the importance of contextual theories. A predicted outcome or its opposite should emerge depending on how and in what the loosely coupled system is embedded. The preceding oppositions also suggest a fairly self-contained research program. Suppose a researcher starts with the first point made, as loose coupling increases the system should contain a greater number of anachronistic practices. Loosely coupled systems should be conspicuous for their cultural lags. Initially, one would like to know whether that is plausible or not. But then one would want to examine in more fine-grained detail whether those anachronistic practices that are retained hinder the system or impose structure and absorb uncertainty thereby producing certain economies in responding. Similar embellishment and elaboration is possible for each function with the result that rich networks of propositions become visible. What is especially attractive about these networks is that there is little precedent for them in the organizational literature. Despite this, these propositions contain a great deal of face validity when they are used as filters to look at educational organizations. When compared, for example, with the bureaucratic template mentioned in the introduction, the template associated with loosely coupled systems seems to take the observer into more interesting territory and prods him or her to ask more interesting questions.

METHODOLOGY AND LOOSE COUPLING

An initial warning to researchers: the empirical observation of unpredictability is insufficient evidence for concluding that the elements in a system are loosely coupled. Buried in that caveat are a host of methodological intricacies. While there is ample reason to believe that loosely coupled systems can be seen and examined, it is also possible that the appearance of loose coupling will be nothing more than a testimonial to bad methodology. In psychology, for example, it has been argued that the chronic failure to predict behavior from attitudes is due to measurement error and not to the unrelatedness of these two events. Attitudes are said to be loosely coupled with behavior but it may be that this conclusion is an artifact produced because attitudes assessed by time-independent and context-independent measures are being used to predict behaviors that are time and context dependent. If both attitudes and behaviors were assessed with equivalent measures, then tight coupling might be the rule.

Any research agenda must be concerned with fleshing out the imagery of loose coupling – a task requiring a considerable amount of conceptual work to solve a few specific and rather tricky methodological problems before one can investigate loose coupling.

By definition, if one goes into an organization and watches which parts affect which other parts, he or she will see the tightly coupled parts and the parts that vary the most. Those parts which vary slightly, infrequently, and aperiodically will be less visible. Notice, for example, that interaction data – who speaks to whom about what – are unlikely to reveal loose couplings. These are the most visible and obvious couplings and by the arguments developed in this paper perhaps some of the least crucial to understand what is going on in the organization.

An implied theme in this paper is that people tend to over-rationalize their activities and to attribute greater meaning, predictability, and coupling among them than in fact they have. If members tend to overrationalize their activity then their descriptions will not suggest which portions of that activity are loosely and tightly coupled. One might, in fact, even use the presence of apparent overrationalization as a potential clue that myth making, uncertainty, and loose coupling have been spotted.

J.G. March has argued that loose coupling can be spotted and examined only if one uses methodology that highlights and preserves rich detail about context. The necessity for a contextual methodology seems to arise, interestingly enough, from inside organization theory. The implied model involves cognitive limits on rationality and man as a single channel information processor. The basic methodological point is that if one wishes to observe loose coupling, then he has to see both what is and is not being done. The general idea is that time spent on one activity is time spent away from a second activity. A contextually sensitive methodology would record both the fact that some people are in one place generating events and the fact that these same people are thereby absent from some other place. The rule of thumb would be that a tight coupling in one part of the system can occur only if there is loose coupling in another part of the system. The problem that finite attention creates for a researcher is that if some outcome is observed for the organization, then it will not be obvious whether the outcome is due to activity in the tightly coupled sector or to inactivity in the loosely coupled sector. That is a provocative problem of interpretation. But the researcher should be forewarned that there are probably a finite number of tight couplings that can occur at any moment, that tight couplings in one place imply loose couplings elsewhere, and that it may be the **pattern** of couplings that produces the observed outcomes. Untangling such intricate issues may well require that new tools be developed for contextual understanding and that investigators be willing to substitute nonteleological thinking for teleological thinking.[5]

Another contextually sensitive method is the use of comparative studies. It is the presumption of this methodology that taken-for-granted understandings – one possible "invisible" source of coupling in an otherwise loosely coupled system – are embedded in and contribute to a context. Thus, to see the effects of variations in these understandings one compares contexts that differ in conspicuous and meaningful ways.

Another methodological trap may await the person who tries to study loose coupling. Suppose one provides evidence that a particular goal is loosely coupled to a particular action. He or she says in effect, the person wanted to do this but in fact actually did that, thus, the action and the intention are loosely coupled. Now the problem for the researcher is that he or she may simply have focused on the wrong goal. There may be other goals which fit that particular action better. Perhaps if the researcher were aware of them, then the action and intention would appear to be tightly coupled. Any kind of intention-action, plan-behavior, or means-end depiction of loose coupling may be vulnerable to this sort of problem and an exhaustive listing of goals rather than parsimony should be the rule.

Two other methodological points should be noted. First, there are no good descriptions of the kinds of couplings that can occur among the several elements in educational organizations. Thus, a major initial research question is simply, what does a map of the couplings and elements within an educational organization look like? Second, there appear to be some fairly rich probes that might be used to uncover the nature of coupling within educational organizations. Conceivably, crucial couplings within schools involve the handling of disciplinary issues and social control, the question of how a teacher gets a book for the classroom, and the question of what kinds of innovations need to get clearance by whom. These relatively innocuous questions may be powerful means to learn which portions of a system are tightly and loosely coupled. Obviously these probes would be sampled if there was a full description of possible elements that can be coupled and possible kinds and strengths of couplings. These specific probes suggest, however, in addition that what holds an educational organization together may be a small number of tight couplings in out-of-the-way places.

---

[5] Cf. Steinbeck (1941), chapt. 14.

ILLUSTRATIVE QUESTIONS FOR A RESEARCH AGENDA

Patterns of Loose and Tight Coupling: Certification versus Inspection

Suppose one assumes that education is an intrinsically uninspected and unevaluated activity. If education is intrinsically uninspected and unevaluated then how can one establish that it is occurring? One answer is to define clearly who can and who cannot do it and to whom. In an educational organization this is the activity of certification. It is around the issues of certification and of specifying who the pupils are that tight coupling would be predicted to occur when technology and outcome are unclear.

If one argues that "certification" is the question "who does the work" and "inspection" is the question "how well is the work done," then there can be either loose or tight control over either certification or inspection. Notice that setting the problem up this way suggests the importance of discovering the distribution of tight and loosely coupled systems within any organization. Up to now the phrase loosely coupled systems has been used to capture the fact that events in an organization seem to be temporally related rather than logically related.[6] Now that view is being enriched by arguing that any organization must deal with issues of certification (who does the work) and inspection (how well is the work done). It is further being suggested that in the case of educational organizations there is loose control on the work – the work is intrinsically uninspected and unevaluated or if it is evaluated it is done so infrequently and in a perfunctory manner – but that under these conditions it becomes crucial for the organization to have tight control over who does the work and on whom. This immediately suggests the importance of comparative research in which the other three combinations are examined, the question being, how do these alternative forms grow, adapt, manage their rhetoric and handle their clientele. Thus it would be important to find organizations in which the controls over certification and inspection are both loose, organizations where there is loose control over certification but tight control over inspection, and organizations in which there is tight control both over inspection and over certification. Such comparative research might be conducted among different kinds of educational organizations within a single country (military, private, religious schooling in the United States), between educational and noneducational organizations within the same country (for example, schools versus hospitals versus military versus business organizations) or between countries looking at solutions to the problem of education given different degrees of centralization. As suggested earlier, it may not be the existence or nonexistence of loose coupling that is a crucial determinant of organizational functioning over time but rather the patterning of loose and tight couplings. Comparative studies should answer the question of distribution.

---

6 Cf. Cohen/March (1974).

If, as noted earlier, members within an organization (and researchers) will see and talk clearly about only those regions that are tightly coupled, then this suggests that members of educational organizations should be most explicit and certain when they are discussing issues related to certification for definition and regulation of teachers, pupils, topics, space, and resources. These are presumed to be the crucial issues that are tightly controlled. Increasing vagueness of description should occur when issues of substantive instruction – inspection – are discussed. Thus, those people who primarily manage the instructional business will be most vague in describing what they do, those people who primarily manage the certification rituals will be most explicit. This pattern is predicted **not** on the basis of the activities themselves – certification is easier to describe than inspection – but rather on the basis of the expectation that tightly coupled subsystems are more crucial to the survival of the system and therefore have received more linguistic work in the past and more agreement than is true for loosely coupled elements.

Core Technology and Organizational Form

A common tactic to understand complex organizations is to explore the possibility that the nature of the task being performed determines the shape of the organizational structure. This straightforward tactic raises some interesting puzzles about educational organizations. There are suggestions in the literature that education is a diffuse task, the technology is uncertain.

This first question suggests two alternatives: if the task is diffuse then would not any organizational form whatsoever be equally appropriate **or** should this directly compel a diffuse form of organizational structure? These two alternatives are not identical. The first suggests that if the task is diffuse then any one of a variety of quite specific organizational forms could be imposed on the organization and no differences would be observed. The thrust of the second argument is that there is one and only one organizational form that would fit well when there is a diffuse task, namely, a diffuse organizational form (for instance, an organized anarchy).

The second question asks if the task in an educational organization is diffuse then why do all educational organizations look the way they do, and why do they all look the same? If there is no clear task around which the shape of the organization can be formed then why is it that most educational organizations do have a form and why is it that most of these forms look identical? One possible answer is that the tasks of educational organizations do not constrain the form of the organization but rather this constraint is imposed by the ritual of certification and/or the agreements that are made in and by the environment. If any of these nontask possibilities are genuine alternative explanations, then the general literature on organizations has been insensitive to them.

One is therefore forced to ask the question, is it the case within educational organizations that the technology is unclear? So far it has been argued that loose coupling in educational organizations is partly the result of uncertain technology. If uncertain technology does not generate loose coupling then researchers must look elsewhere for the origin of these bonds.

Making Sense in/of Loosely Coupled Worlds

What kinds of information do loosely coupled systems provide members around which they can organize meanings, that is, what can one use in order to make sense of such fleeting structures? (By definition loosely coupled events are modestly predictable at best.) There is a rather barren structure that can be observed, reported on, and retrospected in order to make any sense. Given the ambiguity of loosely coupled structures, this suggests that there may be increased pressure on members to construct or negotiate some kind of social reality they can live with. Therefore, under conditions of loose coupling one should see considerable effort devoted to constructing social reality, a great amount of face work and linguistic work, numerous myths[7] and in general one should find a considerable amount of effort being devoted to punctuating this loosely coupled world and connecting it in some way in which it can be made sensible. Loosely coupled worlds do not look as if they would provide an individual many resources for sense making – with such little assistance in this task, a predominant activity should involve constructing social realities. Tightly coupled portions of a system should not exhibit nearly this preoccupation with linguistic work and the social construction of reality.

Coupling as a Dependent Variable

As a general rule, any research agenda on loose coupling should devote equal attention to loose coupling as a dependent and independent variable. Most suggestions have treated loose coupling as an independent variable. Less attention has been directed toward loose coupling as a dependent variable with the one exception of the earlier argument that one can afford loose coupling in either certification or inspection but not in both and, therefore, if one can locate a tight coupling for one of these two activities then he can predict as a dependent variable loose coupling for the other one.

Some investigators, however, should view loose coupling consistently as a dependent variable. The prototypic question would be, given prior conditions such as competition for scarce resources, logic built into a task, team teaching, conflict, striving for professionalism, presence of a central ministry of education, tenure, and so forth, what kind of coupling (loose or tight) among what kinds of elements occurs? If an organization faces a scarcity of resources its pattern of couplings should differ from when it faces an expansion of resources (for instance, scarcity leads to stockpiling leads to decoupling). Part of the question here is, what kinds of changes in the environment are the variables of tight and loose coupling sensitive to? In response to what kinds of activities or what kinds of contexts is coupling seen to change and what kinds of environments or situations, when they change, seem to have no effect whatsoever on couplings within an organization? Answers to these questions, which are of vital importance in predicting the outcomes of any interven-

7 Cf. Mitroff/Kilmann (1975).

tion, are most likely to occur if coupling is treated as a dependent variable and the question is, under what conditions will the couplings that emerge be tight or loose?

Assembling Loosely Connected Events

Suppose one assumes that there is nothing in the world except loosely coupled events. This assumption is close to Simon's stable subassemblies and empty world hypothesis and to the idea of cognitive limits on rationality. The imagery is that of numerous clusters of events that are tightly coupled within and loosely coupled between. These larger loosely coupled units would be what researchers usually call organizations. Notice that organizations formed this way are rather unusual kinds of organizations because they are neither tightly connected, nor explicitly bounded, but they are stable. The research question then becomes, how does it happen that loosely coupled events which remain loosely coupled are institutionally held together in one organization which retains few controls over central activities? Stated differently, how does it happen that someone can take a series of loosely coupled events, assemble them into an organization of loosely coupled systems, and the events remain both loosely coupled but the organization itself survives? It is common to observe that large organizations have loosely connected sectors. The questions are, what makes this possible, how does it happen? What the structure in school systems seems to consist of is categories (for example, teacher, pupil, reading) which are linked by understanding and legitimated exogenously (that is, by the world outside the organization). As John Meyer (1975) puts it, "the system works because everyone knows everyone else knows roughly what is to go on... Educational organizations are holding companies containing shares of stock in uninspected activities and subunits which are largely given their meaning, reality, and value in the wider social market." Note the potential fragility of this fabric of legitimacy.

It remains to be seen under what conditions loosely coupled systems are fragile structures because they are shored up by consensual anticipations, retrospections, and understanding that can dissolve and under what conditions they are resilient structures because they contain mutations, localized adaptation, and fewer costs of coordination.

Separate Intending and Acting Components

Intention and action are often loosely coupled within a single individual. Salancik (1975) has suggested some conditions under which dispositions within a single individual may be loosely coupled. These include such suggestions as follows.

(1) If intentions are not clear and unambiguous, then the use of them to select actions which will fulfill the intentions will be imperfect. (2) If the consequences of action are not known, then the use of intention to select action will be imperfect. (3) If the means by which an intention is transformed into an action are not known or in conflict, then the coupling of action to intention will be imperfect. (4) If intentions are not known to a person at the time of selecting an action, then the relationships between action and intention will be imperfect. This may be more common than

expected because this possibility is not allowed by so-called rational models of man. People often have to recall their intentions after they act or reconstruct these intentions, or invent them. (5) If there exists a set of multiple intentions which can determine a set of similar multiple actions, then the ability to detect a relationship between any one intention and any one action is likely to be imperfect. To illustrate, if there is an intention A which implies selecting actions X and Y, and there is also an intention B which implies selecting actions X and Y, then it is possible that under both presence and absence of intention A, action X will be selected. Given these circumstances, an observer will falsely conclude that this relationship is indeterminant.

The preceding list has the potential limitation for organizational inquiry in that it consists of events within a single person. This limitation is not serious **if** the ideas are used as metaphors or if each event is lodged in a different person. For example, one could lodge intention with one person and action with some other person. With this separation, then all of the above conditions may produce loose coupling between these actors but additional conditions also come into play given this geographical separation of intention from action. For example, the simple additional requirement that the intentions must be communicated to the second actor and in such a way that they control his actions, will increase the potential for error and loose coupling. Thus any discussion of separate locations for intention and action within an organization virtually requires that the investigator specify the additional conditions under which the intending component can control the acting component. Aside from the problems of communication and control when intention and action are separated there are at least two additional conditions that could produce loose coupling.

a. If there are several diverse intending components all of whom are dependent on the same actor for implementing action, then the relationship between any one intention and any one action will be imperfect. The teacher in the classroom may well be the prototype of this condition.
b. The process outlined in the preceding item can become even more complicated, and the linkages between intention and action even looser, if the single acting component has intentions of its own.

Intention and action are often split within organizations. This paper suggests that if one were to map the pattern of intention and action components within the organization these would coincide with loosely coupled systems identified by other means. Furthermore, the preceding propositions begin to suggest conditions under which the same components might be at one moment tightly coupled and at the next moment loosely coupled.

CONCLUSION: A STATEMENT OF PRIORITIES

More time should be spent examining the possibility that educational organizations are most usefully viewed as loosely coupled systems. The concept of organizations as loosely coupled systems can have a substantial effect on existing perspectives about organizations. To probe further into the plausibility of that assertion, it is suggested that the following research priorities constitute a reasonable approach to the examination of loosely coupled systems.

1. Develop Conceptual Tools Capable of Preserving Loosely Coupled Systems

It is clear that more conceptual work has to be done before other lines of inquiry on this topic are launched. Much of the blandness in organizational theory these days can be traced to investigators applying impoverished images to organizational settings. If researchers immediately start stalking the elusive loosely coupled system with imperfect language and concepts, they will perpetuate the blandness of organizational theory. To see the importance of and necessity for this conceptual activity the reader should reexamine the 15 different connotations of the phrase "loose coupling" that are uncovered in this paper. They provide 15 alternative explanations for any researcher who claims that some outcome is due to loose coupling.

2. Explicate What Elements Are Available in Educational Organizations for Coupling

This activity has high priority because it is essential to know the practical domain within which the coupling phenomena occur. Since there is the further complication that elements may appear or disappear as a function of context and time, this type of inventory is essential at an early stage of inquiry. An indirect benefit of making this a high priority activity is that it will stem the counterproductive suspicion that "the number of elements in educational organizations is infinite." The reasonable reply to that comment is that if one is precise in defining and drawing boundaries around elements, then the number of elements will be less than imagined. Furthermore, the researcher can reduce the number of relevant elements if he has some theoretical ideas in mind. These theoretical ideas should be one of the outcomes of initial activity devoted to language and concept development (Priority 1).

3. Develop Contextual Methodology

Given favorable outcomes from the preceding two steps, researchers should then be eager to look at complex issues such as patterns of tight and loose coupling keeping in mind that loose coupling creates major problems for the researcher because he is trained and equipped to decipher predictable, tightly coupled worlds. To "see" loosely coupled worlds unconventional methodologies need to be developed and conventional methodologies that are underexploited need to be given more attention. Among the existing tools that should be refined to study loose coupling are comparative studies and longitudinal studies. Among the new tools that should be "invented" because of their potential relevance to loosely coupled systems are nonteleological thinking,[8] concurrence methodology,[9] and Hegelian, Kantian, and Singerian inquiring systems.[10] While these latter methodologies are unconventional within social science, so too is it unconventional to urge that we treat unpredictability (loose coupling) as our topic of interest rather than a nuisance.

4. Promote the Collection of Thorough, Concrete Descriptions of the Coupling Patterns in Actual Educational Organizations

No descriptive studies have been available to show what couplings in what patterns and with what strengths existed in current educational organizations. This oversight should be remedied as soon as possible.

Adequate descriptions should be of great interest to the practitioner who wants to know how his influence attempts will spread and with what intensity. Adequate description should also show practitioners how their organizations may be more sensible and adaptive than they suspect. Thorough descriptions of coupling should show checks and balances, localized controls, stabilizing mechanisms, and subtle feedback loops that keep the organization stable and that would promote its decay if they were tampered with.

The benefits for the researcher of full descriptions are that they would suggest which locations and which questions about loose coupling are most likely to explain sizeable portions of the variance in organizational outcomes. For example, on the basis of good descriptive work, it might be found that both tightly and loosely coupled systems "know" their environments with equal accuracy in which case, the earlier line of theorizing about "thing and medium" would be given a lower priority.

5. Specify the Nature of Core Technology in Educational Organizations

A surprisingly large number of the ideas presented in this paper assume that the typical coupling mechanisms of authority of office and logic of the task do not operate in educational organizations. Inquiry into loosely coupled systems was triggered partly by efforts to discover what **does** accomplish the coupling in school systems. Before the investigation of loose coupling goes too far, it should be established that authority and task are not prominent coupling mechanisms in schools. The assertions that they are not prominent seem to issue from a combination of informal observation, implausibility, wishful thinking, looking at the wrong things, and rather vague

---

[8] Cf. Steinbeck (1941).

[9] Cf. Bateson (1972), pp. 180–201.

[10] Cf. Mitroff (1974).

definitions of core technology and reward structures within education. If these two coupling mechanisms were defined clearly, studied carefully, and found to be weak and/or nonexistent in schools, **then** there would be a powerful justification for proceeding vigorously to study loosely coupled systems. Given the absence of work that definitively discounts these coupling mechanisms in education and given the fact that these two mechanisms have accounted for much of the observed couplings in other kinds of organizations, it seems crucial to look for them in educational organizations in the interest of parsimony.

It should be emphasized that if it **is** found that substantial coupling within educational organizations is due to authority of office and logic of the task, this does not negate the agenda that is sketched out in this paper. Instead, such discoveries would (1) make it even more crucial to look for patterns of coupling to explain outcomes, (2) focus attention on tight and loose couplings within task and authority induced couplings, (3) alert researchers to keep close watch for any coupling mechanisms other than these two, and (4) would direct comparative research toward settings in which these two coupling mechanisms vary in strength and form.

6. Probe Empirically the Ratio of Functions to Dysfunctions Associated with Loose Coupling

Although the word "function" has had a checkered history, it is used here without apology – and without the surplus meanings and ideology that have become attached to it. Earlier several potential benefits of loose coupling were described and these descriptions were balanced by additional suggestions of potential liabilities. If one adopts an evolutionary epistemology, then over time one expects that entities develop a more exquisite fit with their ecological niches. Given that assumption, one then argues that if loosely coupled systems exist and if they have existed for sometime, then they bestow some net advantage to their inhabitants and/or their constituencies. It is not obvious, however, what these advantages are. A set of studies showing how schools benefit and suffer given their structure as loosely coupled systems should do much to improve the quality of thinking devoted to organizational analysis.

7. Discover How Inhabitants Make Sense Out of Loosely Coupled Worlds

Scientists are going to have some big problems when their topic of inquiry becomes low probability couplings, but just as scientists have special problems comprehending loosely coupled worlds so too must the inhabitants of these worlds. It would seem that quite early in a research program on loose coupling, examination of this question should be started since it has direct relevance to those practitioners who must thread their way through such "invisible" worlds and must concern their sensemaking and stories in such a way that they don't bump into each other while doing so.

REFERENCES

Bateson (1972): Bateson, Mary Catherine: *Our Own Metaphor*. New York: Knopf, 1972.

Cohen/March (1974): Cohen, Michael D. and James G. March: *Leadership and Ambiguity*. New York: McGraw-Hill, 1974.

Glassman (1973): Glassman, R.B.: Persistence and Loose Coupling in Living Systems. In: *Behavioral Science*, Vol. 18, 1973, pp. 83–98.

Heider (1959): Heider, Fritz: Thing and Medium. In: *Psychological Issues*, Vol. 1, Monograph 3, 1959, pp. 1–34.

March/Olsen (1975): March, J.G. and J.P. Olsen: *Choice Situations in Loosely Coupled Worlds*. Unpublished manuscript, Stanford University, 1975.

Meyer (1975): Meyer, John W.: *Notes on the Structure of Educational Organizations*. Unpublished manuscript, Stanford University, 1975.

Mitroff (1974): Mitroff, Ian I.: *The Subjective Side of Science*. New York: Elsevier, 1974.

Mitroff/Kilmann (1975): Mitroff, Ian I. and Ralph H. Kilmann: *On Organizational Stones: An Approach to the Design and Analysis of Organizations Through Myths and Stories*. Unpublished manuscript, University of Pittsburgh, 1975.

Salancik (1975): Salancik, Gerald R.: *Notes on Loose Coupling: Linking Intentions to Actions*. Unpublished manuscript, University of Illinois, Urbana-Champaign, 1975.

Simon (1962): Simon, Herbert A.: The Architecture of Complexity. In: *Proceedings of the American Philosophical Society*, Vol. 106, 1962, pp. 467–482.

Steinbeck (1941): Steinbeck, John: *The Log from the Sea of Cortez*. New York: Viking, 1941.

Stephens (1967): Stephens, John M.: *The Process of Schooling*. New York: Holt, Rinehart, and Winston, 1967.

Weick (1974): Weick, Karl E.: Middle Range Theories of Social Systems. In: *Behavioral Science*, Vol. 19, 1974, pp. 357–367.

William G. Ouchi (1980)

MARKETS, BUREAUCRACIES, AND CLANS

The issue of successfully coordinating different activities under circumstances of uncertainty and turbulence is one of the most fundamental and most difficult ones for modern logisticians and designers of supply networks.

The search for concepts and approaches to the rational analysis and the construction of effective logistics systems, consequently, is a high priority for logistics research and management practice. Classical contributions from the fields of Organizational Theory and Institutional Economics are supplying interesting and innovative answers. Some of them have been presented in earlier sections of this volume.

Another essential contribution is William G. Ouchi's (born 1943, professor for Human Resource and Organizational Behavior at University of California, Los Angeles) paper on *Markets, Bureaucracies, and Clans* of 1980. He showed that – beyond the already established insight about the alternative coordination mechanisms of "markets and hierarchies" (cf. Coase (1937) in this volume and Williamson (1975)) – there is a third generic coordination mechanism: the clan. In a bureaucracy the coordinating elements are employment contracts, surveillance and direction; in a market these are the market prices. In a clan however, coordination of activities is effected primarily by common values, beliefs and traditions. It is apparent from Ouchi's discussion that this insight was inspired by observations made at the time of the extraordinarily efficient industrial operations of Japanese companies, such as Toyota (see the reference to Ohno (1978/1988) in this volume).

The concept of "clan" offers an additional, important analytical instrument to the understanding of large-scale coordinated activities, as they are required in modern logistics and supply chain systems.

Reference:

William G. Ouchi: Markets, Bureaucracies, and Clans. In: *Administrative Science Quarterly*, Vol. 25, March 1980, pp. 129–141.

Further reading:

Coase, Ronald H.: The Nature of the Firm. In: *Economica*, Vol. 4, No. 16, 1937, pp. 386–405.

Williamson, Oliver E.: *Markets and Hierarchies: Analysis and Antitrust Implications*. New York: The Free Press, 1975.

P. Klaus and S. Müller (eds.), *The Roots of Logistics*, DOI 10.1007/978-3-642-27922-5_26,

William G. Ouchi: Markets, Bureaucracies, and Clans[1]

Evaluating organizations according to an efficiency criterion would make it possible to predict the form organizations will take under certain conditions. Organization theory has not developed such a criterion because it has lacked a conceptual scheme capable of describing organizational efficiency in sufficiently microscopic terms. The transactions cost approach provides such a framework because it allows us to identify the conditions which give rise to the costs of mediating exchanges between individuals: goal incongruence and performance ambiguity. Different combinations of these causes distinguish three basic mechanisms of mediation or control: markets, which are efficient when performance ambiguity is low and goal incongruence is high; bureaucracies, which are efficient when both goal incongruence and performance ambiguity are moderately high; and clans, which are efficient when goal incongruence is low and performance ambiguity is high.

THE NATURE OF ORGANIZATIONS

What is an organization, and why do organizations exist? Many of us would answer this question by referring to Barnard's (1968) technological imperative, which argues that a formal organization will arise when technological conditions demand physical power, speed, endurance, mechanical adaptation, or continuity beyond the capacity of a single individual.[2] Yet when the stone is too large or the production facility too complex for a single person, what is called for is cooperation, and cooperation need not take the form of a formal organization. Indeed, grain farmers who need a large grain elevator do not form corporations which take over the farms and make the farmers into employees; instead, they form a cooperative to own and operate the elevator.

Others would refer to March/Simon's (1958) argument that an organization will exist so long as it can offer its members inducements which exceed the contributions it asks of them. While this position explains the conditions under which an organization may continue to exist, it does not explain how an organization can create a whole which is so much greater than the sum of its parts that it can give them more than they contribute.

Most of us, however, would refer to Blau/Scott's (1962) definition of a formal organization as a purposive aggregation of individuals who exert concerted effort

---

[1] I am indebted to many colleagues for their constructive criticisms of this paper, particularly to Chris Argyris, Peter Blau, Larry Cummings, Charles Horngren, Joanne Martin, John Meyer, Jerry Porras, Edgar Schein, W. Richard Scott, Arnold Tannenbaum, Richard Walton, and Oliver Williamson.

[2] Cf. Barnard (1968), pp. 27–28.

toward a common and explicitly recognized goal. Yet we can hardly accept this definition whole, suspecting as Simon has that individuals within organizations rarely have a common understanding of goals.[3]

Another point of view on the question of why organizations exist began with an inquiry by Coase (1937) and has recently been developed by Williamson (1975). In this view, an organization such as a corporation exists because it can mediate economic transactions between its members at lower costs than a market mechanism can. Under certain conditions, markets are more efficient because they can mediate without paying the costs of managers, accountants, or personnel departments. Under other conditions, however, a market mechanism becomes so cumbersome that it is less efficient than a bureaucracy. This transactions cost approach explicitly regards efficiency as the fundamental element in determining the nature of organizations.

MARKETS, BUREAUCRACIES, AND CLANS

Transactions costs are a solution to the problem of cooperation in the realm of economic activity. From the perspective of Mayo (1945) and Barnard (1968) the fundamental problem of cooperation stems from the fact that individuals have only partially overlapping goals. Left to their own devices, they pursue incongruent objectives and their efforts are uncoordinated. Any collectivity which has an economic goal must then find a means to control diverse individuals efficiently.

Many helpful ideas have flowed from this definition of the problem of cooperation. Some (e.g. Etzioni (1965) or Weick (1969)) have emphasized the tension between individual autonomy and collective interests which must attend cooperative action, while others (e.g. Simon (1945)) have emphasized the impossibility of achieving a completely cooperative effort. Our interest is in the efficiency with which transactions are carried out between individuals who are engaged in cooperative action.

Cooperative action necessarily involves interdependence between individuals. This interdependence calls for a transaction or exchange in which each individual gives something of value (for example, labor) and receives something of value (for example, money) in return. In a market relationship, the transaction takes place between the two parties and is mediated by a price mechanism in which the existence of a competitive market reassures both parties that the terms of exchange are equitable. In a bureaucratic relationship, each party contributes labor to a corporate body which mediates the relationship by placing a value on each contribution and then compensating it fairly. The perception of equity in this case depends upon a social agreement that the bureaucratic hierarchy has the legitimate authority to provide this mediation. In either case, individuals must regard the transaction as equitable: it

[3] Cf. Simon (1945), pp. 257–278.

must meet the standards of reciprocity which Gouldner (1961) has described as a universal requirement for collective life.

It is this demand for equity which brings on transactions costs. A transactions cost is any activity which is engaged in to satisfy each party to an exchange that the value given and received is in accord with his or her expectations.

Transactions costs arise principally when it is difficult to determine the value of the goods or service. Such difficulties can arise from the underlying nature of the goods or service or from a lack of trust between the parties. When a company is being sold by one corporation to another corporation, for example, it may not be unambiguously clear what the true value of that company is. If firms similar to the company are frequently bought and sold, and if those transactions occur under competitive conditions, then the market process will be accepted as a legitimate estimator of the true value. But if the company is unique, and there is only one potential buyer, then market forces are absent. How will the buyer and seller determine a fair price? They may call upon a third party to estimate the value of the company. Each party may in addition call upon other experts who will assist them in evaluating both the value of the company and the adequacy of the judgment of the third party. Each side may also require an extensive and complete contract which will describe exactly what is being bought and sold. Each of these activities is costly, and all of them are regarded here as transactions costs: they are necessary to create a perception of equity among all parties to the transaction.

This same argument applies to transactions in which a service, such as the labor of an individual, is the object of exchange. If one individual sells his or her services to another, it may be difficult to assess the true value of that labor. In particular, if the labor is to be used in an interdependent technology, one which requires teamwork, it may be difficult to assess the value contributed by one worker as opposed to another, since their joint efforts yield a single outcome in this case, or in a case where it is likely that task requirements will change, then the auditing and complex contracting required to create the perception of equity can become unbearably costly.

We have identified two principal mechanisms for mediating these transactions: a market and a bureaucracy. These alternatives have received the greatest attention from organization theorists (e.g. Barnard (1968); Weber (1968)) and economists (e.g. Coase (1937); Arrow (1974)). However, the paradigm also suggests a third mechanism: If the objectives of individuals are congruent (not mutually exclusive), then the conditions of reciprocity and equity can be met quite differently.

Both Barnard and Mayo pointed out that organizations are difficult to operate because their members do not share a selfless devotion to the same objectives. Mayo (1945) argued that organizations operated more efficiently in preindustrial times, when members typically served an apprenticeship during which they were socialized into accepting the objectives of the craft or organization. Barnard posed the problem thus:

> "A formal system of cooperation requires an objective, a purpose, an aim… It is important to note the complete distinction between the aim of a cooperative effort and that of an individual. Even in the case where a man enlists the aid of

other men to do something which he cannot do alone, such as moving a stone, the objective ceases to be personal."[4]

While Barnard, like Arrow, views markets and bureaucracies as the basic mechanisms for achieving the continued cooperation of these individuals, he also allowed for the possibility of reducing the incongruence of goals in a manner consistent with Mayo's view of the preindustrial organization:

> "An organization can secure the efforts necessary to its existence, then, either by the objective inducement it provides or by changing states of mind. It seems to me improbable that any organization can exist as a practical matter which does not employ both methods in combination."[5]

If the socialization of individuals into an organization is complete, then the basis of reciprocity can be changed. For example, Japanese firms rely to a great extent upon hiring inexperienced workers, socializing them to accept the company's goals as their own, and compensating them according to length of service, number of dependents, and other nonperformance criteria (see Abegglen (1958); Dore (1973); Nakane (1973)). It is not necessary for these organizations to measure performance to control or direct their employees, since the employees' natural (socialized) inclination is to do what is best for the firm. It is also unnecessary to derive explicit, verifiable measures of value added, since rewards are distributed according to nonperformance-related criteria which are relatively inexpensive to determine (length of service and number of dependents can be ascertained at relatively low costs). Thus, industrial organizations can, in some instances, rely to a great extent on socialization as the principal mechanism of mediation or control, and this "clan" form ("clan" conforms to Durkheim's meaning of an organic association which resembles a kin network but may not include blood relations[6]) can be very efficient in mediating transactions between interdependent individuals.

Markets, bureaucracies, and clans are therefore three distinct mechanisms which may be present in differing degrees, in any real organization.[7] Our next objective is to specify the conditions under which the requirements of each form are most efficiently satisfied.

---

4 Cf. Barnard (1968), pp. 42–43.

5 Cf. Barnard (1968), p. 141.

6 Cf. Durkheim (1933), p. 175.

7 In the broader language necessary to encompass both economics and organization theory, an organization may be thought of as any stable pattern of transactions. In this definition, a market is as much an organization as is a bureaucracy or a clan. The only requirement is that, for the purposes of this discussion, we maintain a clear distinction between the idea of "bureaucracy" and the idea of "organization". Bureaucracy as used here refers specifically to the Weberian model, while organization refers to any stable pattern of transactions between individuals or aggregations of individuals.

THE MARKET FAILURES FRAMEWORK

We can approach this question most effectively by examining the markets and hierarchies approach provided by Williamson (1975) which builds upon earlier statements of the problem by Coase (1937) and others (for a more detailed description of the functioning of each mechanism, see Ouchi (1979)).

Market transactions, or exchanges, consist of contractual relationships. Each exchange is governed by one of three types of contractual relations, all of which can be specified completely. That is, because each party is bound only to deliver that which is specified, the contract must specify who must deliver what under every possible state of nature. The simplest form of contract is the "spot" or "sales" contract. This is what occurs when you walk up to a candy counter, ask for a candy bar, and pay the amount the salesperson asks. In such a transaction, all obligations are fulfilled on the spot. However, the spot market contract is, by definition, incapable of dealing with future transactions, and most exchange relationships involve long-term obligations.

A common device for dealing with the future is the "contingent claims contract", a document that specifies all the obligations of each party to an exchange, contingent upon all possible future states of nature. However, given a future that is either complex or uncertain, the bounded rationality of individuals makes it impossible to specify such a contract completely. Leaving such a contract incompletely specified is an alternative, but one that will succeed only if each party can trust the other to interpret the uncertain future in a manner that is acceptable to him. Thus, given uncertainty, bounded rationality, and opportunism, contingent claims contracting will fail.

Instead of trying to anticipate the future in a giant, once-and-for-all contract, why not employ a series of contracts, each one written for a short period within which future events can confidently be foreseen? The problem with such "sequential spot contracting" is that in many exchange relationships, the goods or services exchanged are unique, and the supplier requires specialized knowledge of how to supply the customer best and most efficiently. The supplier acquires this knowledge over time and in doing so gains a "first mover advantage", which enables him to bid more effectively on subsequent contracts than any potential competitor can. Knowing this, potential competitors will not waste their time bidding, thus producing a situation of "small numbers bargaining" or bilateral monopoly, in which there is only one buyer and seller. Under this condition, competitive pressures are absent, and each party will opportunistically claim higher costs or poor quality, whichever is in his or her interest. In order to maintain such an exchange, each party will have to go to considerable expense to audit the costs or performance of the other. If these transactions costs are too high, the market relationship will fail due to the confluence of opportunism with small numbers bargaining, even though the limitations of uncertainty and bounded rationality have been overcome.

Thus, under some conditions no completely contractual market relationship is feasible. TABLE 1 summarizes the conditions which lead to market failure. According to the paradigm, no one of the four conditions can produce market failure, but almost any pairing of them will do so.

The Market Failures Framework *)		
Human factors		**Environmental factors**
Bounded rationality	←——→	Uncertainty/Complexity
Opportunism	←——→	Small numbers
*) Adapted from Williamson (1975, p. 40).		

TABLE 1: THE MARKET FAILURES FRAMEWORK

The idea of market failure is an analytical device. Economists do not agree on a specific set of conditions that constitute the failure of a market; indeed one point of view argues that even monopolistic conditions may be competitive. However, the idea of market failure as expressed by Williamson (1975) is useful as a conceptual framework within which to compare the strengths of markets as opposed to bureaucracies. The technique is to contend that all transactions can be mediated entirely by market relations, and then ask what conditions will cause some of these market mechanisms to fail and be replaced by bureaucratic mechanisms. In this sense, every bureaucratic organization constitutes an example of market failure.

The bureaucratic organization has two principal advantages over the market relationship. First, it uses the employment relation, which is an incomplete contract. In accepting an employment relation, a worker agrees to receive wages in exchange for submitting to the legitimate right of the organization to appoint superior officers who can (1) direct the work activities of the employee from day to day (within some domain or zone of indifference), thus overcoming the problem of dealing with the future all at once and (2) closely monitor the employee's performance, thus minimizing the problem of opportunism.

Second, the bureaucratic organization can create an atmosphere of trust between employees much more readily than a market can between the parties to an exchange. Because members of an organization assume some commonality of purpose, because they learn that long-term relationships will reward good performance and punish poor performance, they develop some goal congruence. This reduces their opportunistic tendencies and thus the need to monitor their performance.

Bureaucracies are also characterized by an emphasis on technical expertise which provides some skill training and some socialization into craft or professional standards. Professionals within a bureaucratic setting thus combine a primary affiliation to a professional body with a career orientation, which increases the sense of affiliation or solidarity with the employer and further reduces goal incongruence.[8]

---

[8] Despite these desirable properties the bureaucratic type has continually been under attack and revision. As Williamson points out, the move from U-form (functional) to M-Form (divisional) organization among many large firms has been motivated by a desire to simulate a capital market within a bureaucratic framework because of its superior efficiency. By regrouping the parts of the organization, it is possible to create sub-entities that are suf-

In summary, the market failures framework argues that markets fail when the costs of completing transactions become unbearable. At that point, the inefficiencies of bureaucratic organization will be preferred to the relatively greater costs of market organization, and exchange relationships move from one domain into the other.

Consider one example. The 10,000 individuals who comprise the workforce of a steel mill could be individual entrepreneurs whose interpersonal transactions are mediated entirely through a network of market or contractual relationships. Each of them could also have a market relation with yet another combine which owned the capital equipment and facilities necessary to produce steel. Yet steel mills are typically bureaucratic in form and each worker is in an employment, not market, relation with the corporation. Market forces have failed because the determination of value contributed by one worker is highly ambiguous in the integrated steelmaking process, which makes the transactions cost attendant upon maintaining a market too high.

EXTENDING THE MARKET FAILURES FRAMEWORK: CLANS

Bureaucracies can fail when the ambiguity of performance evaluation becomes significantly greater than that which brings about market failure. A bureaucratic organization operates fundamentally according to a system of hierarchical surveillance, evaluation, and direction. In such a system, each superior must have a set of standards to which he can compare behavior or output in order to provide control. These standards only indicate the value of an output approximately, and are subject to idiosyncratic interpretation. People perceive them as equitable only as long as they believe that they contain a reasonable amount of performance information. When tasks become highly unique, completely integrated, or ambiguous for other reasons, then even bureaucratic mechanisms fail. Under these conditions, it becomes impossible to evaluate externally the value added by any individual. Any standard which is applied will be by definition arbitrary and therefore inequitable.

If we adopt the view that transactions costs arise from equity considerations, then we can interpret TABLE 1 in a different light. Simon's work on the employment relation[9] shows that TABLE 1 contains some redundancy. He emphasized that under an employment contract, the employer pays a worker a premium over the "spot" price for

---

ficiently autonomous to permit precise measurement and the determination of an effective price mechanism. Although each division may still operate internally as a bureaucracy, the economies which accrue from this partial market solution are often large, offsetting the diseconomies of functional redundancy which often accompany the separation of the organization into divisions.

[9] Cf. Simon (1957), pp. 183–195.

any piece of work. From the point of view of the worker, this "risk premium" compensates him for the likelihood that he will be asked to perform duties which are significantly more distasteful to him than those which are implied in the employment contract. The uncertainty surrounding the likelihood of such tasks and the expectation that the employer will or will not ask them determines the size of the risk premium. If the employee agreed with all the employer's objectives, which is equivalent to completely trusting the employer never to request a distasteful task, then the risk premium would be zero.

The employment relation is relatively efficient when the measurement of performance is ambiguous but the employer's goals are not. In an employment relation, each employee depends on the employer to distribute rewards equitably; if employees do not trust the employer to do so, they will demand contractual protections such as union representation and the transactions cost will rise.

Thus, the critical element in the efficiency of market versus employment relations has to do with (1) the ambiguity of the measurement of individual performance, and (2) the congruence of the employees' and employer's goals. We can now reformulate the transactions cost problem as follows: in order to mediate transactions efficiently, any organizational form must reduce either the ambiguity of performance evaluation or the goal incongruence between parties. Put this way, market relations are efficient when there is little ambiguity over performance, so the parties can tolerate relatively high levels of opportunism or goal incongruence. And bureaucratic relations are efficient when both performance ambiguity and goal incongruence are moderately high.

What form of mediation succeeds by minimizing goal incongruence and tolerating high levels of ambiguity in performance evaluation? Clearly, it is one which embodies a strong form of the employment relation as defined by Simon (1945) which is a relationship in which the risk premium is minimized. The answer is what we have referred to as the clan, which is the obverse of the market relation since it achieves efficiency under the opposite conditions: high performance ambiguity and low opportunism.

Perhaps the clearest exposition of the clan form appears in what Durkheim refers to as the case of organic solidarity and its contrast with contractual relations:

> "For organic solidarity to exist, it is not enough that there be a system of organs necessary to one another, which in a general way feel solidarity, but it is also necessary that the way in which they should come together, if not in every kind of meeting, at least in circumstances which most frequently occur, be predetermined. Otherwise, at every moment new conflicts would have to be equilibrated… It will be said that there are contracts. But, first of all, social relations are not capable of assuming this juridical form… A contract is not self-sufficient, but supposes a regulation which is as extensive and complicated as

> contractual life itself... A contract is only a truce, and very precarious, it suspends hostilities only for a time."[10]

The solidarity to which Durkheim refers contemplates the union of objectives between individuals which stems from their necessary dependence upon one another. In this sense, any occupational group which has organic solidarity may be considered a clan. Thus, a profession, a labor union, or a corporation may be a clan, and the professionalized bureaucracy may be understood as a response to the joint need for efficient transactions within professions (clan) and between professions (bureaucracy). Goal congruity as a central mechanism of control in organizations also appears repeatedly in Barnard (1968):

> "The most intangible and subtle of incentives is that which I have called the condition of communion... It is the feeling of personal comfort in social relations that is sometimes called solidarity, social integration... The need for communion is a basis of informal organization that is essential to the operation of every formal organization."[11]

Descriptions of organizations which display a high degree of goal congruence, typically through relatively complete socialization brought about through high inclusion (Etzioni (1965)) are also found in Lipset/Trow/Coleman ((1956), pp. 79–80), Argyris ((1964), pp. 10 and 175), Selznick (1966), and Clark (1970). In each case, the authors describe the organization as one in which it is difficult to determine individual performance. However, such organizations are not "loosely coupled" nor are they "organized anarchies" simply because they lack market and bureaucratic mechanisms. A clan, as Durkheim points out, provides great regularity of relations and may in fact be more directive than the other, more explicit mechanisms. That clans display a high degree of discipline is emphasized by Kanter (1972) in her study of utopian communities, some of which were successful businesses such as Oneida and Amana. According to Kanter (1972), this discipline was not achieved through contractualism or surveillance but through an extreme form of the belief that individual interests are best served by a complete immersion of each individual in the interests of the whole.[12]

More recently, Ouchi/Jaeger (1978) and Ouchi/Johnson (1978) have reported on modern industrial organizations which closely resemble the clan form. In these organizations, a variety of social mechanisms reduces differences between individual and organizational goals and produces a strong sense of community (see also Van Maanen (1975); Katz (1978)). Where individual and organizational interests overlap to this extent, opportunism is unlikely and equity in rewards can be achieved at a relatively low transactions cost. Moreover, these organizations are typically in technologically advanced or closely integrated industries, where teamwork is common,

---

[10] Cf. Durkheim (1933), p. 365.

[11] Cf. Barnard (1968) p. 148; see also pp. 89, 152, 169, and 273.

[12] Cf. Kanter (1972), p. 41.

technologies change often, and therefore individual performance is highly ambiguous.

When a bureaucracy fails, then due to excessively ambiguous performance evaluation, the sole form of mediation remaining is the clan, which relies upon creating goal congruence. Although clans may employ a system of legitimate authority (often the traditional rather than the rational-legal form), they differ fundamentally from bureaucracies in that they do not require explicit auditing and evaluation. Performance evaluation takes place instead through the kind of subtle reading of signals that is possible among intimate coworkers but which cannot be translated into explicit, verifiable measures. This means that there is sufficient information in a clan to promote learning and effective production, but that information cannot withstand the scrutiny of contractual relations. Thus, any tendency toward opportunism will be destructive, because the close auditing and hard contracting necessary to combat it are not possible in a clan.

If performance evaluation is so ambiguous and goals so incongruent that a clan fails, what then? We can only speculate, but it seems that this final cell may be the case discussed by Meyer/Rowan (1977) in which control is purely ceremonial and symbolic. School systems, like other organizations, do employ a variety of mechanisms. Yet if there is no effective mechanism of mediation between individuals, the perception of equity may be purely superstitious, based on a broad, community-based acceptance of the legitimacy of the institution.

MARKETS, BUREAUCRACIES, AND CLANS: AN OVERVIEW

Having distinguished three mechanisms of intermediation, we can now summarize them and attempt to set out the general conditions under which each form will mediate transactions between individuals most efficiently. TABLE 2 discriminates markets, bureaucracies, and clans along two dimensions: their underlying normative and informational requirements.

An Organizational Failures Framework

Mode of control	**Normative requirements**	**Informational requirements**
Market	Reciprocity	Prices
Bureaucracy	Reciprocity Legitimate authority	Rules
Clan	Reciprocity Legitimate authority Common values and beliefs	Traditions

TABLE 2: AN ORGANIZATIONAL FAILURES FRAMEWORK

Normative requirements refer to the basic social agreements that all members of the transactional network must share if the network is to function efficiently, without undue costs of performance auditing or monitoring. A norm of reciprocity, according to Gouldner (1961) is one of only two social agreements that have been found to be universal among societies across time and cultures (the other is the incest taboo). If no such norm were widely shared, then a potential trader would have to consume so much energy in setting the contractural terms of exchange in advance and in auditing the performance of the other party afterwards that the potential transaction would cost too much. Under such conditions, a division of labor is unthinkable and social existence impossible. Therefore, a norm of reciprocity underlies all exchange mechanisms.

A norm of legitimate authority is critical for two reasons. As discussed above, it permits the assignment of organizational superiors who can, on an ad hoc basis, specify the work assignments of subordinates, thus obviating the need for a contingent claims employment contract which would be either so complex as to be infeasible or so simple as to be too confining or else incomplete. Legitimate authority also permits organizational superiors to audit the performance of subordinates more closely than is possible within a market relationship. In a bureaucracy, legitimate authority will commonly take the "rational/legal" form, whereas in a clan it may take the "traditional" form (cf. Blau/Scott (1962), pp. 27–38). Legitimate authority is not ordinarily created within the organization but is maintained by other institutions such as the church or the educational system (Weber (1947); Blau/Scott (1962); Barnard (1968), pp. 161–184). While the legitimacy of a particular organization may be greater or smaller as a result of its managerial practices, it is fundamentally maintained within a society generally.

Common values and beliefs provide the harmony of interests that erase the possibility of opportunistic behavior. If all members of the organization have been exposed to an apprenticeship or other socialization period, then they will share personal goals that are compatible with the goals of the organization. In this condition, auditing of performance is unnecessary except for educational purposes, since no member will attempt to depart from organizational goals.

A norm of reciprocity is universal, legitimate authority is accepted, though in varying degree, in most formal organizations, and common values and beliefs are relatively rare in formal organizations. Etzioni (1965) has described this last form of control as being common only to "total organizations" such as the military and mental hospitals, and Light (1972) describes its role in ethnically bound exchange relationships. However, we have also noted that a partially complete form of socialization, accompanied by market or bureaucratic mechanisms, may be effective across a wider range of organizations. Mayo (1945) contended that instability of employment, which upsets the long socialization period necessary, is the chief enemy of the development of this form of control.

The informational prerequisites of each form of control are prices, rules, and traditions. Prices are a highly sophisticated form of information for decision making. However, correct prices are difficult to arrive at, particularly when technological interdependence, novelty, or other forms of ambiguity obscure the boundary between tasks or individuals. Rules, by comparison, are relatively crude informational devices. A rule is specific to a problem, and therefore it takes a large number of rules to control organizational responses. A decision maker must know the structure of the rules in order to apply the correct one in any given situation. Moreover, an organization can never specify a set of rules that will cover all possible contingencies. Instead, it specifies a smaller set of rules which cover routine decisions, and refers exceptions up the hierarchy where policymakers can invent rules as needed. As Galbraith (1973) has pointed out, under conditions of uncertainty or complexity the number of exceptions becomes so great that the hierarchy becomes overloaded and the quality of decision making suffers.

Traditions are implicit rather than explicit rules that govern behavior. Because traditions are not specified, they are not easily accessible, and a new member will not be able to function effectively until he or she has spent a number of years learning them (Van Maanen/Schein (1978)). In terms of the precision of the performance evaluation they permit, traditions may be the crudest informational prerequisite, since they are ordinarily stated in a general way which must be interpreted in a particular situation. On the other hand, the set of traditions in a formal organization may produce a unified, although implicit philosophy or point of view, functionally equivalent to a theory about how that organization should work. A member who grasps such an essential theory can deduce from it an appropriate rule to govern any possible decision, thus producing a very elegant and complete form of control. Alternatively, a disruption of the socialization process will inhibit the passing on of traditions and bring about organizational inefficiency

SOME CONCLUDING THOUGHTS

Under conditions of extreme uncertainty and opportunism, transactions cost may rise. Indeed, Denison (1978) has observed that net productivity declined in the United States between 1965 and 1975 due to changes in "the industrial and human environment within which business must operate".[13] According to Denison, output per unit of input has declined for two reasons: 78 percent of the decline is due to increased costs of air, water, and safety on the job, and the remaining 22 percent is attributable to increased needs for surveillance of potentially dishonest employees, customers, contractors, and thieves. The resources put into improvements in air, water, and safety are not a net loss to society although they may reduce corporate profitability. The increased need for surveillance in business, however, may represent the fact that the cost of monitoring transactions has risen. Mayo (1945) might have predicted this change as an inevitable result of the instability which accompanies industrialization. In our framework, we could advance the following explanation: exchange relationships are generally subject to so much informational ambiguity that they can never be governed completely by markets. Consequently, they have

[13] Cf. Denison (1978), p. 21.

been supplemented through cultural, clan mechanisms. As instability, heterogeneity, and mobility have intensified in the United States, however, the effectiveness of these cultural mechanisms has been vitiated and bureaucratic mechanisms of surveillance and control have increased. Although bureaucratic surveillance may be the optimal strategy under present social conditions, it is nonetheless true that the United States is devoting more of its resources to transactional matters than it did ten years ago, and that represents a net decline in its welfare.

The degree of uncertainty and opportunism that characterize American society may be such that no mechanisms of control ever function very well. We have already observed that the conditions necessary for a pure market, bureaucracy, or clan are rare. Even a combination of these control mechanisms may be insufficient in many cases, however. In organizations using new technologies or in the public sector, the rate of change, instability of employment, or ambiguity of performance evaluation may simply overwhelm all rational control attempts.

In these cases, exchange becomes institutionalized. Meyer/Rowan's (1977) central thesis is that school systems, by their nature, evade any form of rational control. They have no effective price mechanism, no effective bureaucratic control, and no internally consistent cultures.[14] Thus school systems (as distinguished from education, which need not be done by large organizations) continue to grow and survive because the objectives which they are believed to pursue have been accepted as necessary by society. Since rational control is not feasible within the school, no one knows whether it is actually pursuing these goals, but an institutionalized organization (the church is another example) need not give evidence of performance.[15]

All work organizations are institutionalized in the sense that fundamental purposes of all viable organizations must mesh at least somewhat with broad social values (Parsons/Shils (1951)). This institutionalization permits organizations to survive even under conditions that severely limit their capacity for rational control. Ultimately, organizational failure occurs only when society deems the basic objectives of the organization unworthy of continued support.

What is an organization? An organization, in our sense, is any stable pattern of transactions between individuals or aggregations of individuals. Our framework can thus be applied to the analysis of relationships between individuals or between subunits within a corporation, or to transactions between firms in an economy. Why do organizations exist? In our sense, all patterned transactions are organized, and thus all stable exchanges in a society are organized. When we ask "why do organizations exist," we usually mean to ask "why do bureaucratic organizations exist," and the answer is clear. Bureaucratic organizations exist because, under certain specifiable conditions, they are the most efficient means for an equitable mediation of transactions between parties. In a similar manner, market and clan organizations exist because each of them, under certain conditions, offers the lowest transactions cost.

[14] See also Meyer et al. (1978).

[15] See also Ouchi (1977), pp. 97–98.

REFERENCES

Abegglen (1958): Abegglen, James C.: *The Japanese Factory: Aspects of its Social Organization*. Glencoe, IL: Free Press, 1958.

Argyris (1964): Argyris. Chris: *Integrating the Individual and the Organization*. New York: Wiley, 1964.

Arrow (1974): Arrow, Kenneth J.: *The Limits of Organization*. New York: Norton, 1974.

Barnard (1968): Barnard, Chester I.: *The Functions of the Executive*, 30th Anniversary ed. Cambridge: Harvard, 1968.

Blau/Scott (1962): Blau, Peter M. and W. Richard Scott: *Formal Organizations*. San Francisco: Scott, Foreman, 1962.

Clark (1970): Clark, Button R.: *The Distinctive College*. Antioch, Reed, and Swarthmore. Chicago: Aldline, 1970.

Coase (1937): Coase, Ronald H.: The Nature of the Firm. In. *Economica*, New Series, Vol. 4, 1937, pp. 386–405.

Denison (1978): Denison, Edward F.: *Effects of Selected Changes in the Institutional and Human Environment upon Output per Unit of Input*. Brookings General Series Reprint No. 335. Washington: Brookings, 1978.

Dore (1973): Dore, Ronald: *British Factory – Japanese Factory*. Berkeley: University of California, 1973.

Durkheim (1933): Durkheim, Emile: *The Division of Labor in Society*. G. Simpson, trans. New York: Free Press, 1933.

Etzioni (1965): Etzioni, Amitai: Organizational Control Structure. In: James G. Marc (ed.), *Handbook of Organizations*. Chicago: Rand McNally, 1965, pp. 650–677.

Galbraith (1973): Galbraith, Jay: *Designing Complex Organizations*. Reading/MA: Addison-Wesley, 1973.

Gouldner (1961): Gouldner, Alvin W.: The Norm of Reciprocity. In: *American Sociological Review*, Vol. 25, 1961, pp. 161–179.

Kantar (1972): Kantar, Rosabeth Moss: *Commitment and Community*. Cambridge: Harvard, 1972.

Katz (1978): Katz, Ralph: Job Longevity as a Situational Factor in Job Satisfaction. In: *Administrative Science Quarterly*, Vol. 23, 1978, pp. 204–223.

Light (1972): Light, Ivan H.: *Ethnic Enterprise in America*. Berkeley: University of California, 1972.

Lipset (1956): Lipset, Seymour M., Martin A. Trow, and James S. Coleman: *Union Democracy*. Glencoe, IL: Free Press, 1956.

March/Simon (1958): March, James G. and Herbert A. Simon: *Organizations*. New York: Wiley, 1958.

Mayo (1945): Mayo, Elton: *The Social Problems of an Industrial Civilization*. Boston: Division of Research, Graduate School of Business Administration, Harvard University, 1945.

Meyer/Rowan (1977): Meyer, John W. and Brian Rowan: Formal Structure as Myth and Ceremony. *American Journal of Sociology*, Vol. 83, 1977, pp. 340–363.

Meyer et al. (1978): Meyer, John W., W. Richard Scott, Sally Cole, and Jo-Ann K. Intili: Instructional Dissensus and Institutional Consensus in Schools. In Marshall W. Meyer and Associates (Eds.): *Environments and Organizations*, San Francisco: Jossey-Bass, 1978, pp. 233–263.

Nakane (1973): Nakane, Chie: *Japanese Society*, rev. ed. Middlesex. England: Penguin, 1973.

Ouchi (1977): Ouchi, William G.: The Relationship between Organizational Structure and Organizational Control. In: *Administrative Science Quarterly*, Vol. 22, 1977, pp. 95–113.

Ouchi (1979): Ouchi, William G.: A Conceptual Framework for the Design of Organizational Control Mechanisms. In: *Management Science*, Vol. 25, 1979, pp. 833–848.

Ouchi/Jaeger (1978): Ouchi, William G. and Alfred M. Jaeger: Type Z Organization: Stability in the Midst of Mobility. In: *Academy of Management Review*, Vol. 3, 1978, pp. 305–314.

Ouchi/Johnson (1978): Ouchi, William G. and Jerry B. Johnson: Types of Organizational Control and their Relationship to Emotional Well-being. In: *Administrative Science Quarterly*, Vol. 23, 1978, pp. 293–317.

Parsons/Shils (1951): Parsons, Talcott and Edward A. Shils: Values, Motives, and Systems of Action. In: Talcott Parsons and Edward A. Shils (Eds.): *Toward a General Theory of Action*. Cambridge: Harvard, 1951, pp. 47–275.

Selznick (1966): Selznick, Philip: *TVA and the Grass Roots* (orig. ed., 1949). New York: Harper Torchbooks, 1966.

Simon (1945): Simon, Herbert A.: *Administrative Behavior*. New York: Free Press, 1945.

Simon (1957): Simon, Herbert A.: *Models of Man*. New York: Wiley, 1957.

Van Maanen (1975): Van Maanen, John: Police Socialization: A Longitudinal Examination of Job Attitudes in an Urban Police Department. In: *Administrative Science Quarterly*, Vol. 20, 1975, pp. 207–228.

Van Maanen/Schein (1978): Van Maanen, John and Edgar H. Schein: *Toward a Theory of Organizational Socialization*. Manuscript, Sloan School of Industrial Administration, Massachusetts Institute of Technology, 1978.

Weber (1925): Weber, Max: *The Theory of Social and Economic Organization* (orig. ed., 1925). A. M. Henderson and T. Parsons, trans. New York: Free Press, 1947.

Weber (1968): Weber, Max: *Economy and Society* (orig. ed. 1925). G. Roth and C. Wittich, (Eds.), New York: Bedminster Press, 1968.

Weick (1969): Weick, Karl E.: *The Social Psychology of Organizing*. Reading, MA: Addison-Wesley, 1969.

Williamson (1975): Williamson, Oliver E.: *Markets and Hierarchies: Analysis and Antitrust Implications*. New York: Free Press, 1975.

The manufacturer's authorised representative in the EU is Springer Nature Customer Service Centre GmbH, Europaplatz 3, 69115 Heidelberg, Germany. If you have any concerns regarding our products, please contact ProductSafety@springernature.com

Printed and bound by CPI Group (UK) Ltd, Croydon, CR0 4YY
01/10/2026
02225976-0020